Rubber-Toughened Plastics

D0208246

ADVANCES IN CHEMISTRY SERIES **222**

Rubber-Toughened Plastics

C. Keith Riew, EDITOR
BF Goodrich Company

Developed from a symposium sponsored by
the Division of Polymeric Materials:
Science and Engineering
of the American Chemical Society,
at the 194th National Meeting
of the American Chemical Society,
New Orleans, Louisiana
August 30–September 4, 1987

American Chemical Society, Washington, DC 1989

Library of Congress Cataloging-in-Publication Data

Rubber-toughened plastics.

(Advances in chemistry series, ISSN 0065–2393)
"Developed from a symposium sponsored by the
Division of Polymeric Materials: Science and
Engineering of the American Chemical Society, New
Orleans, Louisiana, August 30–September 4, 1987."

Bibliography: p.
Includes indexes.

1. Plastics—Additives—Congresses. 2. Rubber—
Congresses.

I. Riew, C. Keith. II. American Chemical Society.
Division of Polymeric Materials: Science and
Engineering. III. American Chemical Society.
Meeting (194th: 1987: New Orleans, La.). IV. Series.

QD1.A355   no. 222   [TP1142]   540s   89–14938
ISBN 0-8412-1488-3   [668.4′22]

Copyright © 1989

American Chemical Society

PRINTED IN THE UNITED STATES OF AMERICA

Advances in Chemistry Series

M. Joan Comstock, *Series Editor*

1989 ACS Books Advisory Board

FOREWORD

The ADVANCES IN CHEMISTRY SERIES was founded in 1949 by the American Chemical Society as an outlet for symposia and collections of data in special areas of topical interest that could not be accommodated in the Society's journals. It provides a medium for symposia that would otherwise be fragmented because their papers would be distributed among several journals or not published at all. Papers are reviewed critically according to ACS editorial standards and receive the careful attention and processing characteristic of ACS publications. Volumes in the ADVANCES IN CHEMISTRY SERIES maintain the integrity of the symposia on which they are based; however, verbatim reproductions of previously published papers are not accepted. Papers may include reports of research as well as reviews, because symposia may embrace both types of presentation.

ABOUT THE EDITOR

C. KEITH RIEW received his M.S. and Ph.D. degrees in organic chemistry from Wayne State University, Detroit, Michigan, and his B.S. degree in chemistry from Seoul National University, Seoul, Korea. He is an R&D Fellow at the BF Goodrich Company, Research and Development Center, Brecksville, Ohio. He has served as research scientist or R&D manager for New Products Research and Development, BF Goodrich Company, for more than 21 years.

He has presented more than 20 technical papers and owns more than 20 patents on the synthesis and application of telechelic polymers, including rubber-toughened or rubber-modified thermosets and hydrophilic polymers. His latest research is in the synthesis, characterization, engineering, and performance evaluation of impact modifiers for thermosets and engineering thermoplastics.

His research interests include correlating dynamic thermomechanical properties to the failure mechanisms of impact-modified thermosets and thermoplastics. In 1983 he coorganized the first international symposium on "Rubber-Modified Thermoset Resins" at the 186th National Meeting of the American Chemical Society, Division of Polymeric Materials: Science and Engineering, Washington, DC, August 28–September 2, 1983. In 1984 he coedited *Rubber-Modified Thermoset Resins,* Advances in Chemistry Series, Volume 208, American Chemical Society, Washington, DC.

In 1987 he organized the second international symposium on "Rubber-Toughened Plastics" at the 194th National Meeting of the American Chemical Society, Division of Polymeric Materials: Science and Engineering, New Orleans, Louisiana, August 31–September 4, 1987.

Dedication

A Tribute to John A. Manson
1928–1988

Following his education at McMaster University and the University of Michigan and an industrial position with the Air Reduction Company, John Manson joined the faculty of Lehigh University in 1967. There he enjoyed an incredibly rich technical career. He authored more than 160 scientific papers in diverse fields that involved interactions with faculty from four different academic departments and two research centers at Lehigh University.

In 1983 he was corecipient of Lehigh University's Libsch Award for outstanding research achievement. His interactions with international colleagues from England, France, Italy, Columbia, Czechoslovakia, and his homeland, Canada, are truly impressive. John Manson coauthored two books, several critical reviews, and numerous encyclopedia articles that represent seminal contributions to the fields of polymer composites, acid–base interactions, polymer fatigue, polymer blends and composites, and polymer concrete.

A day with John Manson may be likened to a stroll through a grove of giant redwoods. The air was always fresh and clear from his extraordinary intellectual honesty and respect for the views of his students and colleagues. His logic and thought processes were as strong and true as the trunks of these great specimens of nature. John's tree will grow no more, but its growth rings will bear permanent witness to his many contributions to so many of us. It will serve us well to periodically review his record as we rededicate our efforts to train the minds of our students and push back the frontiers of the unknown.

R.W. Hertzberg
Department of Materials Science and Engineering
Lehigh University
Bethlehem, PA 18015-3195

ix

Professor Manson fought hard against his cancer. He was writing an article entitled "Fatigue Crack Propagation in Rubber-Modified Epoxies and Polyacetal" for this book when he passed away. I truly regret that I could not include Professor Manson's chapter in this book.

C. Keith Riew

CONTENTS

CHEMISTRY AND PHYSICS

PREFACE

T HE DEVELOPMENT OF RUBBER-TOUGHENED THERMOSET and thermo-plastic resins is an important contribution to the commercial polymer industry. A small amount of discrete rubber particles in a glassy plastic can greatly improve the crack and impact resistance of normally brittle plastic. Generally, this improvement should be accomplished without significant deterioration of the thermomechanical properties inherent to the unmodified original polymer.

Unlike the modification of thermoplastics, which often requires only the simple physical blending of a particular elastomeric modifier, the modification of thermoset resins is achieved almost exclusively through chemistry and is a challenge to the expertise of the polymer chemist. A useful final product with improved durability and applicability is a major concern of materials scientists and engineers who are responsible for answering this challenge.

Rubber-toughened plastics can be used as they are or as matrix resins for composites to produce structural parts. After years of predicting that high-performance plastics materials would become big business, most scientists believe the time has finally come. Part of this new demand for plastics comes from the aerospace and automobile industries, which need light-weight, high-temperature resistant, and most importantly, high-impact and crack-resistant materials for many types of applications.

The toughening of plastics has been pursued by scientists from many disciplines, including polymer chemistry, physics, rheology, engineering, and materials science. Both science and technology can be applied to create products that will be acceptable to various industrial and consumer groups because of greatly improved performance and simplified design, combined with reduced cost. The field has become important not only for specialists involved in industry, but also for those scientists who discover and develop principles of materials science and technology in the universities. These researchers are focusing on quantitative mecha-nisms of toughening to predict product life expectancy. This endeavor, in turn, should enable us to develop new polymers or to tailor existing polymers to meet new requirements.

The 1983 symposium on "Rubber-Modified Thermoset Resins" covered only thermosetting plastics. The 1988 symposium on "Rubber-Toughened Plastics" covered both thermosets and thermoplastics, and thus appealed to a broader audience. The major themes of this book, which is based on the 1988 symposium, are fracture mechanics and failure mechanisms versus toughness of plastics, morphology and mechanical properties versus toughness of plastics, and chemistry and physics of toughening. Our goal is to present reviews and reports by world-famous scientists on the latest advances in toughening science and technology, development of theoretical and practical approaches to toughness improvements for both thermosets and thermoplastics, and improvement of fracture toughness (i.e., crack and impact resistance of plastic parts, composites, adhesives, and coating materials).

I am indebted to each contributor to the book. I wish to thank the Division of Polymeric Materials: Science and Engineering of the American Chemical Society for sponsoring the symposium and this book. I also acknowledge the Division and the Petroleum Research Fund of the American Chemical Society for providing grants for the support of participating foreign speakers from nonprofit organizations. The BF Goodrich Company also provided administrative support. Finally, I thank Mary Ellen Stoll for her secretarial contribution and my wife, Hyunsoo Kim, for all her other help.

C. Keith Riew
BF Goodrich Company
Research and Development Center
Brecksville, OH 44141

February 1988

DEVELOPMENT OF RUBBER-TOUGHENED PLASTICS

Origin and Early Development of Rubber-Toughened Plastics

Raymond B. Seymour

Department of Polymer Science, University of Southern Mississippi, Hattiesburg, MS 39406–0076

The use of brittle polymers, such as polyvinyl chloride (PVC) and polystyrene (PS), was limited prior to the development of rubber-toughened polymers in the 1930s and 1940s. PVC has been toughened by the addition of small amounts of acrylonitrile rubber (NBR) and other elastomeric materials. Over 1.5 million tons of rubber-toughened PVC is produced annually in the United States. The preferred toughening additive for PS is a styrene–butadiene rubber (SBR). Copolymers of styrene, such as styrene–acrylonitrile and styrene–maleic anhydride, have been toughened by the addition of NBR. The low-temperature resistance of these composites has been improved by grafting the polymers onto the elastomer. PS is also toughened by the addition of styrene–butadiene block copolymers. The art of toughening plastics by the addition of elastomers has been extended to include acrylics, polyolefins, polyesters, acetals, nylon, and thermosets. The impact resistance of epoxy resins has been enhanced by the addition of acrylonitrile–butadiene elastomeric copolymers.

NATURE WAS PARTICULARLY GENEROUS in supplying *Hevea* elastomers, tough balata and gutta percha, and moderately tough shellac, as well as strong collagen, silk, cotton, and wood. However, the first synthetic polymers were brittle and certainly could not be characterized as tough. Toughness may be defined as resistance to impact (i.e., low-impact polymers have Izod impact strengths greater than 105 J/m).

The brittleness of oleoresinous polymers was masked by using them as thin films on solid substrates. The brittleness of ebonite, which was produced by the reaction of a large amount of sulfur with *Hevea braziliensis*, was reduced by decreasing the amount of sulfur reactant and thus decreasing the cross-linked density and permitting elasticity in the principal sections between the cross-links.

Synthetic Polymers

Cellulose nitrate, which was produced by Schönbein (*1*) in the 1840s, was used as a film by Parkes (*2*) in the 1850s. This product could not be classified as a plastic, which, by definition, must be capable of being made into articles by molding or forming. This intractable macromolecule was converted to a moldable plastic by the Hyatt brothers (*3*), who added camphor to serve as a flexibilizing agent or plasticizer.

Casein plastics (Galalith), which were hardened by condensation of the casein with formaldehyde, were brittle but became more ductile under high humidity. Glyptal polyester resins were produced by W. Smith (*4*), who condensed glycerol and phthalic anhydride. Although these polymers were brittle, they could be used as coatings on rigid surfaces. Tougher polyester resins were obtained by Kienle and Hovey (*5*), who incorporated unsaturated vegetable oils in glyptal resins and produced alkyd resins.

Polyvinyl Chloride

Both polystyrene (PS) (*6*) and polyvinyl chloride (PVC) (*7*) were produced in the laboratory in the 1830s, but neither was manufactured commercially until a century later. PVC was a brittle polymer that, when heated at processing temperatures, decomposed to produce HCl and a colored diene polymer. Hence, it is not surprising that Klatte, who patented the production of vinyl chloride in 1917 (*8, 9*), stated in 1930 (*10*) that "no hopes are being considered for the technical development of PVC."

Nevertheless, industrial chemists who had acquired some knowledge of polymer structure made modifications that ensured commercialization of vinyl chloride polymers, which are often called by the general term "vinyls". Clarke (*11*) of Eastman patented diphenyl phthalate plasticizers, Wilkie (*12*) of U.S. Industrial Alcohols patented diethyl phthalate plasticizers, and chemists (*13, 14*) at du Pont patented dibutyl phthalate as a plasticizer for PVC. Brous and Semon (*15*) of Goodrich and Wick (*16*) of I. G. Farben Industries are credited with the commercialization of plasticized PVC, which Goodrich called Koroseal.

Reid (*17*) of Carbide and Carbon Chemicals, Voss and Dickhauser (*18*) of I. G. Farben Industries, Lawson (*19*) of du Pont, and Seymour (*20*) of Goodyear recognized that copolymerization would reduce the intractibility of PVC. They produced commercial copolymers of vinyl chloride with vinyl acetate (Vinylite), vinyl alkyl ethers (Igelite), and vinylidene chloride (Pliovic).

During the 1940s, while chemists in the American PVC industry continued to emphasize the use of plasticizers for toughening PVC, German chemists (10) directed their efforts toward the development of improved heat stabilizers and processing aids for rigid PVC. Although this rigid heat-stabilized PVC was brittle, it was thermally formable, heat weldable, and resistant to corrosives. Hence, it was acceptable as a construction material and is still being used for sheet and pipe.

This rigid PVC was also toughened by the incorporation of acrylonitrile–butadiene copolymer elastomers (NBR) (21). However, this improvement in toughness was noted only when the difference in solubility parameters (δ) of the PVC (δ 19.4 MPa$^{1/2}$) and the additive was 0.4–0.8 MPa$^{1/2}$ (22). Thus, when added to PVC at a concentration of 5 parts per hundred, the following notched Izod impact values, in foot pounds per inch notch, were obtained: butadiene diethyl fumarate (δ 17.2 MPa$^{1/2}$, impact 15.0), butadiene–methylisopropenyl ketone (δ 17.4 MPa$^{1/2}$, impact 15.9), and NBR (δ 18.7 MPa$^{1/2}$, impact 17.4).

A toughened PVC called Lucovyl–H4010 was produced and patented in 1959 (23). This commercial product was obtained by the copolymerization of butadiene and acrylonitrile in a PVC latex. With a refractive index similar to PVC, this product was readily dispersed in PVC to provide a two-phase system with very small rubber particles (0.1 μm).

Nakamura (24) toughened PVC by blending with high-density polyethylene (HDPE), and this effect was enhanced by cross-linking with the addition of peroxy compounds. He was able to reduce the brittle temperature of PVC by as much as 50 °C by cross-linking with 5–15 phr (parts per hundred parts of resin) of HDPE.

However, these elastomeric impact modifiers have been replaced, to a large extent, by acrylonitrile–butadiene–styrene (ABS) graft copolymers, methacrylate–butadiene–styrene (MBS) terpolymers, chlorinated polyethylene (CPE) (Hostalite Z), polyacrylates (Kydex), polyurethanes (Nythene), and ethylene–vinyl acetate (EVA)–PVC graft copolymers (Sumigraft) (24–26).

The events leading up to the commercial use of acrylics as impact modifiers in PVC have been reported (27). The difference in solubility between PVC and MBS (methacrylate–butadiene–styrene copolymers) is sufficient to ensure a two-phase system. However, because of somewhat similar indices of refraction, clear PVC products can be produced.

NBR–PVC blends are available from Alpha, Goodrich, and Uniroyal (28). Improved compatibility has been obtained by reducing the molecular weight of NBR and by the addition of compatibilizing agents (29). Reviews on rubber-toughened PVC are available (30, 31). Over 3.5 million tons of PVC is produced annually in the United States, and over 50% of this production is rubber-toughened PVC.

Polystyrene

Ostromislensky (32), while employed by Naugatuck Chemical Company, patented a process for making rubber-toughened polystyrene by polymerizing a solution of *Hevea* rubber in styrene monomer. Amos reviewed the Dow development of high-impact polystyrene (HIPS) (33), which was marketed in 1948 (34). Seymour (35) produced HIPS in the early 1940s, and the patent for this pioneer rubber-toughened polystyrene was assigned to Monsanto in 1951 (35). However, Monsanto did not market this product until some years later.

Boyer produced HIPS, which was called Styralloy, by blending emulsions of styrene–butadiene rubber (SBR) and polystyrene (36). The Monsanto process was similar, except that different ratios of butadiene and styrene were blended with polystyrene. Hayward of Dow also produced HIPS by the polymerization of a solution of SBR and styrene in a 10-gal can. The addition of soybean oil reduced the production of incompatible domains that result from cross-linked polymer in this HIPS (37–40).

Acrylonitrile–Butadiene–Styrene

Daly (41) produced acrylonitrile–butadiene–styrene (ABS) (Kralastic) compositions in 1952 by melt-blending copolymers of styrene and acrylonitrile (SAN) with NBR.

ABS was produced by solution polymerization, in which the SAN was grafted in a shell-like manner to the NBR of polybutadiene (PB). ABS has also been produced by blending emulsion copolymers of SAN and NBR and by emulsion polymerization of the three monomers. This process generates a product with higher impact resistance than the solution process. Much of the ABS used today is a graft copolymer that Childers (42) produced by emulsion copolymerization techniques in 1958. Some ABS is also produced by a suspension polymerization process (43).

ABS, under the trade names of Marbon and Cycolac, has been marketed by Borg Warner since 1953 (44) and by General Electric since 1988. Dow produced ABS with low acrylonitrile (AN) content, which was called the "poor man's ABS", in the 1960s (36). Davenport also produced ABS in 1939 (45), but basic patents were issued to Calvert in 1959 and 1966 (46).

Polybutadiene and ethylene–propylene–diene–monomer rubber (EPDM) are also used as additives in place of NBR. However, because unsaturated ABS polymers are not weather resistant, acrylic elastomers and chlorinated polyethylene have been used in place of NBR. These products, called acrylic–styrene–acrylonitrile (ASA) (Geloy) (47) and acrylonitrile–chlorinated polyethylene–styrene (ACS) (48) are produced in much smaller quantities than ABS, which has an annual production of 600,000 tons in the United States. Clear ABS has been produced by the copolymerization of methyl methacrylate with other monomers used in ABS (*49*). ABS, with improved resistance to elevated temperatures, has been obtained by replacing the styrene monomer with α-methylstyrene (*50*) and by blending with polycarbonate.

The toughening of acrylonitrile copolymers by elastomers has been extended to copolymers of styrene, acrylonitrile, and maleic anhydride (Cadon) (*51–54*). The resistance of ABS-type plastics to weathering has been improved by substituting olefinic elastomers (OSA) (*55*). Styrene–butadiene–styrene (SBS) block copolymers and ABS terpolymers (Blendex) have also been used to toughen several other polymers, such as PVC and polycarbonate (PC) (*56*).

Styrene–Diene Block Copolymers

The original latex blends of PS and elastomers were simple mixtures, but HIPS produced by polymerization of styrene, in which the elastomer is dissolved, is actually a graft copolymer. Hence, after Szwarc's development in anionic copolymerization (*57*), it appeared logical to produce blocks of styrene and dienes.

Diblock, radial-branched copolymers of styrene and butadiene (SB) (Solprene, K-resins) were produced by Phillips (*58*); triblock copolymers of styrene–isoprene and styrene (SIS) (Kraton) were introduced by Shell in the 1960s (*59, 60*). The Shell product made from butadiene is harder than that made from isoprene block copolymers (*61, 62*). Both types of Kraton were hydrogenated in 1972 to produce more weather-resistant polymers. The properties of these block copolymers may be modified by the addition of elastomers, such as ethylene–propylene copolymers (EPDM).

Acrylics

The acrylic impact modifiers used for PVC have been modified to ensure matching indices of refraction and used for rubber-toughened polymethyl methacrylate (PMMA) (*63*). About 75,000 tons of rubber-toughened acrylic plastics is used annually in the United States.

Polyolefins

Elastomeric polyolefins, such as polyisobutylene, butyl rubber, and ethylene–propylene copolymers (EPM, EPDM), have been used as elastomeric components of many rubber-toughened plastics (*64*). A polyolefin thermoplastic elastomeric blend (PTR–1000) was marketed by Uniroyal in 1972 (*65*). Allied Chemical produced a butyl–rubber grafted polyethylene (ET Polymer, Paxon) in 1970 (*66*). In 1974, competitive PTRs were introduced by du Pont, Goodrich, Hercules, and Exxon under the trade names of Somel,

Telcar, and Profax (67). These elastomers have also been used to improve the properties of other polyolefins.

The toughness of polypropylene has been enhanced by block copolymerization with ethylene by using Ziegler-type catalysts, such as γ-titanium trichloride and diethylaluminum chloride in heptane, with hydrogen as a chain-transfer agent (68). Optimum toughness is observed when the ethylene–propylene block is a random elastic copolymer (69).

Polyesters

The first all-synthetic fiber was a flexible aliphatic polyester produced by Carothers and Hill (70) in 1932. The stiffer aromatic polyester fiber (PET) (Terylene), produced by Whinfield and Dickson (71) in 1940, was suitable for textiles but extremely difficult to mold. Polybutylene terephthalate (PBT), which is more readily moldable, was produced commercially for use as a high-performance polymer by Celanese (Celanex) in 1968 (72).

The 300,000 tons of PET used for blow-molding soft drink bottles is the largest nonfibrous use of this polyester. However, over 60,000 tons of PET and PBT are used annually as engineering plastics. Rubber-toughened PET (Rynite) was produced by Dreyrup of du Pont in 1978, Celanese (Grafite, Gaftuf, Duraloy), Mobag (Pocan), GE (Valox), and BASF (Ultradur) (73).

Aromatic polymers of carbonic acid (Lexan, Merlon) were synthesized by Schnell (74) and Fox (75) of Bayer and G.E., respectively, in the mid-1950s. These clear amorphous thermoplastics were blended with PBT and elastomers (Xenoy), with polyurethanes (Texin) and silicones.

Block copolymers of stiff and flexible polyester (Hytrel), which were introduced by du Pont in 1972, are classified as thermoplastic elastomers (76).

Acetals

Staudinger et al. (77) produced polymers of formaldehyde (POM, Acetal) in 1859 and 1922, respectively. However, these thermally stable acetal polymers (Delrin) were not commercially available until the late 1950s (78). Rubber-toughened acetal polymers with notched Izod impact-resistance values as high as 1000 J/m (18.7 ft lb per in. of notch) were introduced by du Pont (79), Celanese (Duraloy), and BASF (Ultraform) in the early 1980s.

This improved toughness, which is coupled with improved resistance to crack propagation, may be related to nonuniform crack failure (80).

Nylons

Nylon-66, which was synthesized by Carothers (70) in the mid-1930s, was the first high-performance polymer. It is now used at an annual rate of

250,000 tons by the American plastics industry. A rubber-toughened "super-tough nylon" was introduced commercially by du Pont in 1976 (*81*). Nylon-6 has also been toughened by the addition of elastomers (Ultramid, Capron, Nycoa, and Crilon). EPDM has been used to toughen Nylon-66, to produce a plastic with a notched impact resistance of at least 160 J/m.

Improvement in the notched impact resistance of nylon may be related to improved resistance to fatigue in notched specimens. The effect has not been observed in unnotched specimens, in which crack formation is the principal action.

Thermosets

High-impact phenolic molding compounds were produced in the 1940s by adding liquid polysulfides (Thiokol LP-2) or NBR to the novolac prepolymers before advancing and molding (*82*). This toughening technique was extended to epoxy resin formulations by Sultan and McGarry in the mid-1960s (*83*). Goodrich and Goodyear produce specific additives (Hycar, ATBN, and CTBN) for toughening epoxy polymers (*84*). This liquid rubber oligomer, which is added at a concentration of 10 parts per 100 parts of epoxy resin, forms a homogenous solution, but phase separation occurs during the curing step. Toughened plastics are obtained by the use of CTBN with methylol, epoxy, phenol, and carboxyl and terminal groups (*85*). Phase separation of the rubber-toughened prepolymer can be prevented by the addition of ABS (*86*). The rubber particle size and extent of crazing in CTBN-modified epoxy resins may be important (*87*). However, the tensile yield strength is directly proportional to the rate of strain and to the rubber content (*88*).

Improved toughening of epoxy resins has been noted by the use of an excess of amine-curing agent (50%), by the use of chain extenders of bisphenol A, and by the use of high-molecular-weight cross-linking agents such as polyoxypropyleneamines (*89*).

The CTBN–epoxy prepolymer mixture is homogeneous, but CTBN precipitates as a discrete phase (*90*) as the resin is cured. Moderately active curing agents (such as piperidine) provide better phase separation than more active agents (such as triethylenetetramine) (*85*). Toughness is also enhanced by the addition of bisphenol A (*90*).

Mechanism of Toughening of Plastics

Researchers who investigated the stress–strain behavior of HIPS in the early 1940s observed that, unlike polystyrene, which exhibited a brittle fracture with a 1–2% elongation at the yield point, HIPS elongated as much as 40% before breaking. HIPS and other rubber-toughened plastics undergo stress whitening, and this energy-absorbing process contributes to an improvement in impact strength as the temperature is increased (*91*).

There is an optimum size for the discrete rubber particles in each system. The toughening effect is related to the size of these particles (92), which are effective at temperatures above their glass-transition temperatures (T_g). The rubber particles must be at least as large as the cracks, 300–500 nm (93).

In 1950 Burchdahl (94) observed prominent secondary loss peaks of mechanically dampened HIPS below room temperature. Mertz (95) showed in 1956 that the impact energy absorbed by HIPS was the sum of the energy required to fracture the PS and the rubbery particles. Mertz also postulated that stress whitening was caused by scattered light from the microcracks and that void formation, which resulted from opening of the cracks, permitted large strain.

Bucknall (96) upgraded the Mertz microcrack concept in 1965 by using optical microscopy to relate stress whitening to multiple crazes that preceded the cracks. By subsequent investigations with electron microscopy, Matsuo (97) showed that the spherical rubber particles in ABS became spheroidal as branched crazes formed in the plastic matrix.

Newman (98) used optical microscopy in 1965 to show that toughening could be related to shear yielding in the matrix. Subsequently, Bucknall (99) showed that crazing and shear yielding could occur simultaneously in rubber-toughened plastics.

T_g, morphology, interfacial adhesion, particle size, and volume fraction of the elastomeric components affect mechanical properties of the resinous composite. The maximum increase in toughness is noted when the T_g of the elastomer is above that of the temperature used, but the exact effect of morphology of the particles is questionable. Although good interfacial adhesion is essential, the need for specific particle diameter size and particle distribution has been questioned. However, there is general agreement that toughness is a function of the volume fraction of the elastomer.

Bucknall (100) has used the Eyring model for flow (101) to relate the kinetics of deformation in rubber-toughened plastics to multiple crazing and shear yielding in the vicinity of the crack tip. Li (102) and Bucknall (103–106) have provided additional information on these failure phenomena.

Conclusions

The toughening of brittle general-purpose plastics, such as polystyrene and polyvinyl chloride, has contributed considerably to the growth of the plastics industry. More important, the toughening of high-performance plastics, such as acetals, aromatic polyesters, and nylon, has provided second-generation materials that can outperform most classic construction materials. With continued advances in the understanding of the science of toughening, new toughening materials should be introduced that will permit the solution of many heretofore unsolvable problems and ensure the continued growth of the modern plastics industry.

References

1. Schönbein, C. F. *Philos. Mag.* **1847,** *31,* 7.
2. Parkes, A. *J. Soc. Arts* **1865,** *14,* 81.
3. Hyatt, V. W.; Hyatt, I. V. U.S. Patent 105 338, 1870.
4. Smith, W. *J. Soc. Chem. Ind. London* **1901,** *20,* 1075.
5. Kienle, R. H.; Hovey, A. B. *J. Chem. Soc.* **1929,** *51,* 509.
6. Simon, E. *Annalen* **1839,** *31,* 265.
7. Regnault, V. *Annalen* **1835,** *115,* 63.
8. Klatte, F. Ger. Patent 281 877, 1913.
9. Klatte, F. U.S. Patent 1 241 738, 1917.
10. Kauffman, M. *The Chemistry and Industrial Production of Polyvinyl Chloride;* Gordon & Breach: New York, 1969.
11. Clarke, H. T. U.S. Patent 1 398 939, 1920.
12. Wilkie, H. F. U.S. Patent 1 449 159, 1922.
13. Lawson, W. E. Br. Patent 379 292, 1930.
14. Wickert, J. N. Br. Patent 408 969, 1931.
15. Brous, S. L.; Semon, W. *Ind. Eng. Chem.* **1935,** *27,* 667.
16. Wick, G. U.S. Patent 2 191 056, 1935.
17. Reid, E. W. U.S. Patent 1 935 517, 1928.
18. Voss, A.; Dickhauser, E. U.S. Patent 2 012 177, 1928.
19. Lawson, W. E. Br. Patent 319 588, 1928.
20. Seymour, R. B. U.S. Patent 2 328 743, 1943.
21. Kulich, D. M.; Kelley, P. D.; Pace, J. E. *Encyclopedia of Polymer Science and Technology;* Wiley: New York, 1985; Vol. 1, p 338.
22. Seymour, R. B.; Stahl, G. A. *Macromolecular Solutions;* Pergamon: Elmsford, NY, 1985.
23. Semon, W. A.; Stahl, G. A. In *History of Polymer Science and Technology;* Seymour, R. B., Ed.; Dekker: New York, 1982.
24. Nakamura, Y. *J. Polym. Sci., Polym. Chem. Ed.* **1978,** *16,* 1981, 2055.
25. Riddle, E. H. *Monomeric Acrylic Esters;* Van Nostrand Reinhold: New York, 1954.
26. Jennings, G. B. U.S. Patent 2 646 417, 1953.
27. Dunkelberger, D. L. In *Polymeric Composites: Their Origin and Development;* Seymour, R. B.; Deanin, R. D., Eds.; VNU Science: Ultrecht, Holland, 1987.
28. Hopkins, H. P. *SPE J.* **1960,** *16,* 304.
29. Giles, H. F. *Mod. Plast.* **1986,** *63,* 105.
30. Bucknall, C. B. *Toughened Plastics;* Applied Science: London, 1977.
31. Carlson, A. W.; Jones, T. A.; Martin, J. L. *Mod. Plast.* **1957,** *44,* 155.
32. Phillips, R.; Exler, D. S.; Heiberger, C. A. *SPE J.* **1965,** *21,* 135.
33. Ostromislensky, I. I. U.S. Patent 1 613 673, 1927.
34. Amos, J. L. *Polym. Eng. Sci.* **1974,** *14,* 1.
35. Seymour, R. B. U.S. Patent 2 574 438, 1952.
36. Boyer, R. In *History of Polymer Science and Technology;* Seymour, R. B., Ed.; Dekker: New York, 1982; Chapter 19.
37. Hayward, R. M.; Elly, J. U.S. Patent 2 668 806, 1954.
38. Morris, E. D.; Griess, G. A. U.S. Patent 2 606 103, 1952.
39. Amos, J. L.; McIntire, O. R.; McCurdy, J. L. U.S. Patent 2 644 692, 1954.
40. Stein, A.; Walter, R. L. U.S. Patent 2 862 909, 1958.
41. Daly, L. E. U.S. Patent 2 439 202, 1948.
42. Childers, C. W.; Fisk, C. F. U.S. Patent 2 820 773, 1958.
43. Lee, L. H. U.S. Patent 3 238 275, 1966.
44. Zahn, E. *Appl. Polym. Symp.* **1969,** *11,* 219.

45. Davenport, M. E.; Hubbard, L. W.; Pettit, M. R. Br. Patent 32 549, 1949.
46. Calvert, W. C. U.S. Patent 2 908 991, 1959; 3 238 275, 1966.
47. Ogawa, M.; Takezoe, S. *Jpn. Plast. Age* **1973**, *11*, 39.
48. Pavelich, W. A. In *High Performance Polymers: Their Origin and Development;* Seymour, R. B.; Kirschenbaum, G. S., Eds.; Elsevier: New York, 1986; Chapter 13.
49. Frazier, W. *Chem. Ind.* **1966, 1399.**
50. Basdekis, C. H. *ABS;* Van Nostrand Reinhold: New York, 1964.
51. Seymour, R. B.; Kispersky, J. P. U.S. Patent 2 439 277, 1948.
52. Lee, Y. C.; Trementozzi, G. A. U.S. Patent 4 2411 695, 1982.
53. ARCO U.S. Patent 4 325 037, 4 336 345, 1982.
54. Hall, W. J.; Kruse, R. L.; Mendelson, R. A.; Trementozzi, Q. A. In *The Effects of Hostile Environments on Coatings and Plastics;* Garner, D. P.; Stahl, G. A., Eds.; ACS Symposium Series 229; American Chemical Society: Washington, DC, 1983; Chapter 5.
55. Ziska, J. L. *Mod. Plast.* **1986**, *63*, 8.
56. Grawboski, T. S.; Irvin, H. H. U.S. Patent 3 953 800, 1962; 3 130 177, 1964.
57. Szwarc, M.; Levy, M.; Mulkovich, R. *J. Am. Chem. Soc.* **1956**, *78*, 2656.
58. Phillips Chemical Co. Technical Bulletin 19430, 1980.
59. Haws, J. R.; Wright, R. E. In *Handbook of Thermoplastic Elastomers;* Balker, B. M., Ed.; Van Nostrand Reinhold: New York, 1979; Chapter 3.
60. Porter, L. M. U.S. Patent 3 149 182, 1964.
61. Holden, G.; Bishop, E. T.; Legge, N. R. *J. Polym. Sci. Part C* **1969**, *26*, 37.
62. Crouch, W. W.; Short, J. N. *Rubber Plast. Age* **1961**, *42*, 276.
63. Bauer, R. G.; Pierson, R. M.; Mast, W. C.; Bletso, N. C.; Shepherd, L. In *Multicomponent Polymer Systems;* Advances in Chemistry 99; American Chemical Society: Washington, DC, 1971; p 251.
64. Tornqvist, G. M. In *History of Polyolefins;* Seymour, R. B.; Cheng, T., Eds.; Reidel: Dordrecht, Holland, 1986; Chapter 9.
65. Morris, H. L. In *Handbook of Thermoplastic Elastomers;* Balker, B. M., Ed.; Van Nostrand Reinhold: New York, 1979; Chapter 2.
66. Severini, F.; Mariani, E.; Pagliari, A.; Cerri, E. In *Multicomponent Polymer Systems;* Advances in Chemistry 99; American Chemical Society: Washington, DC, 1971; p 260.
67. Heegs, T. G. In *Block Copolymers;* Allport, D. C.; James, W. H., Eds.; Applied Science: London, 1973; Chapter 8.
68. Baldwin, F. P.; Ver Strate, G. *Rubber Chem. Technol.* **1972**, *45*, 709.
69. Han, C. D.; Kin, Y. W.; Chen, S. V. *J. Appl. Polym. Sci.* **1975**, *19*, 284.
70. Carothers, W. H.; Hill, J. W. *J. Am Chem. Soc.* **1932**, *54*, 1579.
71. Whinfield, J. R.; Dickson, J. T. Br. Patent 578,079, 1941.
72. Seymour, R. B. *Polymers for Engineering Applications;* American Society for Metals: Metals Park, OH, 1987.
73. Dreyrup, E. J. U.S. Patent 2 015 013 4, 1978.
74. Schnell, H. *Angew Chem.* **1956**, *68*, 633.
75. Fox, D.; Christopher, W. *Polycarbonates;* Van Nostrand Reinhold: New York, 1964.
76. Dreyrup, E. J. In *High Performance Polymers: Their Origin and Development;* Seymour, R. B.; Kirschenbaum, G. S., Eds.; Elsevier: New York, 1986.
77. Staudinger, H.; Staudinger, M.; Sauter, E. *Z. Phys. Chem.* **1937**, *B33*, 403.
78. MacDonald, R. N. U.S. Patent 2 768 994, 1956.
79. Walling, C.; Brown, F.; Bartz, K. U.S. Patent 3 027 352, 1961.
80. Flexman, E. A. *Mod. Plast.* **1985**, *62*, 72.

81. Kohan, M. I. In *High Performance Polymers: Their Origin and Development;* Seymour, R. B.; Kirschenbaum, G. S., Eds.; Elsevier: New York, 1986; Chapter 4.
82. Kinloch, A. J.; Shaw, A. J.; Tod, D. A. In *Rubber-Modified Thermoset Resins;* Riew, C. K.; Gillham, J. K., Eds.; Advances in Chemistry 208; American Chemical Society: Washington, DC, 1984; p 101.
83. Sultan, J. N.; McGarry, F. J. *Polym. Eng. Sci* **1973,** *13,* 29.
84. *Toughness and Brittleness of Plastics;* Deanin, R. D.; Crugnola, A. M., Eds.; Advances in Chemistry 154; American Chemical Society: Washington, DC, 1976.
85. Rowe, E. H.; Siebert, G. R.; Drake, R. S. *Mod. Plast.* **1978,** *49,* 110.
86. Batzer, H.; Zaher, S. A. *J. Appl. Polym. Sci.* **1975,** *19,* 585.
87. Manzione, L. T.; Gillham, J. K.; McPherson, C. A. *J. Appl. Polym. Sci.* **1981,** *26,* 889.
88. Beck, R. H.; Gratch, S.; Newman, S.; Rusch, K. C. *J. Polym. Sci., Polym. Lett. Ed.* **1968,** *6,* 707.
89. Morgan, J. R.; King, F. M.; Walkup, C. M. *Polymer* **1984,** *24,* 375.
90. Meeks, A. C. *Polymer* **1974,** *15,* 675.
91. Manson, J. A.; Sperling, L. *Polymer Blends and Composites;* Plenum New York, 1983.
92. Riew, C. K.; Rowe, E. H.; Siebert, A. R. *Org. Coat. Plast. Chem.* **1974,** *34,* 353.
93. Parsons, C. F.; Suck, E. L., Jr. In *Multicomponent Polymer Systems;* Advances in Chemistry 99; American Chemical Society: Washington, DC, 1971; p 340.
94. Burchdahl, R.; Nielsen, L. E. *J. Appl. Phys.* **1950,** *21,* 482.
95. Mertz, E. H.; Claver, G. C.; Baer, M. *J. Polym. Sci.* **1950,** *21,* 482.
96. Bucknall, C. B.; Smith, R. R. *Polymer* **1965,** *6,* 437.
97. Matsuo, M. *Polymer* **1966,** *7,* 421.
98. Newman, S.; Strella, S. *J. Appl. Polym. Sci.* **1965,** *9,* 2297.
99. Bucknall, C. B.; Glayton, D.; Keast, W. E. *J. Mater. Sci.* **1972,** *7,* 1443.
100. Bucknall, C. B. *Rubber Chem. Technol.* **1987,** *60,* 35.
101. Eyring, H. *J. Chem. Phys.* **1936,** *4,* 283.
102. Li, J. C. M.; Pampillo, C. A.; Davis, L. A. In *Deformation and Fracture of High Polymers;* Kausch, H. H.; Hassell, J. A.; Haffee, R. I., Eds.; Plenum: New York, 1973.
103. Bucknall, C. B.; Drinkwater, I. C. *J. Mater. Sci.* **1973,** *8,* 1800.
104. Bucknall, C. B.; Page, C. J.; Young, V. O. In *Toughness and Brittleness of Plastics;* Deanin, R. D.; Crugnola, A. M., Eds.; Advances in Chemistry 154; American Chemical Society: Washington, DC, 1976; p 179.
105. Bucknall, C. B.; Partridge, I. K.; Ward, M. W. *J. Mater. Sci.* **1983,** *19,* 2064.
106. Bucknall, C. B.; Page, C. J. *J. Mater. Sci.* **1982,** *17,* 808.

RECEIVED for review February 11, 1988. ACCEPTED revised manuscript September 6, 1988.

Rubber-Toughened Styrene Polymers

A Review

Adolf Echte

Kunststofflaboratorium, BASF Aktiengesellschaft, D–6700 Ludwigshafen, Federal Republic of Germany

This chapter surveys the possible ways to influence the molecular and morphological structures of rubber-toughened styrene polymers in industrial production processes. Two important types of production processes are used in industry. The first is polymerization of styrene monomer or its mixture with other monomers in the presence of dissolved rubber. The second is a graft polymerization of the monomer(s) onto preformed rubber in emulsion. In both cases, the structure of the dispersed rubber phase depends to a great extent on the process conditions. Transparent and very tough styrene–butadiene copolymers produced by anionic polymerization have recently become of interest. The most important characteristics of these two-phase or multiphase products are the molecular weight of the matrix; phase–volume ratio; type of particle, particle size, and size distribution; interfacial bonding; and rubber cross-linking density. The proper combination of all these properties has resulted in numerous novelties in high-impact polystyrene and acrylonitrile– butadiene–styrene copolymers, and it will continue to do so in the future.

WHEN THE DEMAND FOR STYRENE MONOMER, acrylonitrile, and styrene–butadiene rubber (SBR) declined drastically in the late 1940s, the rubber modification of polystyrene (PS) and styrene–acrylonitrile copolymers (SAN) became of industrial interest. Since then, the so-called high-impact polystyrene (HIPS) and the rubber-toughened SAN (acrylonitrile–butadiene–styrene, ABS) have gained a broad market. In 1985 the consumption of HIPS passed 3×10^6 tons per year (t/a), and ABS reached 1.5 $\times 10^6$ t/a. Applications for HIPS are mainly packaging (United States, 30%;

Western Europe, 34%), appliances (United States, 8%; Western Europe, 8%), housewares (United States, 5%; Western Europe, 8%), and toys (United States, 6%; Western Europe, 3%); there are also many minor applications. The percentages are related to total polystyrene consumption, including expendable polystyrene (EPS), because it is not possible to differentiate the grades properly (1). ABS has been used mainly in automobile manufacture (United States, 18%; Western Europe, 25%; Japan, 19%), appliances (United States, 17%; Western Europe, 19%; Japan, 29%), electronics (United States, 9%; Western Europe, 17%), and pipes and fittings (United States, 15%; Western Europe, 2%), with a great number of minor applications.

Polystyrene and SAN are transparent and brittle thermoplastic materials. The addition of rubber increases impact strength considerably, as can be seen from the stress–strain curves determined by an instrumental impact pendulum (Figure 1).

SBR was replaced long ago by polybutadiene as a toughening agent in HIPS and ABS. Polybutadiene has a much lower glass-transition temperature (–85 °C) than SBR (–50 °C) (Figure 2) and therefore gives a much better low-temperature impact strength. This strength is required mainly in refrigerators, and to a lesser extent in automotive and some other applications.

HIPS and ABS are two-phase materials. Buchdahl and Nielsen (2) showed this by dynamic mechanical testing of PS–SBR mixtures in 1950. Two phases also emerge clearly from TEM images made from microtome cut films stained by osmium tetraoxide, according to Kato (3).

Such pictures even proved the dispersed phase to be of a two-phase nature, especially for HIPS and ABS made in bulk (Figure 3). "Bulk" as used here means polymerization of styrene in the presence of dissolved rubber. It comprises the continuous mass process, as well as the discontinuous mass suspension process to be discussed later. The particles contain relatively large inclusions of matrix material. Because of their appearance

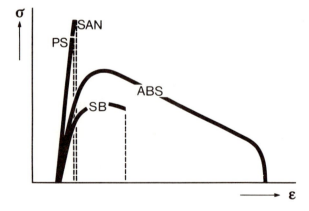

Figure 1. Impact resistance of styrene polymers. SB is styrene–butadiene HIPS.

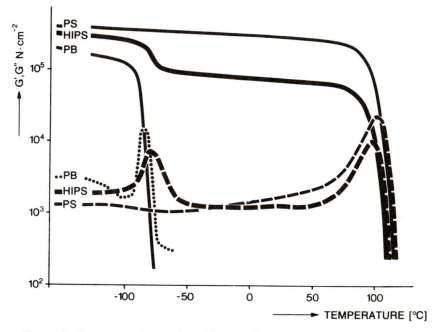

Figure 2. Storage modulus G' and loss modulus G'' of HIPS and its two components.

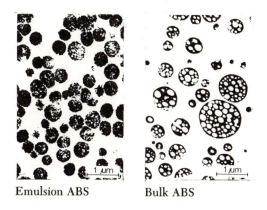

Emulsion ABS   Bulk ABS

Figure 3. Morphology of ABS polymerized in emulsion or in bulk. (Reproduced with permission from ref. 4. Copyright 1979 Verlag Chemie.)

in the images, we term this type of particles "cell particles". The inclusions enlarge the share of the rubbery phase far beyond the calculated rubber content. Therefore, as well as for other reasons, HIPS is made only in bulk.

ABS, however, is produced either in bulk or in emulsion. Most producers prefer the emulsion process to the bulk, generally because the optimum rubber particle size is much smaller and can be adjusted more easily by emulsion polymerization.

In two-phase polymer systems, the technical properties do not depend solely on the properties of the pure components. In a complex way, numerous parameters affect the technical properties of the final products (List 1).

These molecular and morphological parameters can be brought into a scheme that may help the practitioner to choose the kind of operation necessary for running a process according to the specifications of the products (Chart 1).

Technical properties can be influenced during production processes by various suitable measures to be discussed later. In the cases of HIPS and

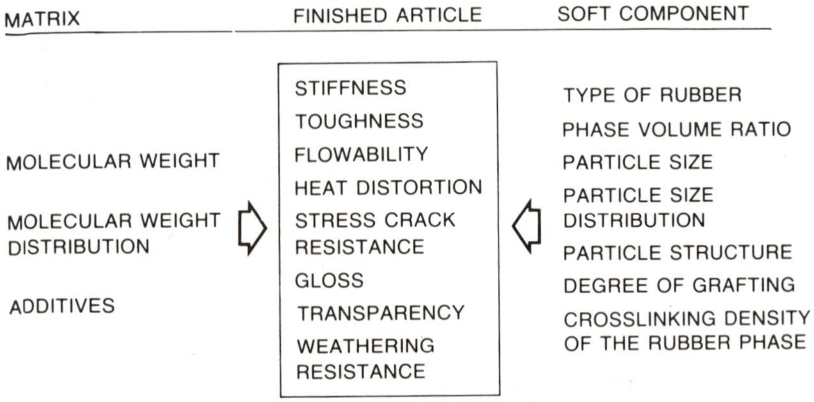

MATRIX	FINISHED ARTICLE	SOFT COMPONENT
MOLECULAR WEIGHT MOLECULAR WEIGHT DISTRIBUTION ADDITIVES	STIFFNESS TOUGHNESS FLOWABILITY HEAT DISTORTION STRESS CRACK RESISTANCE GLOSS TRANSPARENCY WEATHERING RESISTANCE	TYPE OF RUBBER PHASE VOLUME RATIO PARTICLE SIZE PARTICLE SIZE DISTRIBUTION PARTICLE STRUCTURE DEGREE OF GRAFTING CROSSLINKING DENSITY OF THE RUBBER PHASE

List 1. *Molecular and morphological parameters that influence technical properties.*

	Modulus	impact-strength	heat distortion temperat.	solvent crazing resistance	flow properties	Gloss and/or transpar.
Phase Volume Ratio	↘	↗	↘	↗	↘	↘
Particle Size and-Distribution	↘	⌢	—	↗	↗	↘
Degree of Crosslinking	↗	⌢	↗	↘	↗	↗
Matrix Molecular Weight	—	↗	—	↗	↘	—
Lubricant concentration	↘	↗	↘	↘	↗	↗

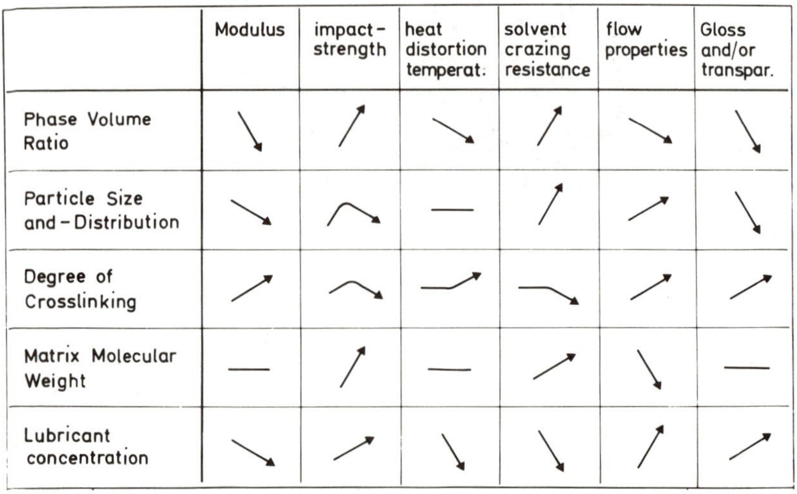

Chart 1. *Functional trends to adjust technical properties of HIPS.*

ABS, a measure that will improve one property affects practically all the others. Therefore, product development for market demand is always a question of optimization. Optimization requires the ability to adjust the different parameters separately, as far as possible without consequences for other parameters.

Production of HIPS or ABS in Bulk

In principle, the bulk process consists of polymerizing styrene (or styrene–acrylonitrile) in the presence of dissolved rubber. Polystyrene is formed in the homogeneous rubber solution. The system then separates into two phases after a few percent of conversion, because the two polymer solutions are incompatible (Figure 4).

The monomer is distributed between the two solvents, rubber and polystyrene. Because the amount of polystyrene increases with conversion and the amount of rubber remains constant, the polystyrene phase increases in volume at the expense of the rubber phase. When the phase–volume ratio approaches unity, phase inversion begins, and the rubber phase is distributed within the surrounding polystyrene phase (6, 7). The viscosity of the system increases in the same way, and the rubber-phase droplets become fixed in size at roughly 30–35% conversion (8). Conversion is then carried to a high level, and the residual monomers and any solvent are removed.

Phase inversion can be observed with a phase-contrast microscope (9) and by viscosity measurements (5, 10) (Figure 5). Phase inversion occurs in

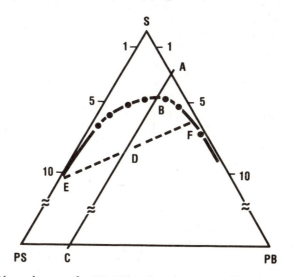

Figure 4. Phase diagram for PS–PB–styrene monomer. Key: A, starting point; B, phase-separation point; C, final composition; D, phase-inversion point; E, composition of the polystyrene phase at phase inversion; F, composition of the polybutadiene phase at phase inversion. (Reproduced with permission from ref. 5. Copyright 1983 Hanser.)

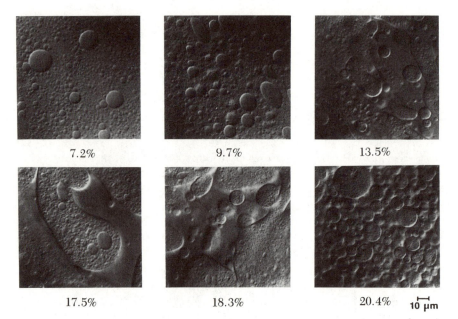

Figure 5. Prepolymerization process. Polymerization of styrene in the presence of 6.4% dissolved polybutadiene rubber. Samples were taken at solids contents indicated.

a rather broad conversion range. Under the microscope, it can be seen that the rubber phase itself is not homogeneous. Phase inversion occurs between a polystyrene solution and a subemulsion of polystyrene droplets in rubber solution. This subemulsion is the origin of the heterogeneous structure of the final rubber particles. Viscosity passes through a maximum–minimum with increasing conversion and shows a relatively broad transition span (Figure 6).

This curvature may be quantitatively approximated by an equation put forward by Freeguard and Karmarkar (11) for the viscosity of a two-phase oil-in-oil emulsion.

$$\log \frac{\eta_E}{\eta_c} = 2.5 \frac{\eta_d - 0.505 \, \eta_c}{\eta_d + \eta_c} (\phi_c + \phi_c^{5/3} + \phi_c^{11/3}) \qquad (1)$$

where η is viscosity, ϕ is volume fraction, E is emulsion, c is continuous phase, and d is dispersed phase. A more recent version was formulated by Song Zhiqiang and co-workers (12). This equation in the original paper was misprinted. It should read as indicated here.

$$\frac{1}{\eta_E} = \frac{A \, (1 - \phi_{PS})}{\eta_R} + \frac{B \, \phi_{PS}}{\eta_{PS}} \qquad (2)$$

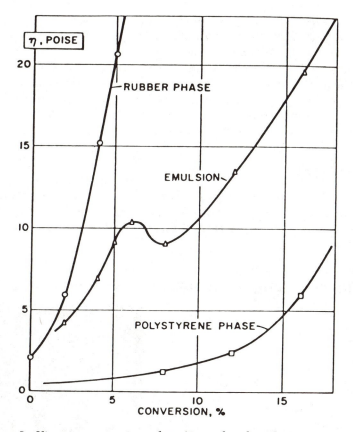

Figure 6. Viscosity–conversion plot. (Reproduced with permission from ref. 10. Copyright 1969 Wiley.)

where R is rubber, PS is polystyrene, and *A* and *B* are constants. Equations 1 and 2 are both derived from the power law for pseudoplastic liquids. In this reaction pathway, the molecular and morphological parameters of the final products must be adjusted to achieve the set of technical properties desired.

Molecular Weight of the Matrix

Matrix molecular weight, which affects mainly mechanics and melt flow, is usually regulated by process conditions (temperature, initiator concentration, and solvent content) and by the addition of a suitable chain-transfer agent. *tert*-Dodecylmercaptan is commonly used for this purpose. Its transfer constant is roughly 4, and it is, therefore, effective for the whole course of the reaction.

Mercaptans with primarily bound thiol groups are even more effective (transfer constant >20), but they exhaust at an early stage of conversion.

Otherwise, chain-transfer agents with transfer constants of about 0.1 (e.g., dimeric α-methylstyrene or terpinolene) become effective only when an advanced stage of conversion is reached.

An increasing molecular weight means a higher melt viscosity. The higher the melt viscosity, the higher the molecular orientation in injection-molded parts. Orientation imposes anisotropic behavior (Figure 7) that may be of greater influence on the properties than the molecular and morphological parameters in the isotropic state.

Type of Rubber

In most cases the rubber type is polybutadiene. Its glass-transition temperature depends on the microstructure (Table I). In practice it is –85 °C, sufficient for good low-temperature impact behavior. Because of its lack of

Orientation

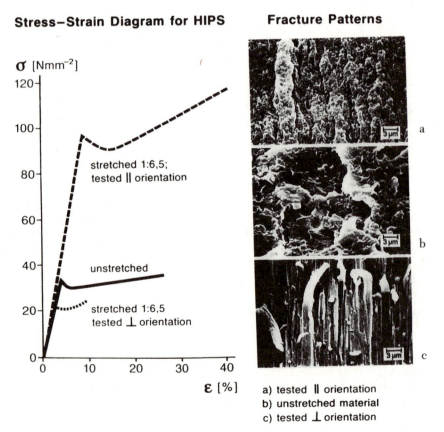

Figure 7. Influence of orientation on mechanical properties.

**Table I. Microstructure and Glass-Transition
Temperature of Polybutadiene**

$$\text{~~CH}_2\diagdown\!\!{}_{\displaystyle C=C}\!\!{}^{\diagup CH_2-CH_2}\diagdown\!\!{}_{C=C}\!\!{}^{\diagup H}\quad {}^{H}\diagdown\!\!{}_{C}\!\!{}^{\diagup CH=CH_2}$$

Structure	Components (%)			T_g (°C)
	1,4-cis	1,4-trans	1,2-Vinyl	
High-*cis*	98	1	1	−105
Medium-*cis*	35	55	10	−85
Emulsion	24	54	22	−75

crystallinity, medium-*cis*-polybutadiene has slight advantages over high-*cis*-polybutadiene. The melting point of high-*cis*-polybutadiene is roughly −40 °C, and this negatively affects the low-temperature impact behavior of the final HIPS.

The solution viscosity of the rubber used is important for the adjustment of the particle size in the reaction course, as will be discussed later. It depends on the molecular weight and on the degree of branching. Most polybutadienes used commercially have molecular weights of between 180,000 and 260,000 (viscosity average). They are usually long-chain-branched to suppress cold flow. Other rubbers occasionally used are SBR and EPDM (ethylene–propylene–diene–monomer rubber), the latter yielding products with improved weatherability (*13*).

Phase–Volume Ratio

Phase–volume ratio is influenced by the rubber content, by the number and size of occlusions, and by the degree of grafting. The early work of Cigna (*14*) indicates that the phase–volume ratio, as well as the rubber content, is decisive for the properties of a rubber-toughened styrene polymer. This finding was later confirmed by Bucknall et al. in a series of papers (*15*).

Grafting at an early stage of polymerization turns the rubber solution in styrene into an emulsion of polystyrene solution in styrene dispersed in the rubber solution. This emulsion is stable enough that the polystyrene solution droplets will coalesce only partially, and the phase inversion takes place between a polystyrene solution and the unchanged emulsion. Thus, the dispersed rubber phase still contains inclusions of polystyrene solution. This step is how the inhomogeneity of the final rubber particles begins (*7*). The more polystyrene solution is trapped inside the polybutadiene (PB) phase after the phase inversion, the higher is the phase–volume ratio, PB/PS. Therefore the phase–volume ratio increases when grafting increases in the very early stages of the reaction, provided that the particle size is kept constant.

The number and size of the occlusions depend on the reaction conditions.

If the reaction is carried out continuously in tower reactors or in a reactor cascade, as is usually the case, then the reaction pathway is not basically different from a discontinuous run, and the particles contain a large number of big inclusions in both cases (Figure 8A).

If, however, the reaction is run in a continuous stirred tank reactor (CSTR) at elevated conversions, a relatively diluted rubber solution (6–10% solids) enters a reaction mass of some 60% solids, with polystyrene solution as the continuous phase. The entering feed solution is dispersed immediately in the reaction mass, losing monomer rapidly to it. This loss results in little grafting and only a few small inclusions within the particles, depending additionally on the residence time of the individual particles in the segregated reactor. Consequently, the rubber efficiency (i.e., the ratio of the whole particle mass to its polybutadiene content) becomes poor (Figure 8B). Therefore, continuous industrial processes are run in reactor cascades or tower reactors.

Rubber-Particle Size

The rubber-particle size spectrum comes into existence within the small conversion range between phase inversion and the high-viscosity stage. In this range, it is governed by three main influences: shear stress brought about by the agitator speed, the viscosity ratio of the two phases, and the interfacial tension between them. This dependence can be seen, at least qualitatively, from Flumerfelt's equation (16), which was derived from di-

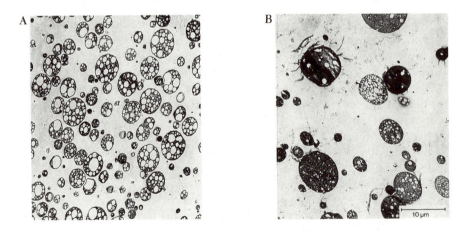

Figure 8. Rubber efficiency of cell particles. Particles were obtained by polymerizing a solution of polybutadiene in styrene: A, in a CSTR at 60% conversion; B, in a series of tower reactors, with sample taken at 60% conversion. Both samples were dried in vacuo at 213 °C and 1 mbar, cut, stained with OsO₄, and tested by transmission electron microscopy (TEM).

mensional analysis of viscoelastic fluids under shear stress and experimentally confirmed by means of several two-phase systems.

$$D_c = \frac{K \lambda \sigma \eta_d^{\beta-1}}{\eta_c^{\beta}} = K' \frac{\sigma}{\tau} \left(\frac{\eta_d}{\eta_c}\right)^{\beta-1} \tag{3}$$

$$\left[\tau \propto \frac{\eta_c}{\lambda}\right]$$

where D_c is the critical drop breakup diameter; K, K', and β are constants; λ is the time constant; σ is interfacial tension; η_d is the viscosity of the dispersed phase; η_c is the viscosity of the continuous phase; and τ is shear stress.

The influence of agitator speed on particle size, crucial for the break-through of the process in the early 1950s (*17*), has been shown in various publications (*9, 18, 19*). Particle size decreases with increasing agitator speed (Figure 9). At high shear the particle size levels out, and in the case of bulk ABS, Ide and Sasaki (*19*) observed reagglomeration at very high shear.

The influence of the viscosity ratio with respect to drop breakup in a system of two immiscible fluids was studied by Rumscheidt and Mason (*20*) and Karam and Bellinger (*21*) in the 1960s (Figure 10). Rumscheidt and Mason demonstrated the mechanism of drop breakup at different levels of viscosity ratio, and Karam and Bellinger showed that the critical deformation *D* required for a drop to split off is at a minimum when the viscosity ratio approaches unity. This situation holds good qualitatively for polymerizing styrene–rubber solutions, but the quantitative correlations are much more complex because of the viscoelasticity of the two phases, discontinuity of shear, mass-transport phenomena, and other influences.

In practice, in the phase-inversion regime the viscosity of the polybutadiene phase exceeds that of the polystyrene phase by a factor of 20 or more at 25 °C (*5*). This ratio, however, comes back to lower values because the rubber phase is itself an oil-in-oil emulsion. The influence of the viscosity ratio can be demonstrated by changing the phase viscosities separately. This step can easily be done by using rubbers of different molecular weights at a constant polystyrene molecular weight or by changing the polystyrene molecular weight at a constant polybutadiene molecular weight (*9*) (Figure 11).

The third parameter for adjusting particle size is interfacial tension. It can be varied by suitable surfactants in this oil-in-oil emulsion, as has been shown by Molau (*6*). Graft copolymers and block copolymers of styrene and butadiene act as such surfactants; an increase in grafting or the addition of block copolymers in the early stages of polymerization reduces the particle size (Figure 12).

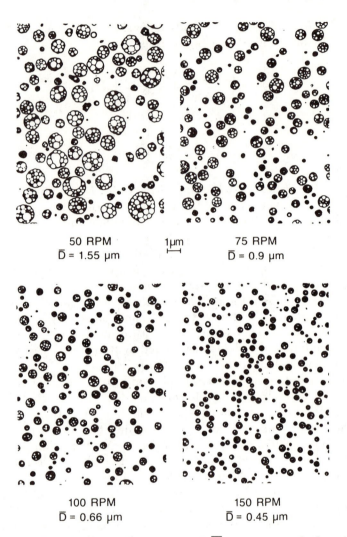

50 RPM
$\overline{D}$ = 1.55 µm

1µm

75 RPM
$\overline{D}$ = 0.9 µm

100 RPM
$\overline{D}$ = 0.66 µm

150 RPM
$\overline{D}$ = 0.45 µm

Figure 9. Particle size and agitator speed. $\overline{D}$ is mean particle diameter.

Grafting is generally considered to consist of primary radical attack on the rubber backbone. Most authors describe this reaction as hydrogen abstraction from the allylic position in the polybutadiene molecule (22–24), but recently arguments were put forward that grafting might be copolymerization (25). Nevertheless, grafting-active initiators like benzoyl peroxide greatly promote the reaction, whereas 2,2'-azobisisobutyronitrile (AIBN) does not (23). Rosen (26) has calculated the maximum possible degree of grafting f_{max} as a function of the conversion x (Figure 13). This calculation assumes that grafting occurs only in the rubber solution and that every

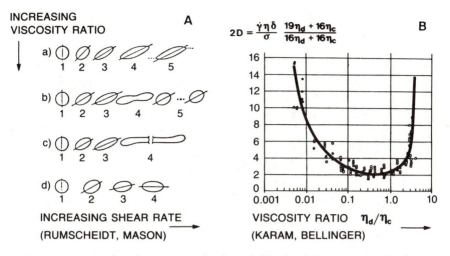

INCREASING
VISCOSITY RATIO

$$2D = \frac{\dot{\gamma}\eta\delta}{\sigma} \frac{19\eta_d + 16\eta_c}{16\eta_d + 16\eta_c}$$

INCREASING SHEAR RATE
(RUMSCHEIDT, MASON)

VISCOSITY RATIO η_d/η_c
(KARAM, BELLINGER)

Figure 10. Drop breakup in a simple shear field. The deformation D of a drop is given by the ratio (L–B)/(L+B), L being the major axis and B the minor axis of the ellipsoidally deformed droplet in the shear field. Key: $\dot{\gamma}$ is shear rate; η is viscosity; δ is Sauter diameter of the original drop; and σ is interfacial tension. Subscripts d and c mean discontinuous and continuous, respectively. (A: Reproduced with permission from ref. 20. Copyright 1961 Academic. B: Reproduced from ref. 21. Copyright 1968 American Chemical Society.)

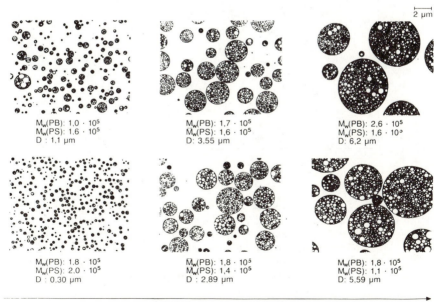

INCREASING VISCOSITY RATIO

Figure 11. Particle size and viscosity ratio. $\overline{D}$ is mean particle diameter.

INFLUENCE OF ADDED S-B BLOCKCOPOLYMER ON PARTICLE SIZE

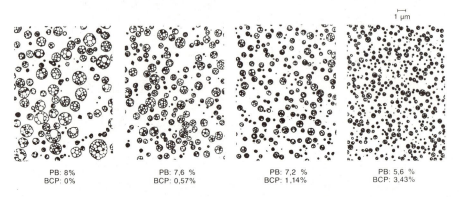

| PB: 8% | PB: 7,6 % | PB: 7,2 % | PB: 5,6 % |
| BCP: 0% | BCP: 0,57% | BCP: 1,14% | BCP: 3,43% |

Figure 12. Particle size and interfacial tension. Interfacial tension is supposed to drop by increasing block copolymer content.

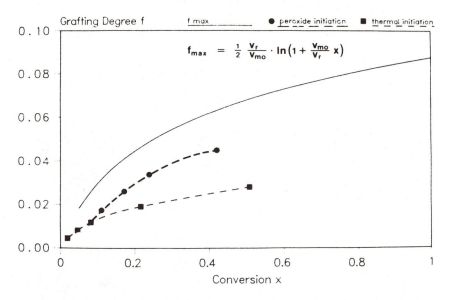

Figure 13. Maximum degree of grafting versus conversion. V_r is the volume of rubber solution and V_{mo} is the volume of monomer at $x = 0$. Experimental work was run discontinuously with 8% polybutadiene in styrene at 115 °C (thermal polymerization) and 80 °C (peroxide initiation).

growing chain in this phase results in grafting. This upper limit is far from being attained, as the comparison with the experiment shows.

In bulk ABS, furthermore, grafting is affected by preferential solubility of styrene in the rubber phase and of acrylonitrile in the matrix phase (27, 28). This preferential solubility cuts down the degree of grafting and affects a shift in composition between the grafts and the matrix copolymer.

Molau (29) has further shown that the block length of these surfactants must surpass a minimum value in the order of magnitude of the critical entanglement chain length to be able to make the two phases compatible.

Structure of Particles

Block copolymers, furthermore, open up the possibility of change in the internal structure of the particles. Polymerization of styrene in the presence of preformed styrene–butadiene di-, tri-, or multiblock copolymers produces a multitude of different particle structures (9, 30–32) (Figure 14).

These different structures are related to the morphological setup of the block copolymer used. Block copolymers exhibit crystallike domain arrays whose structure depends on their composition (Figure 15). Following Molau's scheme (33), the domain structure of A–B block copolymers shifts with increasing B content from spheres of B in A-matrix via cylinders of B in A-matrix to alternating lamellae of A and B followed by the inverse structures.

If styrene is polymerized in the presence of dissolved block copolymers, their domains are preserved in solution in the form of micelles when the concentration exceeds a critical value of about 9% (comparable to the concentration where the individual molecular coils in the solution come into contact). These micelles somehow retain the structure of the parent block copolymer. Additional grafting onto the polybutadiene part of the block

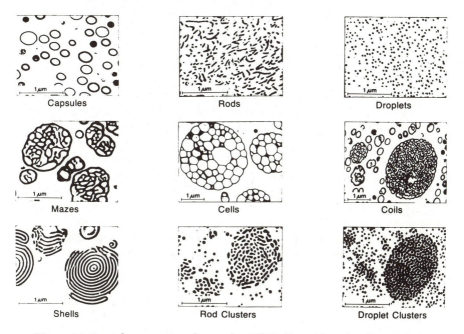

Figure 14. Particle structures observed in HIPS. (Reproduced with permission from ref. 32. Copyright 1982 Hanser.)

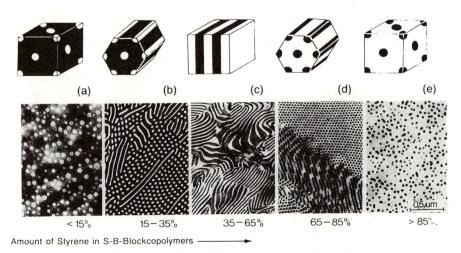

Figure 15. Block copolymer structures, schematic and observed. (Reproduced
with permission from ref. 4. Copyright 1979 Verlag Chemie.)

copolymers leads to a higher polystyrene content in the ordered micelles.
If this polystyrene content exceeds a certain critical limit, the structure of
the micelles changes according to the corresponding structural change in
the parent block copolymer, as if it had had a higher polystyrene content
from the beginning. If this change happens before phase inversion, the type
of particles generated afterward corresponds to the new micelle structure,
or, in other words, to the block copolymer structure that belongs to this
higher polystyrene content (Figure 16). This scheme could be sustained by
comparing polymerizations with simulations where grafting had been totally
excluded (9).

The reaction pathway can be monitored by viscosity–conversion measurements, as has been shown for polybutadiene. Structural change within
the continuous rubber phase, as well as phase inversion, both manifest
themselves in the sigmoidal shape of the viscosity–conversion curve (Figure
17). In the case of block copolymers, structural changes in the rubber phase
are superimposed on the shape of the viscosity–conversion plot. The corresponding changes in structure can be made visible by electron micrographs
taken of cast films.

If styrene–butadiene–styrene (SBS) triblock copolymers are used instead of the diblocks, the process runs basically the same way (Figure 18).
There is one difference from the proceedings with SB diblocks: when these
SBS triblocks contain much polybutadiene, very large particles that can
hardly be split by the agitator are formed.

Finally, in the case of butadiene–styrene–butadiene (BSB) triblock
copolymers, onionlike shell particles or rod clusters or droplet clusters
are obtained (Figure 19). These forms probably result from the ability of triblock copolymers to bridge the gap between two similar adjacent domains.

Amount of Block-styrene in S-B-Copolymers

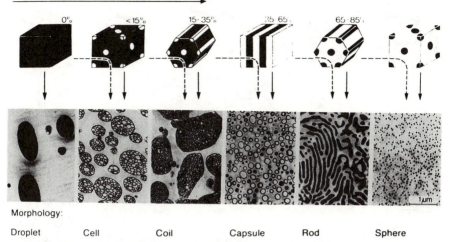

Morphology:

Droplet   Cell   Coil   Capsule   Rod   Sphere

Figure 16. Particle structures from SB diblock copolymers. Particle structure related to domain structure: solid lines, without grafting; dotted lines, with grafting before phase inversion. (Reproduced with permission from ref. 4. Copyright 1979 Verlag Chemie.)

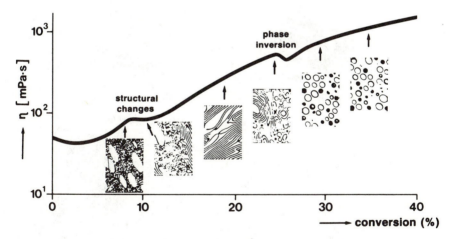

Figure 17. Formation of capsule particles. Initial block copolymer concentration, 12%; temperature, 123 °C, 0.05% tert-dodecylmercaptan; viscosity taken from the torque of the agitator calibrated with liquids of known viscosity. (Reproduced with permission from ref. 34. Copyright 1981 Verlag Chemie.)

Bridging prevents these domains from separating during phase inversion (Figure 20).

These findings were investigated further by simulated polymerizations (i.e., solution blendings) where no grafting reaction was possible (30). As a result, it can be established that if grafting doesn't interfere, the type of

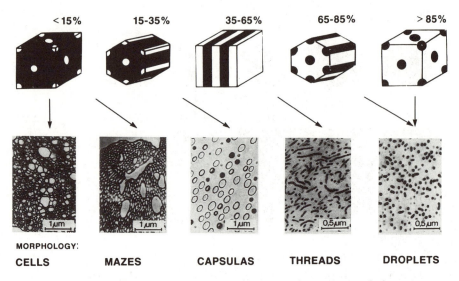

MORPHOLOGY:

CELLS   **MAZES**   **CAPSULAS**   **THREADS**   **DROPLETS**

Figure 18. Particle structures from SBS triblock copolymers. Particle formation during polymerization of styrene solutions.

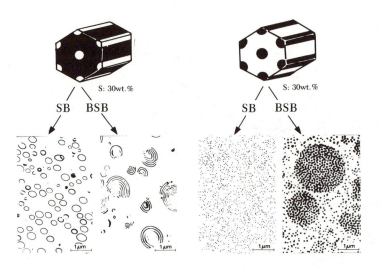

Figure 19. Particle structure from BSB triblock copolymers.

particles obtained corresponds to that of the basic block copolymer. Grafting shifts its primary composition to higher styrene content (Figure 21), and this shift brings about a change in the type of rubber particles.

This concept holds good if the change in structure occurs before particle formation; in other words, before phase inversion (Figure 22). After this stage, in general, the particle morphology is stable against further grafting

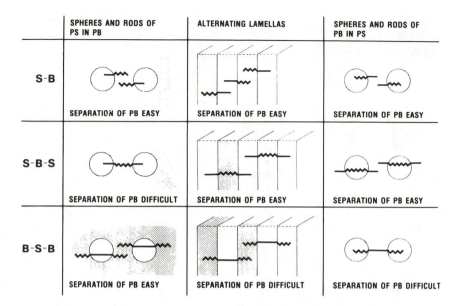

Figure 20. Particle formation. The bridging ability of triblocks across the phases renders the separation of domains difficult when the phase to disperse or to dilute is bridged by the middle block of the triblock. (Reproduced with permission from ref. 30. Copyright 1980 Hüthig and Wepf.)

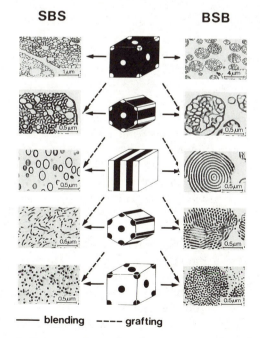

Figure 21. Particle formation by graft polymerization or by solution blending. (Reproduced with permission from ref. 30. Copyright 1980 Hüthig and Wepf.)

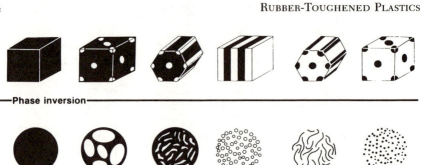

Figure 22. Block copolymer and rubber particles.

and subsequent structural change because of the increasing viscosity of the system (Figure 23). If, however, phase inversion takes place just before the block copolymer micelles change their structure according to their altered composition, this change may still take place after phase inversion inside the just-formed particle, provided the viscosity is low enough. The result is disintegration of the primary cell particle to open labyrinths, capsules, or even rod clusters.

Rubber Cross-Link Density

Rubber cross-link density is a further important factor. Rubber particles have to withstand the forces imposed by the different shrinkage of rubber and matrix when the melt cools to ambient temperatures. This situation builds up a triaxial stress field around the particles that is said to participate in craze nucleation under load. To prevent voiding inside the particles during solidification of the matrix, the rubber has to be cross-linked. To retain its elasticity and not to impair its glass-transition temperature and loss modulus, cross-linking must not be too intensive. The dependence of the swelling index of polybutadiene as a measure of its cross-link density on thermal treatment was elaborated by McCreedy and Keskkula (35) (Figure 24).

Rubber particles in HIPS or ABS from the bulk processes are heterogeneous. Therefore, swelling in good solvents for both components is only a rough measure of the degree of cross-linking. Nevertheless, this method has been used up to now for practical reasons. Recently, Karam and Tien (36) put forward a quantitative analysis of the swelling behavior of a heterogeneous gel on the basis of the modified Flory–Rehner equation. They succeeded in describing the cross-link density in these gels quantitatively when the proportion of occlusions is known.

Because cross-linking is thermally activated, most cross-links are made in the devolatilization step of the continuous processes. This step is time- and temperature-dependent, and the process conditions have to be adapted to the swelling index desired.

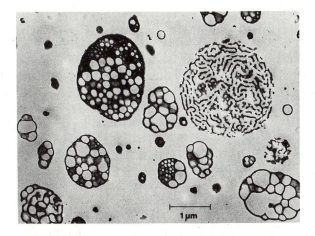

Figure 23. *Structural changes inside preformed particles when the critical grafting degree is obtained not before but after phase inversion and particle formation.*

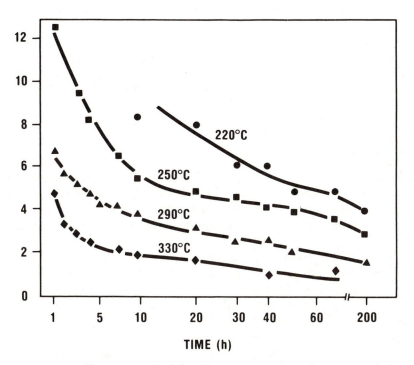

Figure 24. *Swelling index of polybutadiene versus time and temperature. (Reproduced with permission from ref. 35. Copyright 1979 Butterworth.)*

Additives

Common additives in HIPS are lubricants and antioxidants. The lubricant of choice is mineral oil. Alternatives (such as butyl stearate and phthalic esters) were not able to keep up with this inexpensive and effective lubricant. As an inert material, the mineral oil is added in most cases to the polymerization feed. It is usually present in amounts of 2–3% by weight. In some specialties, the amount of mineral oil may be lower (for better heat distortion stability) or higher (for easy flow). It lowers the glass-transition temperature of polystyrene by 3 °C per weight percent. The viscosity of HIPS is exactly correlated to the logarithmic blending rule (37):

$$\ln \eta = (1 - c) \ln \eta_p + c \ln \eta_{oil} \tag{4}$$

where η is viscosity, c is concentration of oil, and subscript p indicates polymers. With this equation we can easily calculate the relative influences of both components.

Antioxidants are usually taken from the group of hindered phenols. Several types are in common use. They protect HIPS efficiently against oxidative degradation, but no stabilizer is able to make HIPS (as well as ABS) sufficiently stable for long-term outdoor applications (carbon black being the exception).

Occasionally, further additives are used in the resins for special applications. The most important are colors and pigments. Others are UV stabilizers, antistatic agents, external lubricants, and mold-release agents. They all affect the properties to a certain extent, and their influence has to be tested in each case.

Industrial Processes

Two kinds of processes to produce HIPS are currently in industrial use, both invented and developed in the early 1950s. One of them is run discontinuously, the other continuously. Both must allow for the adjustment of all the product parameters: grafting mainly before phase inversion, particle-size adjustment just after phase inversion, matrix molecular weight over the whole range of conversion, and cross-linking in the final stage of the reaction.

The discontinuous process was elaborated by Monsanto (38). The facility consists of a prepolymerization vessel where mass polymerization is run in the presence of a graft-active radical initiator (e.g., dibenzoyl peroxide, ref. 21), and a second vessel for the suspension polymerization step, followed by equipment for finishing (Figure 25).

In the prepolymerization stage, grafting creates sufficient emulsifying material and provides occlusion precursors. After phase inversion, particle size and structure are fixed by the agitator speed, by the phase–viscosity

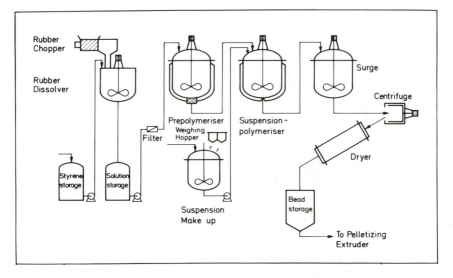

Figure 25. HIPS: Mass suspension process. (Reproduced from ref. 39. Copyright 1981 American Chemical Society.)

ratio given by the respective molecular weights and concentrations of the polymers involved, and by the graft copolymer concentration.

When the particles have been fixed in size and structure, the highly viscous reaction mass is dispersed in water with suspending agents. Polymerization is carried through nearly to completion by means of an oil-soluble high-temperature initiator and by stepwise increased temperatures. The final conversion and temperature determine rubber cross-linking.

The continuous process was elaborated by Dow in the early 1950s (*40*) and was surveyed by Platzer (*41*). In this process, polymerization is performed in bulk in a series of tower reactors (Figure 26).

In the first reactor, grafting and phase inversion are followed by particle-size adjustment. In the second and third tower reactors, polymerization is taken to 80–85% conversion, and the volatiles are degassed from the reaction mass after it is superheated by flashing into a vacuum chamber. The polymer melt is then forwarded to granulation. In this section (devolatilization and melt transport) the rubbery phase cross-links, and each producer has a proprietary set of conditions to get optimum quality.

In the polymerization step, there are a great variety of reactor lines, details of which have been published. Most of these reactors are variants of the original three-tower process, suitable for optimizing grafting and particle-size adjustment (*32*) (Figure 27).

Thus producers are able to adapt their product mix to market requirements and to launch new products for new applications. Commercial HIPS commodities have matrix molecular weights ($\overline{M}_w$) of between 170,000 and

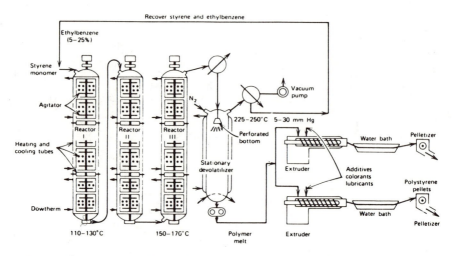

Figure 26. Continuous bulk polymerization of styrene in the presence of dissolved rubber in a series of towers. A mixture of diphenyl and diphenyl oxide is circulated through grids of pipes with slow agitation for temperature control. (Reproduced from ref. 41. Copyright 1970 American Chemical Society.)

220,000 and a dispersity index $\overline{M}_w/\overline{M}_n$ of from 2.5 to 3.0. The rubber used preferentially is medium-*cis*-polybutadiene. The phase–volume ratio, based on a 6–10% rubber content, is normally between 25 and 35%, taken as gel content. The mean particle size is normally set at 2.5 μm in diameter to obtain the optimum impact strength (*42–44*). The particles are most frequently of the cell type. The swelling index is normally between 9.0 and 12.0, corresponding to the molecular weight of polybutadiene segments between cross-links of 10,000–30,000 (*36*). A common additive is mineral oil in the concentration range of 2–3%, the residual monomer content being less than 500 ppm.

This old rubber-modified polystyrene, however, is still a subject of novelties. For instance, BASF has launched several series of new products in the past 12 years. One of them includes products with improved Frigen-resistance that contain very large particles in combination with a high phase–volume ratio and some other adapted properties (*45*). In Europe they have taken a considerable share of the market in refrigerators from ABS. Another series recently launched consists of products with excellent gloss combined with high toughness. These materials are based on very small capsule particles, with a mean diameter of about 0.3 μm and a small amount of cell particles (*46*). Products with pure capsule particle morphology are translucent and have excellent gloss, but only medium impact strength.

In addition, products with excellent Izod toughness and fair gloss, as well as those with fair Izod toughness and very good gloss, have been commercialized with good success (Figure 28). All these glossy impact products are apt to make inroads on diverse ABS applications.

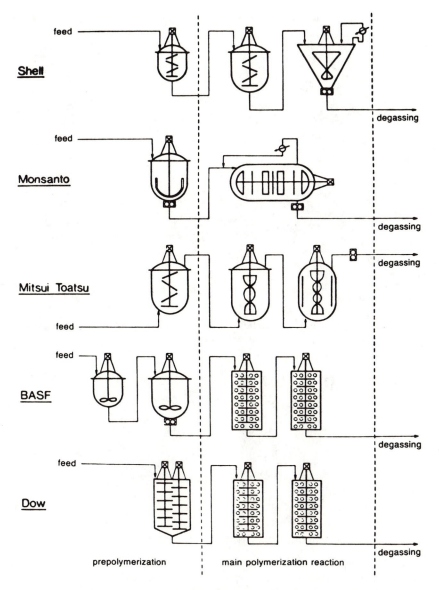

Figure 27. *Reactor cascades for HIPS production. (Reproduced with permission from ref. 32. Copyright 1982 Hanser.)*

ABS: Properties and Production

ABS (rubber-toughened acrylonitrile–butadiene–styrene copolymer) is a very versatile thermoplastic material for a broad range of applications. It can be made by bulk polymerization, and both discontinuous mass suspension and continuous bulk processes are in use (Figure 29). But the most common

Morphology, Izod Impact Strength and Gloss

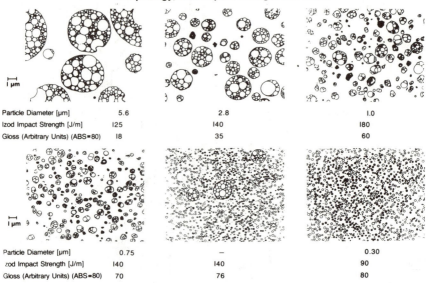

Particle Diameter [µm]	5.6	2.8	1.0
Izod Impact Strength [J/m]	125	140	180
Gloss (Arbitrary Units) (ABS=80)	18	35	60

Particle Diameter [µm]	0.75	–	0.30
zod Impact Strength [J/m]	140	140	90
Gloss (Arbitrary Units) (ABS=80)	70	76	80

Figure 28. Structure and properties of commercial HIPS-types, including the recent developments.

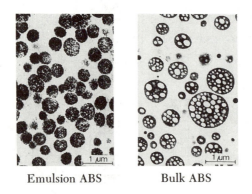

Emulsion ABS   Bulk ABS

Figure 29. Morphology of ABS polymerized in emulsion or bulk. (Reproduced with permission from ref. 47. Copyright 1981 Council of Scientific and Industrial Research.)

ABS process is emulsion polymerization, at least for the rubbery phase. Emulsion polymerization provides products with much smaller particles than the continuous bulk process.

The emulsion process is run discontinuously in several steps (32) (Figure 30). First, a polybutadiene latex is produced. Butadiene polymerizes rather slowly. Small particles are made to reduce the cycle period and, in addition, the reaction is not run to completion. Because these particles are not of the

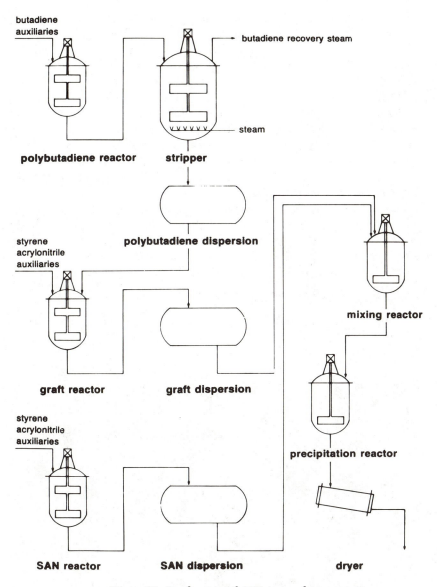

Figure 30. Production of ABS in emulsion.

optimum size, a subsequent agglomeration step is necessary. This step is possible by the addition of suitable lattices with hydrophilic groups (48) or by a gradual decrease of the pH value by means of acetic anhydride (49). A grafting step follows particle agglomeration to the required size. Without the addition of extra emulsifier, but with additional initiator together with the matrix monomers, a seeded polymerization is started to cover the particles with a matrix polymer shell. The graft dispersion (Figure 31) is mixed with a SAN dispersion in suitable amounts, and the latex mixture is then precipitated by electrolytes, washed, dried, and processed to granules.

The properties of these materials depend on the same molecular and morphological parameters as HIPS properties. Matrix molecular weight, rubber type, phase–volume ratio, rubber-particle size, particle structure, cross-link density, and additive content govern the technical properties.

But there are some very important differences. Styrene–acrylonitrile copolymer is a polar matrix and therefore much more cohesive than polystyrene. This cohesiveness is considered to be the reason for the higher toughness of ABS, as is evident from the fact that ABS deforms not only by crazing but also by shear yielding (50, 51), which is activated by smaller particles than are usually present in HIPS.

Furthermore, the matrix properties are influenced by composition, as well as molecular weight. The composition of both matrix and graft shell should be matched so that they will remain compatible. Deviations in acrylonitrile content of more than 5% cause incompatibility in SAN (52).

Molecular and Morphological Parameters

Matrix and Molecular Weight.   Matrix composition and molecular weight are determined by feed composition and reaction conditions. Styrene

Figure 31. ABS graft dispersion.

and acrylonitrile have an azeotropic composition at 25% (by weight) acrylonitrile because of their copolymerization parameters of $r_1 = 0.4$ and $r_2 = 0.04$, respectively, in bulk polymerization. In emulsion polymerization, however, the azeotropic composition is 28.5% (by weight) acrylonitrile (53). Because the emulsion polymerization of the matrix and that of the graft shell are run discontinuously, the composition of the polymer is confined to a relatively narrow range of 28.5 ± 5% to avoid a shift in the chemical composition. In addition to this condition, the relatively high solubility of acrylonitrile in water and the low solubility of styrene may lead to a shift in composition of the two monomers at the reaction site. This shift is particularly likely in the grafting reaction, in which it is superimposed by the different solubility of the two monomers in the primary rubber particles (27).

Rubber Type.   The rubber type used in ABS is polybutadiene in most cases. It is made by emulsion polymerization, usually at 60 °C with potassium persulfate as initiator and sodium stearate, rosin acid salts, or alkyl sulfonates as emulsifiers. The microstructure of the polybutadiene depends somewhat on the reaction temperature, and the cross-link density depends on the final conversion. Both parameters influence the glass-transition temperature.

Polybutadiene is susceptible to oxygen attack. In finished articles, this mainly affects the polybutadiene particles in the surface and causes cavitation that acts as notches and causes bleaching (Figure 32). According to Stabenow and Haaf (54), clusters are the preferred path of oxygen attack; the cavitation causes flaws in the surface and makes the material brittle. Cavitation is hampered by closed graft shells around every particle, which prevent clustering.

Acrylate rubbers and EPDM are sometimes taken to replace polybutadiene in order to obtain better weatherability (Figure 33). Acrylate rubbers

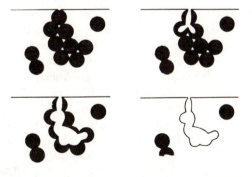

Ozone − attack on ABS with Particle Clusters
(schematic)

Figure 32. Surface cavitation by oxygen attack. (Reproduced with permission from ref. 54. Copyright 1973 Hüthig and Wepf.)

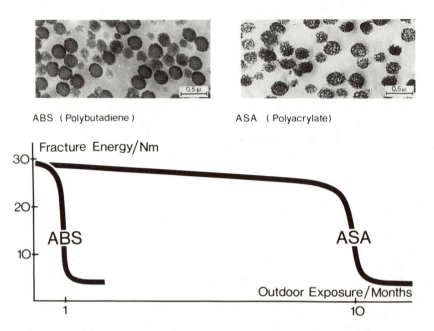

ABS (Polybutadiene)   ASA (Polyacrylate)

Figure 33. Rubber type and weatherability. (Reproduced with permission from ref. 47. Copyright 1981 Council of Scientific and Industrial Research.)

are made in emulsion polymerization similar to polybutadiene and subsequently treated analogously to get acrylonitrile–styrene–acrylic ester copolymer (ASA); EPDM is used only in a bulk precipitation process. This process provides products with irregularly shaped rubber particles, acrylonitrile–(ethylene–propylene–terpolymer)–styrene copolymer (AES) (Figure 34).

Because the rubbery and rigid phases of emulsion-polymerized ABS are made in separate reactors, the phase–volume ratio is determined by the blend ratio of the two components. As in HIPS, the phase–volume ratio has the greatest influence on impact strength. In contrast to HIPS, the rubber particles contain only a few or are not filled with matrix material occlusions. The total amount of polybutadiene in ABS polymers is therefore three or four times as high as in corresponding HIPS types (55). Impact strength increases with increasing rubber content (Figure 35). But, as in the case of HIPS, interfacial bonding by grafts and particle size are most important. In a series with a constant particle size and a constant degree of grafting, impact strength increases with the phase–volume ratio (56, 57).

Young's modulus and yield stress decrease with the rubber content. The decrease in yield stress fits the Ishai–Cohen model very well, as was shown by Pavan and co-workers (58) (Figure 36).

Rubber Particle Size.   In emulsion polymerization the rubber particle size is determined by a suitable choice of emulsifier concentration and

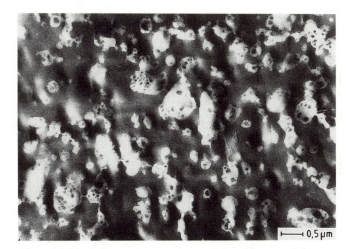

Figure 34. AES particle structure.

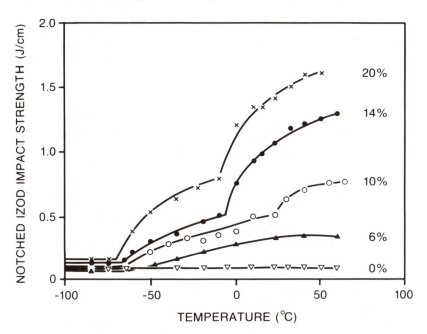

Figure 35. Impact strength of ABS versus temperature and rubber content. (Reproduced with permission from ref. 56. Copyright 1977 Applied Science.)

mode of operation (monomer-to-water ratio, temperature). In principle, big rubber particles are made by reduced emulsifier concentration, by monomer feed, or by seeded polymerization. In practice, however, it is preferable to produce small particles within reasonable cycle times. The final particle size is adjusted afterward by agglomeration techniques; this approach leads to a

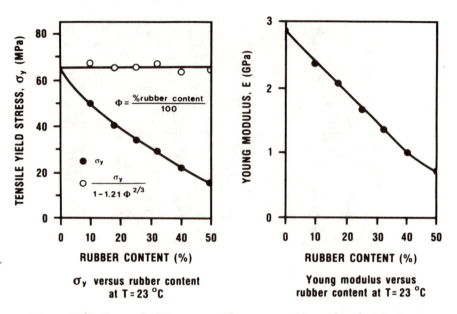

Figure 36. Stiffness of ABS versus rubber content. (Reproduced with permission from ref. 58. Copyright 1985 Plastics and Rubber Institute.)

bimodal particle-size distribution of unagglomerated and agglomerated particles (Figure 37).

In ABS polymers, the optimum particle size is much smaller than in HIPS. Because of the polar character of the matrix, the material tends to deform by shear yielding at low deformation rates. At higher deformation rates, crazes are nucleated (50). Very small particles make the ABS resins stiff, glossy, and less tough; big particles impair stiffness and gloss, but improve toughness (Figure 38). The optimum mean particle size for a good combination of toughness, stiffness, and surface gloss is about 0.3–0.5 μm (47). Measurements of specimens processed under identical conditions show that gloss can be related exponentially to particle size (59).

Structure of Particles. The particle structure of emulsion-polymerized ABS is of the core-shell type (Figure 39). Particles of polybutadiene must be tied in the surrounding matrix by a graft shell to provide good coupling of the two phases. The more closely covered the particle, the better the coupling and the more uniform the distribution in the matrix. Because the grafting is nearly complete (scarcely any free SAN polymer present after the graft reaction), the proportion of grafted SAN in the interface depends on the total amount of glassy monomer added and on the distribution of the grafted polymer between internal and external grafts. The proportion of graft copolymer included rises with increasing particle size and decreases with its cross-link density (47).

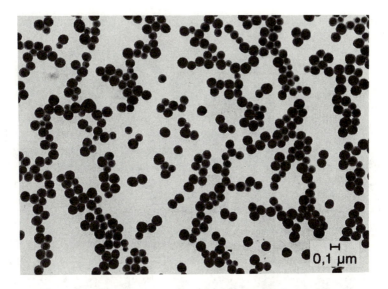

Figure 37. Bimodal particle-size distribution in ABS by agglomeration.

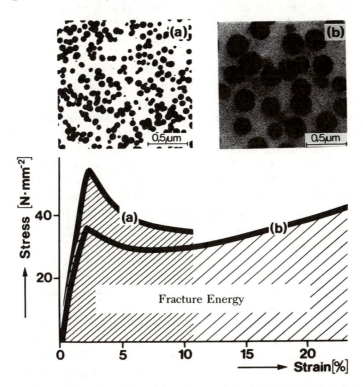

Figure 38. Particle size and mechanical properties of ABS. (Reproduced with permission from ref. 47. Copyright 1981 Council of Scientific and Industrial Research.)

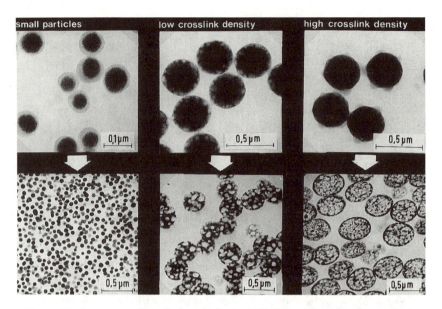

Figure 39. Structure of grafted particles in ABS. (Reproduced with permission from ref. 47. Copyright 1981 Council of Scientific and Industrial Research.)

Emulsion-grafted small particles exhibit a closed shell of graft copolymer on their surface, and no inclusions are visible in the images either before or after the mixing process. Bigger particles with low cross-link density have hardly any detectable graft shell, and the mixture with matrix polymer shows that the particles contain many large inclusions. If the cross-link density is enhanced, the graft shell becomes more clearly marked, but a considerable part of matrix polymer is still located inside the particle.

If the size and cross-link density are constant, the amount of grafts is of crucial importance for the mechanical properties (Figure 40). If the particles are not completely covered by the graft shell, they will tend to agglomerate to large and irregularly shaped clusters during the mixing process, and the impact strength of the final ABS will depend on the mixing conditions.

If the primary particles are small (0.1 μm in diameter), an increase in grafting will first reduce the cluster dimensions to the optimum value. Consequently, impact strength increases. If grafting is increased further, particles will be prevented from clustering and impact strength will drop again. With larger particles, less than 50% grafting is sufficient to obtain very tough materials, and products with bimodal particle-size distribution are even more advantageous (57). In contrast to HIPS, the internal structure of ABS is of minor influence on the mechanical properties. Most ABS types have more or less solid particles with few and small inclusions, depending on particle size and cross-link density. In ASA, however, grafting occurs mainly in the interior, even in highly cross-linked particles. The acrylate rubber is more

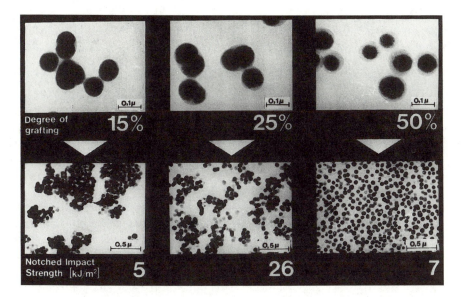

Figure 40. Influence of degree of grafting on ABS structure. (Reproduced with permission from ref. 47. Copyright 1981 Council of Scientific and Industrial Research.)

hydrophilic than the grafted matrix copolymer and therefore tends more to keep in contact with the aqueous milieu.

Cross-Linking.   The cross-link density of rubber depends mainly on the final conversion in the butadiene polymerization. Additionally, the graft reaction and thermal treatment during the compounding step will contribute to the cross-linking. Little has been published on quantitative correlations to technical properties (*60, 61*).

Additives.   There are no peculiarities in additives for ABS beyond those with HIPS. As a lubricant, mineral oil is not suitable because of the differences in polarity and therefore in solubility. The most important lubricants are esters of phthalic acid and adipic acid, fatty acid esters of glycerol, and, in some cases, waxes. External lubricants are calcium or zinc stearate, amides of fatty acids, or polyethylene waxes. Stearylethanolamine acts as an antistatic agent.

ABS contains phenolic antioxidants similar to those used in HIPS; in special cases it also contains UV-absorbers. Flame retardancy is provided by high-bromine aromatics such as octabromodiphenyl ether, octabromodiphenyl, or tetrabromophthalic acid, all in combination with antimony oxide as a synergist (*62*).

The proper choice of the additive system, which depends on the application of the final article, has to allow for additive influences on the whole set of properties of the product in service.

Specialty Styrenic Materials

Styrenic materials are very versatile. They are easy to process and they have a good overall price–performance relationship. But for a series of applications, they fail to meet the requirements. The most important properties involved are transparency (in combination with high impact strength) and heat distortion temperature.

Transparency.   The two-phase nature of impact-grade styrenic materials causes turbidity by light scattering at the dispersed phase. This effect occurs because the particles are often larger than the wavelength of light or their size lies in its order of magnitude, and because they exhibit refractive indices different from those of the matrix (Figure 41). According to Mie's theory, the turbidity coefficient depends on both particle diameter and refractive index difference (63).

If the diameter of the particles is small enough with respect to the wavelength of visible light (some 0.05 part of it), they can be considered

$$I_x / I_o = \exp\left[-(\varepsilon + \tau)x\right]$$

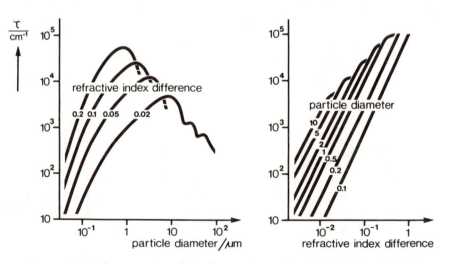

Figure 41. Light scattering dependence on particle size and refractive index difference. Key: τ is the turbidity coefficient; ε is the extinction coefficient; x is sample thickness; and I_o and I_x are the intensity of the entering and leaving beam, respectively. (Reproduced with permission from ref. 4. Copyright 1979 Verlag Chemie.)

optically homogeneous in spite of their real domain structure, and their cumulative refractive index can be calculated by linear interpolation from the components. The refractive index of the composite particle is lower than that of the SAN matrix (~1.57) but higher than that of polybutadiene (1.52). Polymethyl methacrylate (PMMA) is compatible with SAN to a certain degree, and its refractive index (1.49) is lower than that of the particle. It is, therefore, possible to match the refractive indices of the matrix and the grafted rubber by blending the matrix with some PMMA to get transparent materials. This approach holds good for monomodally distributed grafted particles. However, the degree of grafting of bimodally distributed particles depends on their specific surfaces, which is smaller with larger particles. Therefore, the refractive indices of small and big particles become different, and the products turn turbid even if the calculated composition of the grafted rubber matches the matrix refractive index (Figure 42).

Clusters of particles act as scattering centers like single particles. Transparency therefore also depends on the dispersion of the particles in the matrix (i.e., on the processing conditions). As a result, transparent items

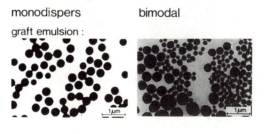

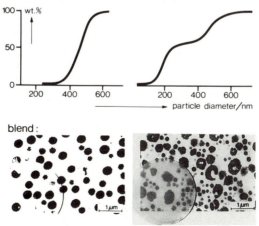

Figure 42. Particle size distribution and transparency. (Reproduced with permission from ref. 4. Copyright 1979 Verlag Chemie.)

may turn turbid on annealing if they first were in a nonequilibrium state of dispersion, and agglomerate later because of the lack of shearing forces when heated again above their glass-transition temperature.

It is possible to calculate the diameter of a sphere optically equivalent to the cluster from a Guinier plot of the intensity of scattered light against the scattering angle (Figure 43).

Scattering intensity is also affected by the spatial arrangement of the scattering centers. Styrene–butadiene block copolymers are highly trans-

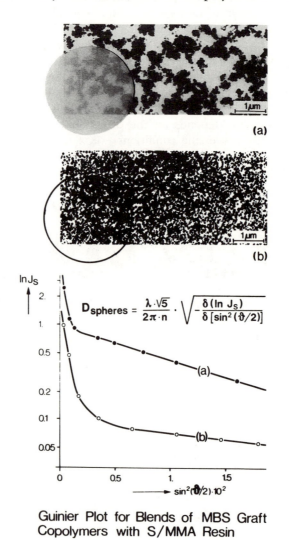

$$D_{spheres} = \frac{\lambda \cdot \sqrt{5}}{2\pi \cdot n} \cdot \sqrt{-\frac{\delta (\ln J_S)}{\delta [\sin^2 (\vartheta/2)]}}$$

Guinier Plot for Blends of MBS Graft
Copolymers with S/MMA Resin

Figure 43. Clustering and transparency. (Reproduced with permission from ref. 4. Copyright 1979 Verlag Chemie.)

parent, although they are separated in domains of different refractive indices (Figure 44). These domains are not small enough to explain the excellent transparency. One can characterize the influence of regularity by a lattice factor L and consider the intensity of scattered light to be a product of the squares of this lattice factor and a particle factor P (*4, 47*).

In the region of visible light, the low lattice factor compensates for the higher particle factor to cause transparency. At very low wavelengths with small-angle X-ray scattering, a steep increase in the lattice factor with a maximum at 0.3° leads to the intensity distribution well known from the regular structure of block copolymers.

Styrene–butadiene block copolymers are not compatible with polystyrene (*64*) if its molecular weight exceeds that of the polystyrene block in the copolymer. Therefore, 1:1 blends of block copolymers with polystyrene are more turbid than the parent materials (Figure 45). Extruder blends with a fine dispersion of the two components show only slight turbidity, which is greatly increased when the sample is annealed. This increase stems from the separation of the frozen-in nonequilibrium state of the extruder blend, as can be shown by electron microscopy. This result proves that the regular arrangement of the domains is decisive for the high transparency of block copolymers (*47*).

Such block copolymers are very interesting products as transparent impact-resistant polystyrenes (Figure 46). The structure is, of course, very different from that of a conventional rubber-modified product.

Star-shaped styrene–butadiene block copolymers were developed to obtain a transparent, impact-resistant, and easily processable product. They have been available on the market since the early 1970s. Symmetrical and asymmetrical structures are described in the patent literature (*65–68*) (Figure 47).

Star-block copolymers are made by several versions of multistep anionic copolymerization. The principal structure is starlike, with polybutadiene or polystyrene in the core and with polystyrene endblocks at all or only some of the polybutadiene branches. These products are highly transparent and very tough, and therefore are a very useful extension of the rubber-modified polystyrene family (Figure 48).

Transparency can also be attained by matching refractive indices. Consequently, a blend of SAN-grafted rubber with a blend matrix of SAN and PMMA of appropriate composition yields transparent ABS types. However, PMMA and SAN are compatible only if the SAN contains 12–22% acrylonitrile (*4*).

Heat Distortion.   Heat distortion is directly related to the glass-transition temperature of the polymer and to the content of additives. Styrenic materials have their glass transition at about 100 °C. Two methods can be used to extend their applicability: copolymerization and blending.

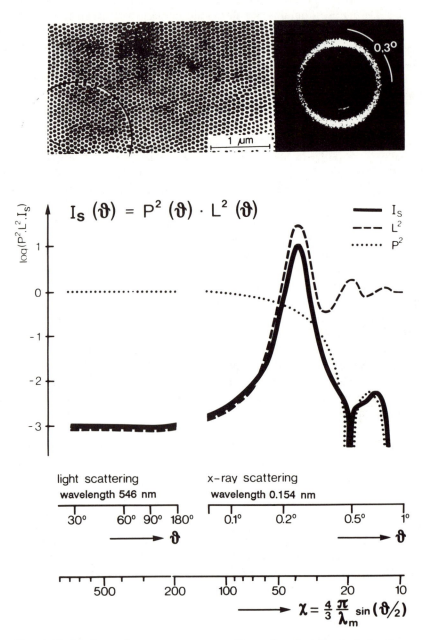

Figure 44. Block copolymer transparency. (Reproduced with permission from ref. 4. Copyright 1979 Verlag Chemie.)

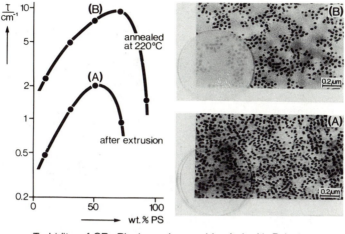

Turbidity of SB-Blockcopolymers blended with Polystyrene
wavelength 500 nm, blending temperature 190 °C

Figure 45. Block copolymer–homopolymer mixtures. (Reproduced with permission from ref. 47. Copyright 1981 Council of Scientific and Industrial Research.)

Copolymerization is confined to only a few comonomers. α-Methylstyrene has found limited application in the ABS field as a partial substitute for styrene, bringing the glass transition up to 115 °C in practical cases. Recently, *p*-methylstyrene was claimed to be able to replace styrene monomer to a considerable extent (69). Poly-*p*-methylstyrene has a glass-transition temperature (T_g) of 108 °C, slightly above the temperature of boiling water. Furthermore, maleic anhydride (MA) is sometimes used as a suitable comonomer for styrene. Rubber-modified types with a glass-transition temperature in the range of 110–150 °C are available according to the MA content. Roughly, the T_g is raised by 3 °C per weight percent of MA in the polymer (70). The impact strength of these polymers is comparable to that of medium-impact polystyrene. The manufacture of these polymers is complicated by the disparity in copolymerization parameters ($r_1 = 0.01$, $r_2 = \sim 0$), which favors alternating copolymerization.

The most versatile method of preparing styrenic materials with a high heat-distortion temperature is blending. Blending can be done with both compatible and incompatible components.

Polystyrene is compatible on a molecular scale with poly-2,6-dimethylphenylene ether (PPE) (71) and also with tetramethylbisphenol-A-polycarbonate (72). The PPE blend has had a good market since the late 1960s, and it is still growing; the polycarbonate blend was of only temporary interest. HIPS–PPE blends have glass-transition temperatures of 100–200 °C (Figure

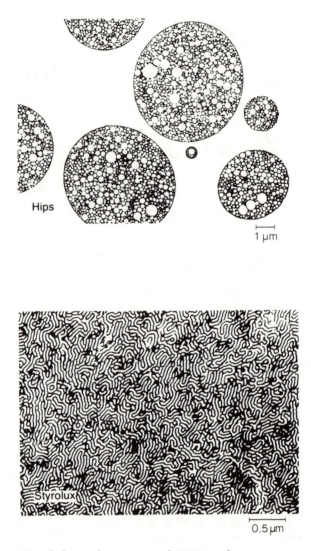

Figure 46. Morphology of conventional HIPS and transparent anionically polymerized grades.

49). In practice, this range is 110–140 °C, because the values are influenced not only by the blend ratio, but also by additives.

These blends exhibit the same morphology as the parent HIPS (i.e., cell particles). Their optimum size is smaller than that found in conventional HIPS, as presented in 1977 by Cooper et al. (73). One of their major benefits is the possibility of providing halogen-free flame-retardant formulations (Figure 50).

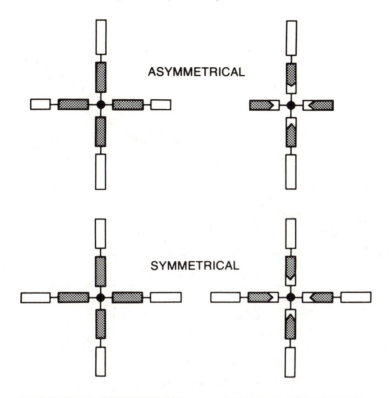

ASYMMETRICAL

SYMMETRICAL

BUTADIENE-CONNECTION   **STYRENE-CONNECTION**

Figure 47. Molecular structure patterns of star-block copolymers.

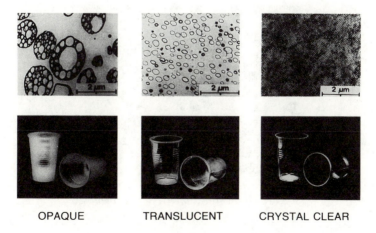

OPAQUE   TRANSLUCENT   CRYSTAL CLEAR

Figure 48. Transparent HIPS application.

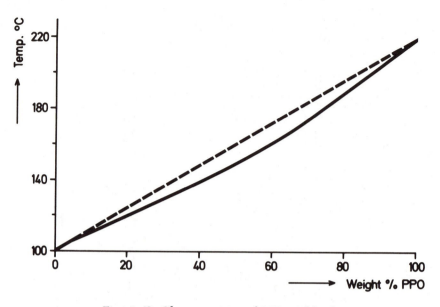

Figure 49. Glass transition of PPE–PS blends.

- high heat resistance
- high impact strength
- high dimensional stability
- high hydrolytic stability
- halogen-free flame retardant agents

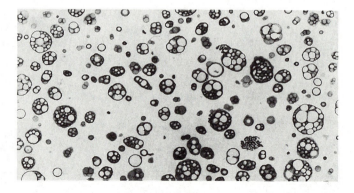

Figure 50. Features of HIPS–PPE blends.

The second blend of commercial interest consists of ABS and polycarbonate (PC). SAN, the matrix material of the ABS, and PC are only semicompatible. These blends exhibit their two-phase nature in electron micrographs and in T_g as well *(74)* (Figure 51).

ABS–PC contains, in addition to PC–SAN, a rubber as a third phase (Figure 52). The properties of these blends depend on their mixing conditions and morphology (Figure 53). If ABS is the continuous phase, the notched Izod impact strength is similar to that of the ABS component. If PC is the continuous phase, impact strength increases considerably. By contrast, tensile strength increases linearly with an increasing PC content *(75)*.

Impact behavior also depends on dispersity and, therefore, on blending and processing conditions. The higher the processing temperature, the coarser the phase distribution *(76)*. If ABS is the continuous phase, a brittle-to-tough transition can be observed at low test temperatures but cannot be seen in the case of a continuous PC phase *(76)*.

Not only ABS but also ASA is a very useful blending component for PC. In addition to the advantages of ABS–PC, it provides a good outdoor performance, light natural color, and good yellowing resistance (Figure 54).

Blending is not confined to the examples mentioned (Figure 55). It has been extended to the whole range of thermoplastic materials, and much more work is going to be done in the field of multiphase polymer blends.

Styrenic materials are full of interesting and challenging novelties in

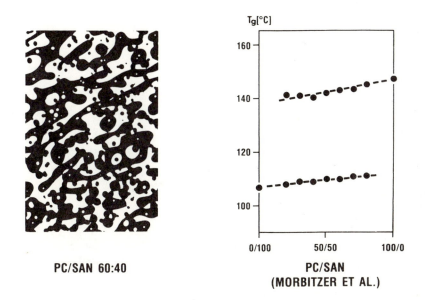

PC/SAN 60:40

PC/SAN
(MORBITZER ET AL.)

Figure 51. Phase structure (left) and glass transition of SAN–PC blends. (Reproduced with permission from ref. 74. Copyright 1985 Hüthig and Wepf.)

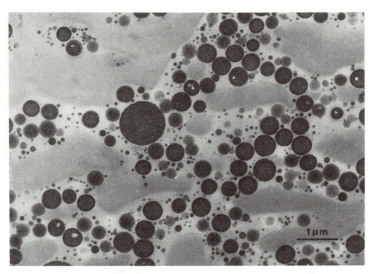

Figure 52. Morphological structure of ABS–PC blends.

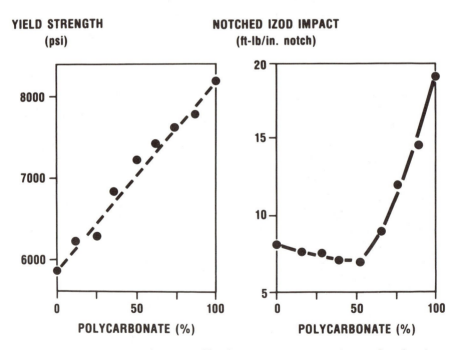

*Figure 53. Properties of ABS–PC blends versus composition. (Reproduced with
permission from ref. 75. Copyright 1984 Wiley.)*

- high resistance to heat deformation
- good rigidity and dimensional stability
- great toughness
- good outdoor performance and resistance to yellowing
- comparatively good resistance to environmental stress cracking
- efficient antistatic
- no problems in processing

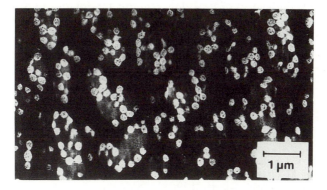

Figure 54. Features of ASA–PC blends.

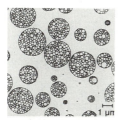

HIPS graft copolymer

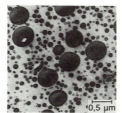

ABS graft copolymer

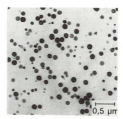

MBS modified PVC

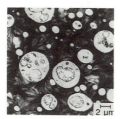

Blend of polyamide
with ABS

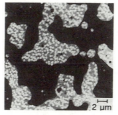

Blend of polyamide
with ASA

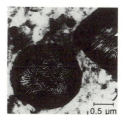

EPDM modified
polypropylene

Figure 55. Multiphase polymer blends; examples.

research and development. Although they are more than 50 years of age, their possibilities are far from being exhausted. We have every reason to look forward to the next decade.

Acknowledgments

This chapter is dedicated to Prof. Dr. Helmut Dörfel on the occasion of his 60th birthday. The author is indebted to G. Heckmann and J. Stabenow for providing the electron micrographs, to H. P. Hofmann and H. D. Schwaben for helpful discussions and kind support, and to B. E. Byrt for reading the manuscript.

References

1. Anonymous *Mod. Plast. Int.* **1986**, *16 (1)*, 24; Anonymous *Mod. Plast. Int.* **1987**, *17 (1)*, 20.
2. Buchdahl, R.; Nielsen, L. E. *J. Appl. Phys.* **1950**, *21*, 482.
3. Kato, K. *J. Electron Microsc.* **1965**, *14*, 219; *Polym. Eng. Sci.* **1967**, *7*, 38.
4. Schmitt, B. J. *Angew. Chem.* **1979**, *91*, 286.
5. Riess, G.; Gaillard, P. In *Polymer Reaction Engineering*; Reichert, K. H.; Geiseler, W., Eds.; Hanser: Munich, 1983; p 221.
6. Molau, G. E. *J. Polym. Sci. Part A-1* **1965**, *12*, 67; ibid. 4235.
7. Molau, G. E.; Keskkula, H. *J. Polym. Sci. Part A-1* **1966**, *4*, 1595.
8. Bender, B. W. *J. Appl. Polym. Sci.* **1965**, *9*, 2887.
9. Echte, A. *Angew. Makromol. Chem.* **1977**, *58/59*, 175.
10. Molau, G. E.; Wittbrodt, W. M.; Meyer, V. E. *J. Appl. Polym. Sci.* **1969**, *13*, 2735.
11. Freeguard, G. F.; Karmarkar, M. *J. Appl. Polym. Sci.* **1971**, *15*, 1649.
12. Zhiqiang, Song; Huigen, Yuan; Zuren, Pen *J. Appl. Polym. Sci.* **1986**, *32*, 3349.
13. Freund, G.; Lederer, M.; Strobel, W. *Angew. Makromol. Chem.* **1977**, *58/59*, 199.
14. Cigna, G. *J. Appl. Polym. Sci.* **1970**, *14*, 1781.
15. Bucknall, C. B.; Cote, F. F. P.; Partridge, I. K. *J. Mater. Sci.* **1986**, *21*, 301; Bucknall, C. B.; Davies, P.; Partridge, I. K. *J. Mater. Sci.* **1986**, *21*, 307; Bucknall, C. B; Davies, P.; Partridge, I. K. *J. Mater. Sci.* **1987**, *22*, 1341.
16. Flumerfelt, R. W. *Ind. Eng. Chem. Fundam.* **1972**, *11*, 312.
17. Amos, J. L. *Polym. Eng. Sci.* **1974**, *14*, 1.
18. Freeguard, G. F. *Br. Polym. J.* **1974**, *6*, 205.
19. Ide, F.; Sasaki, I. *Kobunshi Kagaku* **1970**, *27*, 617.
20. Rumscheidt, F. D.; Mason, S. G. *J. Colloid Sci.* **1961**, *16*, 238.
21. Karam, H. J.; Bellinger, J. C. *Ind. Eng. Chem. Fundam.* **1968**, *7*, 576.
22. Fischer, J. P. *Angew. Makromol. Chem.* **1973**, *33*, 35.
23. Brydon, A.; Burnett, G. M.; Cameron, G. G. *J. Polym. Sci. Polym. Chem. Ed.* **1973**, *11*, 3255; ibid. **1974**, *12*, 1011.
24. Gupta, V. K.; Bhargava, G. S.; Bhattacharyya, K. K. *J. Macromol. Sci. Chem. A* **1981**, *16*, 1107.
25. Hayes, R. A.; Futamura, S. *J. Polym. Sci. Chem. Ed.* **1981**, *19*, 985.
26. Rosen, S. L. *J. Appl. Polym. Sci.* **1973**, *17*, 1805.
27. Locatelli, J. L.; Riess, G. *Makromol. Chem.* **1974**, *175*, 3523.
28. Locatelli, J. L.; Riess, G. *J. Polym. Sci. Polym. Chem. Ed.* **1973**, *11*, 3309.

29. Molau, G. E. *Kolloid Z. Z. Polym.* **1970,** *238,* 4931.
30. Echte, A.; Gausepohl, H.; Lütje, H. *Angew. Makromol. Chem.* **1980,** *90,* 95.
31. Riess, G.; Schlienger, M.; Marti, S. *J. Macromol. Sci. Phys. B* **1980,** *17,* 355.
32. Echte, A. In *Chemische Technologie,* 4th ed.; Harnisch, K.; Steiner, R.; Winnacker, K., Eds.; Hanser: Munich, 1982; Vol. 6, p 373.
33. Molau, G. E. In *Block Polymers;* Aggrawal, S. L., Ed.; Plenum: New York, 1970; p 79.
34. Echte, A.; Haaf, F.; Hambrecht, J. *Angew. Chem. Int. Ed. Engl.* **1981,** *20,* 344.
35. McCreedy, K.; Keskkula, H. *Polymer* **1979,** *20,* 1155.
36. Karam, H. J.; Tien, L. *J. Appl. Polym. Sci.* **1985,** *30,* 1969.
37. Kruse, R. L.; Southern, J. H. *Polym. Eng. Sci.* **1979,** *19,* 815.
38. Stein, R. S.; Walter, R. L. U.S. Patent 28 62 907, 1958, to Monsanto Chemical Company.
39. Pohlemann, H. G.; Echte, A. In *Polymer Science Overview;* Stahl, G. A., Ed.; ACS Symposium Series 175; American Chemical Society: Washington, DC, 1981; p 265.
40. Amos, J. L.; McCurdy, J. L.; McIntryre, O. R. U.S. Patent 26 94 692, 1954, to Dow Chemical Company.
41. Platzer, N. *Ind. Eng. Chem.* **1970,** *62 (1),* 6.
42. Donald, A. M.; Kramer, E. J. *J. Appl. Polym. Sci.* **1982,** *27,* 3729.
43. Bucknall, C. B.; Davies, P.; Partridge, I. K. *J. Mater. Sci.* **1987,** *22,* 1341.
44. Gilbert, D. G.; Donald, A. M. *J. Mater. Sci.* **1986,** *21,* 1819.
45. Mittnacht, H.; Jenne, H.; Bronstert, K.; Lieb, M.; Adler, H. J.; Echte, A. U.S. Patent 41 44 204, 1979, to BASF AG.
46. Echte, A.; Gausepohl, H. U.S. Patent 44 93 922, 1985, to BASF AG.
47. Haaf, F.; Breuer, H.; Echte, A.; Schmitt, B. J.; Stabenow, J. *J. Sci. Ind. Res.* **1981,** *40,* 659.
48. Keppler, H. G.; Wesslau, H.; Stabenow, J. *Angew. Makromol. Chem.* **1968,** *2,* 1.
49. Paster, M. D. U.S. Patent 39 88 010, 1976, to Monsanto Company; U.S. Patent 40 43 955, 1977, to Monsanto Company.
50. Bucknall, C. B.; Drinkwater, I. C. *J. Mater. Sci.* **1973,** *8,* 1800.
51. Bucknall, C. B. *Adv. Polym. Sci.* **1978,** *27,* 121.
52. Molau, G. E. *J. Polym. Sci. Polym. Lett. Ed.* **1965,** *3,* 1007.
53. Lin, Chen-Chong; Ku, Hwar-Ching; Chiu, Wen-Yen *J. Appl. Polym. Sci.* **1981,** *26,* 1327.
54. Stabenow, J.; Haaf, F. *Angew. Makromol. Chem.* **1973,** *29/30,* 1.
55. Bucknall, C. B. *Toughened Plastics;* Applied Science: London, 1977; p 104.
56. Bucknall, C. B. *Toughened Plastics;* Applied Science: London, 1977; p 298.
57. Morbitzer, L.; Kranz, D.; Humme, G.; Ott, K. H. *J. Appl. Polym. Sci.* **1976,** *20,* 2691.
58. Riccò, T.; Rink, M.; Caporusso, S.; Pavan, A. *International Conference; Toughening of Plastics II;* Plastics and Rubber Institute: London, 1985; Vol. 27, pp 2–3.
59. Lednicky, F.; Pelzbauer, Z. *Angew. Makromol. Chem.* **1986,** *141,* 151.
60. Bohn, L. In *Copolymers, Polyblends, and Composites;* Platzer, N. A., Ed.; Advances in Chemistry 142; American Chemical Society: Washington, DC, 1975; p 66; *Angew. Makromol. Chem.* **1971,** *20,* 129.
61. Morbitzer, L.; Ott, K. H.; Schuster, H.; Kranz, D. *Angew. Makromol. Chem.* **1972,** *27,* 57.
62. Brighton, C. A.; Pritchard, G.; Skinner, G. A. *Styrene Polymers Technology and Environmental Aspects;* Applied Science: London, 1979; p 236.
63. Conaghan, B. F.; Rosen, S. L. *Polym. Eng. Sci.* **1972,** *12,* 134.

64. Inoue, T.; Soen, T.; Hashimoto, T.; Kawai, H. *Macromolecules* **1970**, *3*, 87.
65. Zelinski, R. P.; Hsieh, H. L. U.S. Patent 32 81 383, 1966, to Phillips Petroleum Company.
66. Kitchen, A. G.; Szalla, F. J. U.S. Patent 36 39 517, 1972, to Phillips Petroleum Company.
67. Fahrbach, G.; Gerberding, K.; Seiler, E.; Stein, D. U.S. Patent 40 86 298, 1978, to BASF AG.
68. Fahrbach, G.; Gerberding, K.; Seiler, E.; Stein, D. U.S. Patent 41 67 545, 1979, to BASF AG.
69. Keading, W. W.; Young, L. B.; Prapas, A. G. *CHEMTECH* **1982**, *12*, 556.
70. Dean, B. D. *J. Elastomers Plast.* **1985**, *17*, 55.
71. Jelenic, J.; Kirste, R. G.; Oberthür, R. C.; Schmitt-Strecker, S.; Schmitt, B. J. *Makromol. Chem.* **1984**, *185*, 129.
72. Humme, G.; Röhr, H.; Serini, V. *Angew. Makromol. Chem.* **1977**, *58/59*, 85.
73. Lee, G. F.; Katchman, A.; Shank, C. P.; Cooper, G. D. *Polym. Prepr. (Am. Chem. Soc. Div. Polym. Chem.)* **1977**, *18* (*1*), 842.
74. Morbitzer, L.; Kress, H. J.; Lindner, C.; Ott, K. H. *Angew. Makromol. Chem.* **1985**, *132*, 19.
75. Suarez, H.; Barlow, J. W.; Paul, D. R. *J. Appl. Polym. Sci.* **1984**, *29*, 3253.
76. Weber, G.; Schoeps, J. *Angew. Makromol. Chem.* **1985**, *136*, 45.

RECEIVED for review February 11, 1988. ACCEPTED revised manuscript September 7, 1988.

FRACTURE MECHANICS AND FAILURE MECHANISMS

Relationships Between the Microstructure and Fracture Behavior of Rubber-Toughened Thermosetting Polymers

A. J. Kinloch

Imperial College of Science, Technology and Medicine, University of London, Department of Mechanical Engineering, Exhibition Road, London, SW7 2BX, United Kingdom

Thermosetting polymers may be modified by the incorporation of a second phase of dispersed rubbery particles in order to increase greatly the resistance of the material to crack initiation and growth. This chapter discusses the toughening micromechanisms that operate in such materials and the relationships between the microstructure of rubber-toughened thermosetting polymers and their fracture behavior. Finally, a model is proposed to account for the observed structure–property relationships.

POLYMERIZED THERMOSETTING POLYMERS ARE AMORPHOUS and highly cross-linked. This microstructure results in many useful properties for engineering applications, such as high modulus and failure strength, low creep, and good performance at elevated temperatures. However, the structure of thermosetting polymers also leads to one highly undesirable property: They are relatively brittle materials with poor resistance to crack initiation and growth. Incorporation of a second phase of dispersed rubbery particles into thermosetting polymers can greatly increase their toughness without significantly impairing the other desirable engineering properties.

Epoxies, phenolics, polyimides, and polyesters have been modified by the incorporation of a second rubbery phase. Rubber-toughened epoxies

0065–2393/89/0222–0067$07.25/0

have been the most extensively studied of these polymers, thanks to their wide use as structural adhesives and increasing use in fiber composites. The multiphase microstructure of a rubber-modified epoxy can be seen clearly in the transmission electron micrograph of a thin section shown in Figure 1. The rubber employed was a carboxyl-terminated butadiene–acrylonitrile (CTBN) copolymer.

Several general reviews on the science, technology, and applications of rubber-toughened thermosetting polymers (1–4), as well as rubber-toughened thermoplastic polymers (1, 5), are available. Therefore, the aim of this chapter is to discuss in detail our current understanding of the toughening mechanisms and the relationships between the microstructure and fracture behavior. Finally, I will suggest a model to explain the observed micromechanisms and fracture behavior of rubber-toughened thermosetting polymers.

Toughening Mechanisms

Before considering how the microstructure of a rubber-modified thermosetting polymer influences its fracture behavior, we should discuss the mechanisms whereby the presence of rubbery particles increases the toughness of a thermosetting polymer.

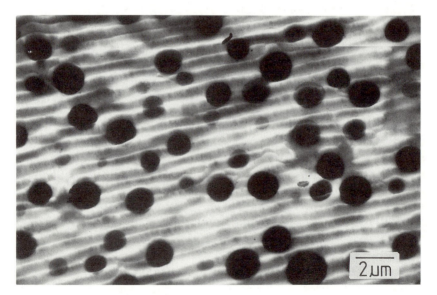

Figure 1. Transmission electron micrograph of a thin section of a rubber-modified epoxy polymer. The section was exposed to osmium tetroxide prior to examination. The selective reactions of the reagent with the unsaturated butadiene group increases the contrast between the rubber particles and the epoxy matrix.

Many different mechanisms have been proposed to explain the greatly improved fracture energy or fracture toughness that may result when a thermosetting polymer possesses a multiphase microstructure of dispersed rubbery particles. Much of the dispute concerns whether the rubbery particles or the thermoset matrix absorbs most of the energy and, if it is the matrix, whether the deformation mechanism involves the formation of crazes. Detailed reviews of these mechanisms have been published (2, 3, 6). Recent work by Kinloch et al. (6, 7) and independently by Yee and Pearson (8) has clearly established that plastic shear-yielding in the matrix is the main source of energy dissipation and increased toughness. Such plastic deformation caused by interactions between the stress field ahead of the crack and rubbery particles occurs to a far greater extent in the matrix of rubber-toughened thermosets than in unmodified polymer. Hence, the first step toward understanding the role of rubbery particles is to consider the stress field around a particle.

Stress Fields Around Rubbery Particles. Goodier (9) derived equations for the stresses in the matrix around an isolated elastic spherical particle embedded in an isotropic elastic matrix, where the matrix is subjected to an applied uniaxial tensile stress remote from the particle. His equations reveal that for a rubbery particle, which typically possesses a considerably lower shear modulus than the matrix, the maximum tensile stress concentration in the matrix resulting from the presence of the particle occurs at the equator of the particle and has a value of about 1.9. Furthermore, assuming that the particle is well bonded to the matrix, the local stress state at this point is one of triaxial tension. This tension arises essentially because of the volume constraint represented by the bulk modulus of the rubbery particle, which is comparable to that of the matrix. The low shear modulus of the rubber particle relative to the thermoset matrix, in spite of its comparable bulk modulus, is a consequence of the Poisson's ratio of the rubber being approximately 0.5; that of the matrix is 0.35. Thus, in contrast to a "hole", which would produce a similar-size stress concentration, the rubbery particle can fully bear its share of the load across the crack front. This ability of rubber particles to bear loads while functioning as stress concentrators can explain the observation that rubbery particles are most effective for toughening the polymer and holes are ineffective.

Broutman and Panizza (10) developed finite-element stress analyses to obtain the stress concentration around rubbery particles when the volume fraction, v_f, of particles is high enough so that the stress fields of the particles interact (i.e., for values of v_f greater than about 0.09). In such cases the maximum stress concentration in the matrix at the equator of the particle may be somewhat higher than for the isolated case. For example, for a volume fraction of 0.2 the maximum stress concentration is about 2.1; for a volume fraction of 0.4 it is about 2.7.

Goodier's analysis (9) also indicates that the stress concentration in the matrix at the particle equator will be present even at temperatures below the glass-transition temperature (T_g) of the rubber. Under such circumstances the shear modulus of the rubbery particle would still be expected to be slightly lower than that of the highly cross-linked thermoset matrix, which is even further below its T_g. However, as the shear moduli will now be much closer in value, the extent of the stress concentration will be decreased somewhat. This relationship explains why rubber-modified thermosets are still somewhat tougher than the unmodified polymer at temperatures well below T_g(rubbery phase).

Matrix Shear Yielding and Particle Cavitation.   The stress field associated with the rubbery particles leads to the initiation of two important deformation processes that can strongly interact, especially at high test temperatures or slow test rates.

One important process is the initiation and growth of multiple localized shear-yield deformations in the matrix. The stress concentrations around the rubbery particles act as initiation sites for the plastic shear deformation. It is not clearly established whether such plastic shear deformation is initiated at the rubber–matrix interface exactly at the equator of the particle (where the maximum tensile stress concentration occurs) or between the equator and the pole of the particle (where the maximum shear stress concentration is likely to occur). However, plastic shear deformation bands are initiated as a result of the stress concentration in the matrix caused by the presence of the rubbery particles. Because there are many particles, considerably more plastic energy dissipation exists in multiphase material than in unmodified polymer. However, plastic deformation is localized through the postyield strain softening of the epoxy matrix and the shear deformations that initiate at one particle but terminate at another. Enhanced plastic deformation in the multiphase material is clearly evident by comparing Figures 2A and 2B and observing the greatly increased surface roughness and drawn ridges of epoxy polymer in Figure 2B. Further, the localized nature of the plastic shear bands, visible as furrows running at angles of approximately 45° to the principal tensile stress (i.e., in the direction of maximum shear stress) is illustrated in Figure 3.

The other major deformation process is cavitation of the rubbery particle. In considering this phenomenon, it is necessary to recall that an overall triaxial stress state (plane strain) usually exists ahead of a crack and produces dilation. Combined with the stresses that are induced in the particle by cooling after cure, dilation causes failure and void formation either in the particle or at the particle–matrix interface. Once formed, the voids grow and so dissipate energy.

However, a more important feature is that such voids can greatly enhance the shear-yielding process. Essentially, they lower the extent of stress triaxiality in the adjacent matrix, which lowers the stress required for shear yielding and so promotes even more extensive plastic shear deformations in

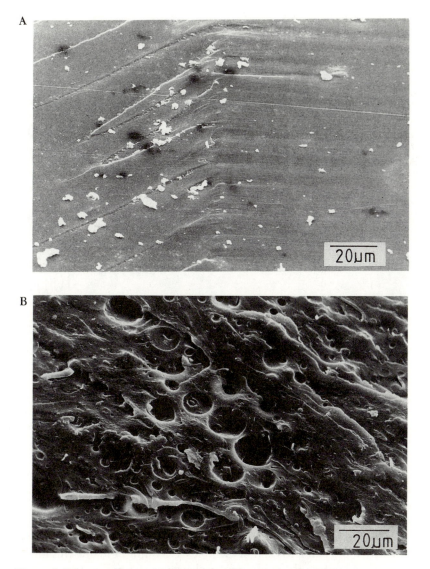

Figure 2. Scanning electron micrographs of fracture surfaces (test temperature, 23 °C). A, Unmodified epoxy polymer; B, Rubber-toughened epoxy polymer [v_f(rubber phase) = 0.14]. (Reproduced from ref. 11. Copyright 1984 American Chemical Society.)

the matrix. However, it appears that such cavitation, and associated relief of the plane-strain constraint, is only effective if it occurs when substantial plastic deformation of the matrix is possible (i.e., cavitation must occur close to the yield point of the matrix). Hence, holes would be expected to be ineffective. In this case they are ineffective because they obviously relieve the overall triaxial stress state at the crack tip from the first instant of loading (i.e., when the material is far from undergoing local plastic deformation).

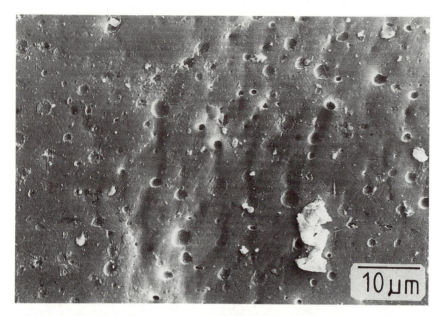

Figure 3. Scanning electron micrograph of surface, normal to the crack plane, of rubber-toughened epoxy. The furrows between the rubber particles, at approximately 45° to the maximum principal tensile stress, are caused by the shear bands that occur at constant volume. (Reproduced with permission from ref. 12. Copyright 1985 Chapman and Hall.)

This cavitation mechanism of the rubbery particles is responsible for the apparent holes in the scanning electron fractographs. Such features are particularly pronounced at higher test temperatures when plastic deformation of the matrix can occur more easily and the rubber particles more readily cavitate (Figure 2B). The higher magnification used in Figure 4 confirms that the holes are filled with rubber; in some instances, an internal void or tear in the rubbery particle can be seen clearly. Other experiments have revealed that such holes become hillocks after the rubbery phase has become swollen with a solvent (Figure 5). Both pieces of evidence clearly indicate that the cavitated rubber is still in the holes as a lining. Finally, this initiation and growth of voids in the rubbery particles gives rise to the stress whitening that is often observed before the crack tip and on the fracture surfaces.

Microstructural Aspects

The microstructural features of rubber-toughened thermosetting polymers that may affect fracture behavior are as follows:

Thermosetting matrix phase
   Cross-link density
   Glass-transition temperature of matrix phase

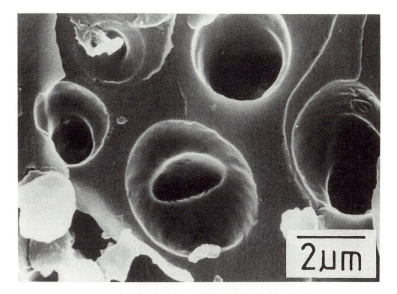

Figure 4. Scanning electron micrograph of fracture surface of rubber-toughened epoxy polymer showing cavitated rubber particles. (Reproduced with permission from ref. 3. Copyright 1986 Elsevier Applied Science Publishers.)

Rubbery dispersed phase
    Volume fraction
    Particle size
    Distribution of particle size
    Intrinsic adhesion across particle–matrix interface
    Morphology of dispersed phase
    Glass-transition temperature of dispersed phase

Microstructural Features of the Matrix Phase

The toughening mechanism outlined highlights the role of the inherent ductility of the matrix in influencing the toughness of the multiphase thermosetting polymer. For example, a decrease in the stress needed for localized shear yielding should obviously assist in increasing the toughness, if all other important microstructural features are unchanged. However, an increase in the inherent ductility can usually only be achieved at the expense of other important properties.

Two interrelated properties may influence the degree of ductility of the matrix: the molecular weight, M_c, between cross-links and the glass-transition temperature, T_g(matrix). These properties and the effects of test temperature and rate will be discussed. Although they are not microstructural features of the matrix, they greatly affect its response.

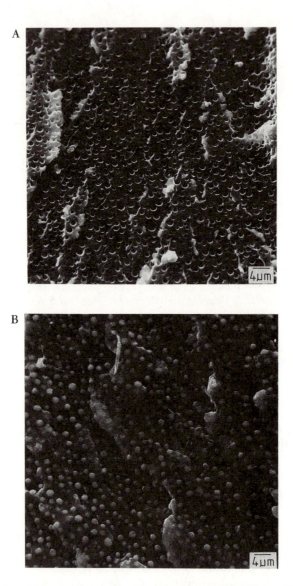

Figure 5. Scanning electron micrographs of fracture surfaces of a rubber-toughened epoxy polymer. A, Before exposure to a solvent; B, After exposure to a solvent that swells the rubbery phase. (Reproduced with permission from ref. 3. Copyright 1986 Elsevier Applied Science Publishers.)

Molecular Weight Between Cross-Links. The effect of molecular weight between cross-links in the thermoset matrix has been investigated by Pearson and Yee (*13*) and Finch et al. (*14*). The former authors studied the toughness of a series of epoxy–CTBN polymers cured with 4,4'-diaminodiphenylsulfone. The molecular weight, M_c, between cross-links in this series was varied by using epoxy resins of different epoxy equivalent weights: the higher the epoxy equivalent weight, the higher is the subsequent value of M_c. Their results are shown in Table I. An increase in M_c produces a small increase in the value of the fracture energy, G_{Ic}, for the unmodified polymers and a dramatic increase in G_{Ic} for the rubber-toughened polymers. These results may be ascribed to an increase in the inherent ductility of the matrix as M_c increases. Greater ductility leads to greater plastic energy dissipation and an increase in G_{Ic}. A greater effect is observed for the rubber-toughened polymer because the rubbery particles initiate a large number of energy-dissipating shear zones.

The more recent studies of Finch et al. (*14*) confirm the increased ductility exhibited as the value of M_c is increased and the resulting increase in G_{Ic} that may be attained. These workers varied the molecular weight between cross-links in the epoxy matrix of an epoxy–CTBN system that was cured with piperidine by changing the cure time and temperature. The microstructure of the dispersed rubbery particles in the modified materials was found not to be significantly changed by the different curing conditions. Plane-strain compression tests on the unmodified polymers clearly revealed that, although the modulus and yield stress were not greatly affected, the maximum plastic strain capability of the matrix was dramatically increased as the value of M_c increased. This finding may be seen from the stress versus strain curves shown in Figure 6. Compression experiments were conducted because the materials fracture before yielding in a standard uniaxial tensile experiment. The increased ductility of the matrix as the cross-link density was decreased was indeed reflected in higher values of G_{Ic}, as shown in Table II.

T_g of the Matrix Phase. The T_g of the matrix phase is obviously affected by the value of M_c, as may be seen from the data in Table II. As

Table I. Effect of Molecular Weight, M_c, Between Cross-Links on the Fracture Energy, G_{Ic} (kJ/m²)

Epoxy Equivalent Weight	Unmodified Polymer[a]	Rubber-Toughened Polymer[b]
172–176	0.16	0.22
475–575	0.20	2.41
1600–2000	0.33	11.5

[a]Epoxy resin; 4,4'-diaminodiphenyl sulfone hardener.
[b]As in footnote *a* but with CTBN.

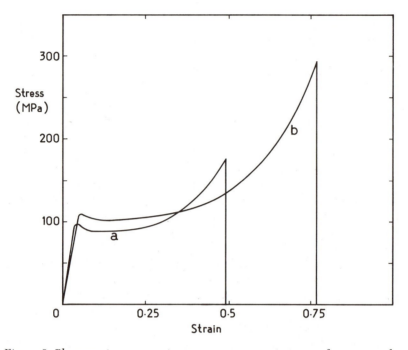

Figure 6. Plane-strain compression stress versus strain curves for epoxy poly-mers having different values of molecular weight, M_c, between cross-links. Epoxy resin; piperidine hardener. a, Cured for 16 h at 120 °C, M_c = 610 g/mol; b, Cured for 6 h at 160 °C, M_c = 4640 g/mol. (Reproduced with per-mission from ref. 14. Copyright 1987 Butterworth.)

Table II. Effect of Varying the Molecular Weight, M_c, Between Cross-Links on the Fracture Energy, G_{Ic}, of an Epoxy–CTBN–Piperidine Material

Cure Conditions	M_c (g/mol)[a]	T_g(matrix) (°C)	G_{Ic} (kJ/m²)
16 h at 120 °C	610	100	2.23
6 h at 160 °C	4640	89	5.90

[a]Average value from ref. 14.

the degree of cross-linking is reduced (i.e., M_c is increased) T_g usually falls. However, this gain in matrix ductility at the expense of a decrease in T_g is often unacceptable, especially if the rubber-toughened polymer is to be exposed to moisture. In this situation the T_g will fall because the material absorbs and is plasticized by water.

The value of the T_g is influenced by other parameters, such as the degree of interchain attraction. Low interchain attractive forces, which may give an increase in matrix ductility, often lead to low T_g values. Thus, increased ductility and toughness are frequently achieved only at the expense of el-evated-temperature properties.

As a corollary to this argument, thermosetting polymers that are ex-
cremely tightly cross-linked and possess high interchain attractive forces (and
hence have high T_g values) are difficult to toughen to a high absolute level.
This condition is evident from work on rubber-toughened epoxies (*11, 15*)
and rubber-toughened bismaleimides (*16, 17*). The need to obtain very tough
thermosets without sacrificing very high glass-transition temperatures is self-
evident and presents an exciting and rewarding challenge.

Test Temperature and Rate.   Although these features are obviously
not microstructural, it is convenient to consider their effects at this point
because their main influence appears to be on the inherent ductility of the
matrix. This influence is illustrated in Figure 7, which shows uniaxial compressive yield stress as a function of test temperature for different epoxy
polymers (*11*). At any particular temperature, the presence of the rubbery
phase decreases the yield stress a little, but the dominating parameter is
the test temperature. The higher the test temperature, the lower is the yield
stress, the greater the plastic strain capability of the epoxy, and the easier
becomes the generation of plastic deformation. This situation is reflected in
the increase of the fracture energy, G_{Ic}, with test temperature, as shown in
Figure 8.

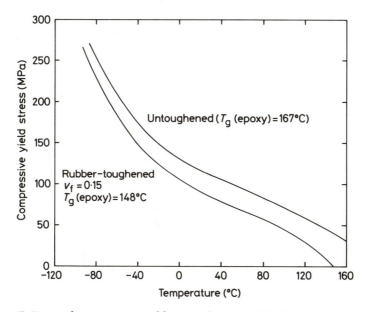

*Figure 7. Uniaxial compression yield stress of an unmodified epoxy and rubber-
toughened epoxy polymers as a function of test temperature. Epoxy resin;
trimethylene glycol p-diaminobenzoate (TMAB) hardener; CTBN rubber. (Reproduced from ref. 11. Copyright 1984 American Chemical Society.)*

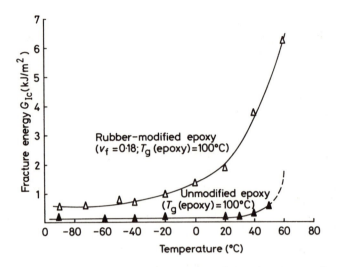

Figure 8. Fracture energy, G_{Ic}, as a function of test temperature for an un-modified and rubber-toughened epoxy polymer. Epoxy resin; piperidine hardener; CTBN rubber. (Reproduced with permission from ref. 20. Copyright 1987 Chapman and Hall.)

Various studies (2, 7, 18, 19) have also revealed that decreasing the rate is equivalent to increasing the temperature. Indeed, workers (19, 20) have found that rate and temperature effects may be interrelated by constructing a master curve of G_{Ic} versus the reduced time of test, t_f/a_T, below the T_g of the epoxy matrix (t_f is the time to failure and a_T is the time–temperature shift factor). The values of a_T were ascertained experimentally by super-imposing yield, stress relaxation, and dynamic mechanical relaxation data. Thus, the effects of rate and temperature on the value of G_{Ic} may be simply combined, as shown in Figure 9. This approach not only enables objective comparison of different polymers by using an intrinsic reference point, but also permits the value of G_{Ic} to be predicted for any given test rate–temperature combination.

Of particular interest in Figures 8 and 9 is the observation that the value of G_{Ic} for the rubber-toughened epoxy increases more rapidly as the T_g of the matrix is approached. This effect arises partly because the yield stress of the matrix is falling as the temperature is increased. More importantly, extensive cavitation of the rubbery particles occurs under these conditions, as may be seen clearly from the intense stress whitening on the fracture surfaces. This cavitation micromechanism will reduce the effective local stress necessary for shear yielding to occur. Thus, cavitation allows even more extensive plastic deformation to result before the crack tip and leads to a high G_{Ic} value.

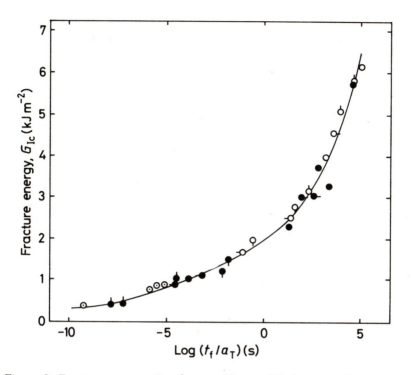

Figure 9. Fracture energy, G_{Ic}, *for a rubber-modified epoxy polymer as a function of* $\log (t_f/a_T)$ *where* t_f *is the time to failure and the shift factor,* a_T, *is determined from yield, stress relaxation, and dynamic mechanical relaxation data. Epoxy resin; piperidine hardener; CTBN rubber.* ○, 60 °C; ◔, 50 °C; ◑, 40 °C; ◓, 37 °C; ◒, 30 °C; ●, 23 °C; ◖, 15 °C; ◗, 0 °C; ◐, –20 °C; ◕, –40 °C; ⊙, –60 °C; Reference temperature, $T_g - 80$ °C = 20 °C. (Reproduced with permission from ref. 20. Copyright 1987 Chapman and Hall.)

Microstructural Features of the Rubbery Phase

Although the microstructural features of the rubbery dispersed phase have a major effect on the toughness of the multiphase polymer, there have been few definitive studies of the influence of such features as volume fraction and particle size. In many studies several crucial features have been changed simultaneously. However, an attempt is made here to establish correlations between the fracture behavior and microstructural features of the rubbery phase.

Volume Fraction. The toughness of a multiphase thermoset generally increases as the volume fraction, v_f, of dispersed rubbery phase is increased, but the modulus and yield strength will usually decrease slightly. Bucknall and Yoshii (*21*) found a linear relation between G_{Ic} and v_f for different epoxy resin hardener systems using a CTBN rubber. However, Kunz et al. (22) found (for both amine-terminated butadiene–acrylonitrile (ATBN) and CTBN modified epoxies) that once a v_f of about 0.1 had been achieved in the polymer, further increases in volume fraction resulted in only comparatively minor increases in the value of G_{Ic}.

More recent work by Kinloch and Hunston (20) has assisted in explaining these apparently conflicting observations and has also shown that the linear relation reported by Bucknall and Yoshii is somewhat fortuitous. Kinloch and Hunston (20) used the concept of time–temperature superpositioning to separate the effects of changing the v_f and the properties of the matrix phase. The effect of the volume fraction of the separated rubbery phase is shown in Figure 10, where G_{Ic} is plotted against the value of v_f at two different values of t_f/a_T. In the case of a unimodal distribution of rubbery particles with a low value of t_f/a_T (i.e., a low test temperature), the relation between G_{Ic} and v_f is such that G_{Ic} increases only very slowly with increasing volume fraction and then reaches a plateau value. At the higher value of t_f/a_T (i.e., a high test temperature) the relation is approximately linear and the rate of increase of G_{Ic} with v_f is far greater than that observed at the low value of t_f/a_T. This result demonstrates that when the epoxy matrix is more ductile, it can respond more readily to the presence of the rubbery particles. Therefore, there is no unique relationship between toughness and the volume fraction of rubbery particles.

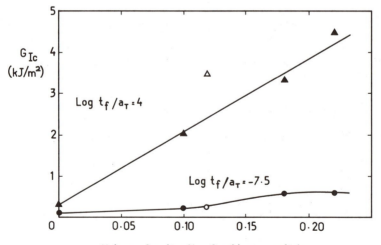

Figure 10. Plots of fracture energy, G_{Ic}, against volume fraction, v_f, of rubbery particles for two test conditions (see Figure 9). Epoxy resin; piperidine hardener; CTBN rubber. Closed points: unimodal distribution; open points: bimodal distribution. (Reproduced with permission from ref. 20. Copyright 1987 Chapman and Hall.)

Finally, the maximum value of v_f that can be achieved via the in situ phase-separation technique is about 0.2–0.3. Attempts to produce higher fractions result in phase inversion and a loss of mechanical properties.

Particle Size. Several studies (23, 24) have suggested that particle size may influence the extent and type of toughening micromechanism, and so affect the measured toughness. For example, Sultan and McGarry (23) proposed that large particles (0.1–1 μm in diameter) cause crazing and are five times more effective in toughening the thermoset than small particles (about 0.01 μm), which initiate shear yielding. However, the evidence for crazing was not convincing, and this mechanism has now been discounted in such thermosets. Furthermore, in these studies the value of v_f was not determined; in order to elucidate the effect of particle size, the value of v_f must be measured and ideally held constant.

In the cases (8, 11, 20, 21) in which this effect has been achieved, particle size does not appear to have a major influence on mechanical properties. However, the average particle diameter was only varied between about 0.5 and 5 μm, and no directly comparable materials have yet been prepared that contain only very small particles (i.e., <0.1 μm). Other workers (20, 22) have prepared toughened epoxy polymers in which small particles (<0.1 μm) were present. However, because these experiments had a bimodal distribution of particle sizes, their results are discussed in the next paragraph.

Particle-Size Distribution. The evidence concerning the influence of particle-size distribution on measured toughness is somewhat contradictory. Kunz et al. (22) prepared a series of toughened-epoxy polymers that contained a bimodal distribution of very small (<0.1 μm) and larger (0.2–1 μm) particles. However, the distribution and volume fraction of rubber were changing together; as the distribution was changing toward a relatively greater number of larger particles, the value of v_f was also increasing. They found that the value of G_{Ic} did not increase with increasing value of v_f as much as had been expected. These observations led to the suggestions that larger particles are not as effective as smaller ones in improving toughness and that the distribution of rubbery particle sizes is a more critical parameter than the overall volume fraction of particles.

However, the work of Kinloch and Hunston (20) indicates that the process of interpreting the effect of particle-size distribution is more complex. At first sight their work also indicates that, at a fixed value of v_f, a bimodal distribution of particle sizes (centered upon maxima of 0.1 and 1.3 μm) can indeed result in higher values of G_{Ic} than a unimodal distribution (mean diameter 1.2 μm). However, as shown in Figure 10, this enhancement of the bimodal distribution was only observed over a certain range of test temperatures and rates, and thus suggested that matrix effects were also playing a role. This suggestion is reinforced by the T_g(matrix) being somewhat different for the unimodal and bimodal materials. Thus, the effect of particle-size distribution has still to be conclusively established.

Interfacial Adhesion. Nearly all of the studies reported in the literature have been concerned with well-bonded rubbery particles, a consequence of the chemical reactivity of the rubber. But poor bonding across the particle–matrix interface has been specifically engineered (11) by using an unreactive liquid rubber. As might be expected, this multiphase epoxy polymer showed virtually no increase in toughness, compared with the unmodified material. Thus, intrinsic adhesion across the particle–matrix interface needs to be sufficiently high to prevent premature particle debonding. Strong interfacial chemical bonds are usually present, of course, when a reactive rubber is used.

Morphology of Dispersed Phase. In only a few studies (8, 21) has the presence of a phase-separated morphology in the rubbery particles been reported. No evidence indicates whether a complex morphology of the particle is desirable or not, but at the very least it should ensure that a relatively high value of v_f is attained.

T_g of the Dispersed Phase. The particles need not be truly rubbery for a multiphase thermoset to be appreciably tougher than the unmodified material. Figures 8 and 9 show quite clearly that the multiphase epoxy polymer is significantly tougher, even when the test temperature is well below T_g(rubbery phase) (i.e., below about –55 °C). Nevertheless, the most substantial increases in G_{Ic} are observed at temperatures above T_g(rubbery phase).

The detailed reasons that may be suggested for this observation have been mentioned in previous discussions. Basically, the very low ratio of shear to bulk modulus of rubbery particles [i.e., when $T_{test} > T_g$(rubbery phase)] means that the particles can not only act as very effective stress initiators for shear yielding in the matrix, but can also bear their share of the triaxial tensile stresses along the crack front. Second, the high shear capability of rubber particles means that they can deform and accommodate the shear strains imposed by the shear bands that they initiate. Third, under certain test conditions, rubber particles may undergo cavitation, relieve the triaxial nature of the stresses acting near the crack tip, and hence enable even further shear yielding to occur in the matrix.

Finally, the multiphase thermosets may include a plastic dispersed phase such as a hydroxyl-terminated polysulfone. The specific attributes of rubbery particles may explain the observation (25) that, although a plastic dispersed phase may increase the toughness of a thermoset, it is usually far less effective in this respect than a rubbery phase.

Modeling Structure–Fracture-Behavior Relationships

To model structure–fracture-behavior relationships successfully, we must quantify the magnitude of the toughening micromechanisms and compare the predicted relationships to the qualitative evidence described. A first step is to consider the model of the "process zone" shown in Figure 11. The process zone is the region around the crack tip where the viscoelastic and plastic energy dissipative processes occur. This figure shows that the rubber particles initiate shear bands, which then propagate. To estimate the relationship between the microstructural parameters and the toughness, we must estimate the volume fraction of plastically sheared matrix material, deduce the energy dissipated by the generation of such bands, and quantify the size of the process zone.

Volume Fraction of Plastically Deformed Material.   If it is assumed that the area of the plastic shear bands initiated by the rubbery particles is proportional to the diametrical cross-sectional area of the particle, whose radius is r_p, and that the shear bands run between rubber particles, then the volume, V_m, of sheared matrix material per particle may be expressed by

$$V_m = \alpha \cdot \pi r_p^2 \cdot d_p \cdot n_s \qquad (1)$$

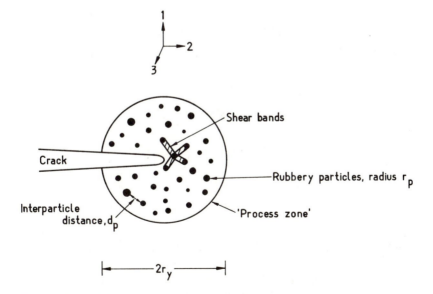

Figure 11. Model of process zone around crack tip showing rubbery particles initiating localized shear bands in matrix material.

where d_p is the interparticle distance and n_s is the average number of shear bands generated per particle. As shear bands will occur on the planes at approximately $\pm 45°$ to the principal stresses "σ_1 and σ_3" and "σ_1 and σ_2", then the value of n_s will be 4 (i.e., each particle will have eight shear bands associated with it, but will share these to give an average of four per particle). The term α is a scaling factor that takes into account the strain field in the process zone; $1 \leq \alpha \leq 0$. Close to the crack tip the plastic strains will be relatively high and $\alpha \to 1$; at the edges of the process zone, where the strains approach the elastic limit, $\alpha \to 0$. Now let

$$\alpha = 1 - \frac{r}{r_y} \tag{2}$$

where r is the radial distance from the crack tip. At $r = 0$, then $\alpha = 1$; at $r = r_y$, then $\alpha = 0$; the mean value of α is 0.5.

The volume fraction of sheared matrix material, V_{fm}, is given by

$$V_{fm} = \alpha \cdot \pi r_p^2 \cdot d_p \cdot n_s \cdot N_v \tag{3}$$

where N_v is the number of rubbery particles per unit volume of material and may be expressed by

$$N_v = \frac{3v_f}{4\pi r_p^3} \tag{4}$$

where v_f is the volume fraction of rubbery particles. But if we assume that the rubbery particles are spherical and arranged in a cubic array, then the value of d_p is given (26) by

$$d_p = r_p \left[\left(\frac{2\pi}{3v_f} \right)^{1/2} - \frac{\pi}{2} \right] \tag{5}$$

Substituting from equations 4 and 5 into equation 3 yields, for $\alpha = 0.5$

$$V_{fm} = \frac{3v_f n_s}{8} \left[\left(\frac{2\pi}{3v_f} \right)^{1/2} - \frac{\pi}{2} \right] \tag{6}$$

Hence, equation 6 gives the volume fraction of plastically deformed material in terms of the microstructural features of the rubber-modified polymer.

Energy Dissipated by Plastic Deformation.   The fracture energy, G_{Ic}, for the rubber-toughened epoxy may be expressed (27) by

$$G_{Ic} = G_{Ieu} + \psi_r \tag{7}$$

where G_{Icu} denotes the fracture energy of the unmodified polymer and ψ_r is the additional plastic deformation dissipated per unit area in the rubber-modified material because of the presence of the rubbery particles.

The term ψ_r may be evaluated from a knowledge of the volume fraction, V_{fm}, of plastically deformed material, and the hysteresis (or loss) strain-energy density, W_d, involved in such deformations. Hence

$$\psi_r = V_{\mathrm{fm}} \int_0^{r_y} W_d(r) \, dr \tag{8}$$

If the polymer is modeled as a perfectly elastic–plastic material, then

$$W_d \approx \tau_y \cdot \gamma(r) \tag{9}$$

where τ_y is the shear yield stress and γ is the plastic strain in the shear band. As discussed earlier, the plastic shear strain, γ, would be expected to be a function of distance, r, from the crack tip and may be expressed by

$$\gamma(r) = \gamma_f - \frac{\gamma_f \cdot r}{r_y} \tag{10}$$

where γ_f is the maximum plastic shear strain that the matrix can undergo prior to fracture. Substituting from equations 9 and 10 into equation 8 gives

$$\psi_r = V_{\mathrm{fm}} \int_0^{r_y} \tau_y \left(\gamma_f - \frac{\gamma_f \cdot r}{r_y} \right) dr \tag{11}$$

Thus

$$\psi_r = \frac{V_{\mathrm{fm}} \cdot \tau_y \gamma_f \cdot r_y}{2} \tag{12}$$

and substituting from equation 6 gives

$$\psi_r = \frac{3 v_f n_s}{16} \left[\left(\frac{2\pi}{3v_f} \right)^{1/2} - \frac{\pi}{2} \right] \cdot \tau_y \gamma_f \cdot r_y \tag{13}$$

The next step is to develop an expression for r_y (i.e., the size of the process zone at fracture).

Estimating the Value of r_y.   The value of r_y may be evaluated, of course, from a knowledge (28) of the fracture energy, G_{Ic}, and the yield stress of the rubber-toughened polymer. However, this approach is not very helpful because the value of G_{Ic} for the rubber-modified material is the parameter that we are attempting to model as a function of microstructure of the multiphase polymer. Therefore, the approach adopted is to estimate the value of r_y for the rubber-modified material relative to the value of the process zone, r_{yu}, for the unmodified polymer. By analogy to an Irwin- or Dugdale-type model (28), the size of the process zone is proportional to the square of the yield stress. Hence, to a first simple approximation, the following relation has been assumed:

$$\frac{r_y}{r_{yu}} \approx k^2 \tag{14}$$

where the ratio k represents the increase in the size of the process zone for the rubber-modified epoxy that arises from the effects of:

- The stress concentration around the equator of the rubbery particle, further raising the local stress ahead of the crack tip to a level such that more extensive plastic yielding of the matrix polymer occurs.

- Cavitation of the rubbery particles, which may occur to a considerable degree under certain circumstances, relieving the triaxial stress state and so decreasing the stress necessary for shear yielding to be initiated.

Both aspects promote more plastic deformation in the matrix, as already discussed qualitatively.

From the stress analysis of Broutman and Panizza (10), the effect of the microstructure of the dispersed rubbery phase on the stress concentrations induced by the particles may be described by

$$k = 1.88 + 0.59v_f + 7.2v_f^2 \tag{15}$$

From the studies of Haward and Owen (29), the effect of cavitated rubbery particles on enabling the plastic deformation to occur more readily may be described by

$$k = 0.76 + 10.95v_f \tag{16}$$

If the data expressed by equations 15 and 16 is extrapolated to $v_f \rightarrow 0$, then the case for a single isolated particle in an infinite matrix is obtained. Obviously, these equations are completely inapplicable and invalid when no rubbery phase at all is present.

When no, or only very limited, cavitation of the rubbery particles occurs, then

$$k(\text{total}) = 1.88 + 0.59v_f + 7.2v_f^2 \tag{17}$$

When cavitation also occurs, then

$$k(\text{total}) = 2.64 + 11.54v_f + 7.2v_f^2 \tag{18}$$

Model Predictions.   From these equations, the energy dissipated in plastic deformations induced by the presence of the rubbery particles when no particle cavitation occurs is given by

$$\psi_r = \frac{3v_f n_s}{16}\left[\left(\frac{2\pi}{3v_f}\right)^{1/2} - \frac{\pi}{2}\right](1.88 + 0.59v_f + 7.2v_f^2)^2 \tau_y \gamma_f \cdot r_{yu} \tag{19}$$

When particle cavitation occurs, the dissipated energy is given by

$$\psi_r = \frac{3v_f n_s}{16}\left[\left(\frac{2\pi}{3v_f}\right)^{1/2} - \frac{\pi}{2}\right](2.64 + 11.54v_f + 7.2v_f^2)^2 \tau_y \gamma_f \cdot r_{yu} \tag{20}$$

These equations can be examined to see whether they agree with experimental observations.

First, they obviously demonstrate that the ability of the matrix to undergo shear yielding is of major importance. Values of τ_y and γ_f are necessary for quantitative predictions, and such data are scarce. However, if the plane-strain compression results shown in Figure 6 are employed, then the data and predictions given in Table III are obtained. As may be seen, predictions from the model described, for which all the parameters may be experimentally assessed, are in good agreement with the experimentally measured values.

Second, the model predicts that, for the rubbery phase, the most important major parameter is the volume fraction, v_f. The particle radius does not appear in the expressions derived from the model. These predictions are in good agreement with the experimental observations described pre-

Table III. Predictions from Model for Matrices with Different M_c Values

M_c (g/mol)	$\sigma_y{}^a$ (MPa)	γ_f	$\tau_y{}^b$ (MPa)	v_f	$G_{Icu}{}^c$ (kJ/m²)	G_{Ic}(model)^d (kJ/m²)	G_{Ic}(experimental) (kJ/m²)
610	96	0.5	38	0.18	0.21	2.7	2.23
4640	111	0.75	44	0.19	0.46	6.9	5.90

NOTE: Epoxy–piperidine–CTBN system; M_c changed by altering the cure schedule (*see* Table II). r_{yu} for unmodified epoxy (M_c = 610) was 23 μm; r_{yu} for unmodified epoxy (M_c = 4640) was 33 μm.
$^a\sigma_y$ and γ_f from plane-strain compression tests.
$^b\tau_y$ calculated from σ_y values, assuming a pressure-dependent yield stress with $\sigma_y = 2\tau_y/(1 - \mu)$ where μ is the pressure-dependent constant and was taken to be 0.2 (30).
$^cG_{Icu}$ is the fracture energy of the unmodified epoxy.
^{d}Extensive cavitation was observed so the values of G_{Ic} for the toughened epoxies were calculated from equations 7 and 20.

viously. The model does not predict any advantage from a having a bimodal, as opposed to a unimodal, distribution of particles. The experimental work that has been conducted in this area has still to yield a conclusive answer to this aspect of the effects of the microstructure of the rubber phase. However, the model does assume good adhesion between the rubbery particle and the matrix.

Third, the model indicates a rate–temperature dependence of the measured fracture energy of the rubber-toughened thermosetting polymer, through the variation of τ_y and γ_f and depending upon whether extensive cavitation accompanies the fracture process. At low temperatures or fast test rates, when very little cavitation is observed, equation 19 is appropriate. However, extensive cavitation may occur at higher temperatures or slow rates, and then equation 20 is more appropriate. Some predictions from the model are shown in Figure 12.

For a test temperature of 20 °C (giving a log t_f/a_T value of 2.5) the data shown in Table III ($M_c = 610$) have been used to predict the relationship between v_f of the rubbery dispersed phase and the resulting fracture energy, G_{Ic}. When the predictions are examined against the experimental data, there is good agreement. At a low test temperature (e.g., about –40 °C, corresponding to a log t_f/a_T value of about –7.5), there are currently no available

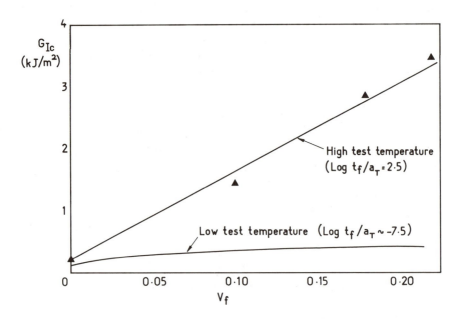

Figure 12. Fracture energy, G_{Ic}, versus volume fraction, v_f, of rubbery particles for two test conditions. The solid lines are predictions from the model and the points are experimental data. Epoxy resin; piperidine hardener; CTBN rubber.

data for the shear properties of the matrix. Therefore, an estimate of these properties has been made (τ_y = 60 MPa and γ_f = 0.2) and used in the calculations. The accuracy of this estimate may somewhat alter the exact values of the G_{Ic} for the rubber-toughened epoxies, but the general relationship shown in Figure 12 for a log t_f/a_T value of about −7.5 would not be changed radically. Comparison of Figure 12 to the experimental results shown in Figure 10 demonstrates that the model does indeed predict the correct form of the G_{Ic} versus v_f relationships for both a high and a low test temperature.

Concluding Remarks

The development of multiphase thermosetting polymers based upon the inclusion of a rubbery second phase has led to major improvements in the toughness of highly cross-linked thermosets. However, detailed quantitative studies to ascertain the relationships between the microstructure and fracture behavior have been undertaken only recently.

This chapter has reviewed the experimental relationships that have been reported between the microstructure and fracture behavior of rubber-toughened thermosets and proposed a quantitative model to explain and predict the observed data. The model still requires several refinements, but does appear to agree, both qualitatively and quantitatively, with the observed microstructure–fracture relationships.

List of Symbols

G_{Ic}	fracture energy
G_{Icu}	fracture energy of unmodified material
M_c	molecular weight between cross-links of the epoxy phase
N_v	number of rubbery particles per unit volume
T_g	glass-transition temperature
V_{fm}	volume fraction of sheared matrix
V_m	volume of sheared matrix
W_d	hysteresis, or loss, strain–energy density
a_T	time–temperature shift factor
d_p	interparticle distance
n_s	average number of shear bands generated per particle
r	distance
r_p	radius of particle
r_y	radius of process zone for rubber-modified polymer
r_{yu}	radius of process zone for unmodified polymer
t_f	time to failure
v_f	volume fraction of particles
α	scaling factor

γ plastic shear strain
γ_f maximum plastic shear strain at failure
μ pressure-dependent constant
σ_y yield stress (from plane-strain compression tests)
τ_y shear yield stress
ψ_r additional plastic energy dissipated per unit area of crack
            growth in rubber-modified material because of the presence
            of rubber particles

References

1. Bucknall, C. B. *Toughened Plastics*; Applied Science: London, 1977.
2. Kinloch, A. J.; Young, R. J. *Fracture Behaviour of Polymers*; Applied Science: London, 1983; p 421.
3. *Structural Adhesives: Developments in Resins and Primers*; Kinloch, A. J., Ed.; Applied Science: London, 1986; p 127.
4. Kinloch, A. J. *Adhesion and Adhesives*; Chapman and Hall: London, 1987.
5. Keskkula, H.; Turley, S. G.; Boyer, R. F. *J. Appl. Polym. Sci.* **1971**, *15*, 351.
6. Kinloch, A. J.; Shaw, S. J.; Hunston, D. L. *Polymer* **1983**, *24*, 1355.
7. Kinloch, A. J.; Shaw, S. J.; Tod, D. A.; Hunston, D. L. *Polymer* **1983**, *24*, 1341.
8. Yee, A. F.; Pearson, R. A. *J. Mater. Sci.* **1986**, *21*, 2462.
9. Goodier, J. N. *Trans. ASME* **1933**, *55*, 39.
10. Broutman, L. J.; Panizza, G. *Int. J. Polym. Mater.* **1971**, *1*, 95.
11. Chan, L. C.; Gillham, J. K.; Kinloch, A. J.; Shaw, S. J. In *Rubber-Modified Thermoset Resins*; Riew, C. K.; Gillham, J. K., Eds.; Advances in Chemistry 208; American Chemical Society: Washington, DC, 1984; p 261.
12. Kinloch, A. J.; Maxwell, D.; Young, R. J. *J. Mater. Sci. Lett.* **1985**, *4*, 1276.
13. Pearson, R. A.; Yee, A. F. *Polym. Mater. Sci. Eng. Prepr., Am. Chem. Soc.* **1983**, *186*, 316.
14. Finch, C. A.; Hashemi, S.; Kinloch, A. J. *Polym. Commun.* **1987**, *28*, 322.
15. Aronhime, M. T.; Peng, X.; Gillham, J. K. *Polym. Mater. Sci. Eng. Prepr., Am. Chem. Soc.* **1984**, *51*, 420.
16. Kinloch, A. J.; Shaw, S. J.; Tod, D. A. In *Rubber-Modified Thermoset Resins*; Riew, C. K.; Gillham, J. K., Eds.; Advances in Chemistry 208; American Chemical Society: Washington, DC, 1984; p 101.
17. Shaw, S. J.; Kinloch, A. J. *Int. J. Adhes. Adhes.* **1985**, *5*, 123.
18. Bitner, J. R.; Rushford, J. L.; Rose, W. S.; Hunston, D. L.; Riew, C. K. *J. Adhes.* **1982**, *13*, 3.
19. Hunston, D. L.; Kinloch, A. J.; Shaw, S. J.; Wang, S. S. *Adhesive Joints*; Mittal, K. L., Ed.; Plenum: New York, 1984; p 789.
20. Kinloch, A. J.; Hunston, D. L. *J. Mater. Sci. Lett.* **1987**, *6*, 137.
21. Bucknall, C. B.; Yoshii, T. *Br. Polym. J.* **1978**, *10*, 53.
22. Kunz, S. C.; Sayre, J. A.; Assink, R. A. *Polymer* **1982**, *23*, 1897.
23. Sultan, J. N.; McGarry, F. J. *Polym. Eng. Sci.* **1973**, *13*, 29.
24. Bascom, W. D.; Ting, R. Y.; Moulton, R. J.; Riew, C. K.; Siebert, A. R. *J. Mater. Sci.* **1981**, *16*, 2657.
25. Bucknall, C. B.; Partridge, I. K. *Polymer* **1983**, *24*, 639.
26. Tyson, W. R. *Acta Metall.* **1963**, *11*, 61.
27. Kinloch, A. J.; Young, R. J. *Fracture Behaviour of Polymers*; Applied Science: London, 1983; p 77.

28. Kinloch, A. J.; Young, R. J. *Fracture Behaviour of Polymers*; Applied Science: London, 1983; p 91.
29. Haward, R. N.; Owen, D. R. J. *J. Mater. Sci.* **1973**, *8*, 1136.
30. Bowden, P. B.; Jukes, J. A. *J. Mater. Sci.* **1972**, *7*, 52.

RECEIVED for review February 11, 1988. ACCEPTED revised manuscript October 4, 1988.

Matrix Ductility and Toughening of Epoxy Resins

G. Levita

Department of Chemical Engineering, Industrial Chemistry, and Materials Science, University of Pisa, Via Diotisalvi 2, 56100 Pisa, Italy

Epoxy resins can be modified by a number of reagents to produce solids that possess a variety of mechanical properties. Some of these reagents lead to the formation of two-phase structures that are much tougher than single-phase epoxies. Toughness depends largely on the physical and geometric properties of the separated phase. However, the properties of the rigid epoxy matrix are equally important, because the plastic processes responsible for toughening take place within the matrix. The ductility of epoxies can be varied in several ways. Some of them and the consequences for fracture resistance are considered in this chapter.

T HE THEORETICAL STRENGTH, σ_t, OF ELASTIC SOLIDS is given (1) by the simple model $\sigma_t \sim (E\gamma / d)^{0.5}$, where E, γ, and d are the elastic modulus, the surface energy, and the interatomic distance, respectively. For rigid polymers σ_t exceeds the experimental values by about one order of magnitude. In the case of epoxy resins the difference between theoretical and experimental values can be significantly lower. The failure stress of resorcinol resins, for instance, is $\sim$160 MPa (2), and the ratio of theoretical to experimental strength is about 2. Such a remarkable performance has been attributed (2) to the high packing efficiency of epoxy networks because of strong segmental interactions that originate from hydrogen bonds. The number of mechanically active molecular segments consequently increases.

The strength of polymers is controlled by the number of macromolecular chains that carry the external load. An increase in this number leads to a lowered concentration of overstressed bonds. Overstressing is reduced when

0065–2393/89/0222–0093$07.50/0

stress-relief mechanisms, such as yielding, come into play. In the case of yielding, stress is transferred from overstressed regions (typically the crack tip) to adjacent zones, where it levels off to the yield value. If postyield deformability exists, an energy sink is also provided. Indeed, the only way to increase fracture resistance is to promote plastic processes. In the case of cross-linked polymers, these processes are severely restricted. As a consequence, epoxy resins are rather brittle. The fracture energy, G_{Ic}, of glassy resins (~100–500 J/m^2) is 10–40 times lower than that of polycarbonate, which ranks midway between brittle and tough polymers.

To activate plastic processes, molecular mobility must be increased. Unfortunately, this increase is accompanied by a decrease of elastic modulus, heat-distortion temperature, and hardness. Such a contradictory situation can be illustrated by the results of Mostovoy and Ripling (3, 4) who found that the fracture energy in amine- and anhydride-cured epoxies decreased as the modulus increased. The molecular mobility of interest here gives rise to conformational transitions such as those proposed by Robertson for yielding (5). Short-range mobility, which is responsible for secondary relaxations (particularly β), can also play a role.

Yielding behavior is of great importance in fracture. Specimen sizing is based on the value of the yield stress, which is controlled by the degree of plastic constraint. Plane stress or plane strain are limit conditions that can induce dramatic transitions from ductile to brittle behavior, as in the case of polycarbonate.

Rubber toughening is the major source of crack-resistant polymers. This technology has been applied to almost every thermoplastic and also to thermosets. All polymer systems consist of a rigid matrix containing a soft small-particle phase whose behavior is supposed to be rubbery. Because the morphologies and elastic constants of all systems are similar, one might expect the basic mechanisms of toughening to be, as a first approximation, the same for all polymers. In rubber-toughened polymers, in order to avoid loss of rigidity, the volume fraction of the dispersed phase is kept to the minimum value necessary for the attainment of a satisfactory level of toughness.

Reinforcing mechanisms must exist within the matrix if they are to be active on a large scale. Thus the properties of the matrix polymer are at least as important as those of soft filler. A great deal of the fracture resistance of rubber-modified polymers comes from the deformation behavior of the matrix. Figure 1 illustrates how the fracture energy of toughened thermoset plastics indeed depends upon the fracture energy of the matrix. All data are relative to blends containing about 10% rubber (mostly carboxyl-terminated butadiene–acrylonitrile copolymer (CTBN)) tested at nonimpact rates. The line in Figure 1 has no particular meaning except to represent an amplification factor of 10.

Although G_{Ic} can occasionally be as high as 1 kJ/m^2, a very high value for cross-linked polymers, most G_{Ic} data for neat resins are grouped around

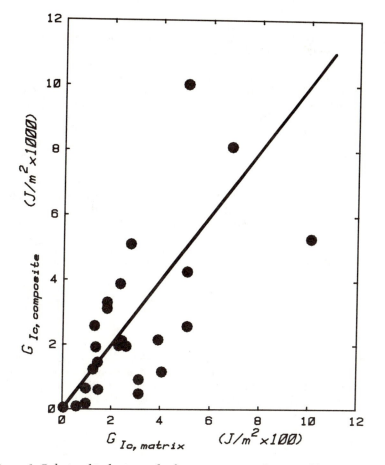

Figure 1. Relationship between the fracture energy of some rubber-toughened thermosets, G_{Ic}(composite), and the fracture energy of the corresponding matrix, G_{Ic}(matrix). Rubber content ~10%. Data from refs. 6–13.

100–300 J/m^2. Different epoxy matrix properties can be obtained by a number of approaches, including modification of cure conditions, use of chain extenders, and use of high-molecular-weight components. The effects of such techniques on fracture behavior are dealt with in this chapter.

Experimental Methods and Materials

A series of commercial diglycidyl ether of bisphenol A (DGEBA) resins of 400–4000 molecular weight (MW) were used. The epoxy prepolymer with the lowest molecular weight (Epon 828, MW 390, Shell) was a liquid at room temperature. The medium- to high-molecular-weight resins—Eposir 7161 (MW 980), 7170-P (MW 1640), and 7180 (MW 3700) by Società Italiana Resine (SIR) were solid (melting point 50, 80, and 110 °C). The hardeners were piperidine (PIP, boiling point 106 °C) and diaminodiphenyl sulfone (DDS, melting point 175 °C). PIP was used at a concentration of 5 phr. DDS was employed with a little excess (5%) over stoichiometry to compensate for its sluggish reactivity.

The reactive rubbers were either amine-terminated butadiene–acrylonitrile copolymer (ATBN) or CTBN (BF Goodrich). ATBN 1300×21 and ATBN 1300×16 were used for the DGEBA–PIP formulations, with acrylonitrile (AN) contents of 10 and 16%, respectively. In the case of DGEBA–DDS formulations, only the CTBN 1300×13 was used (AN 27%). In all blends, the rubber level was 10% by weight. Technical-grade bisphenol A (BPA) was used as the chain extender.

Details about sample preparation and cure conditions of formulations based on the 828 resin can be found in refs. 6–8. High-molecular-weight resins were first melted at 180 °C before adding DDS. The mixtures were then gently stirred until they clarified. The viscous liquids were poured into aluminum trays and cured under vacuum (to ensure outgassing) at 180 °C for 12 h. Rubber-modified formulations were obtained by adding CTBN after the hardener dissolved into the resin.

Fracture tests were carried out with a three-point bend assembly. Samples had span-to-width ratios of 4 or 8. The critical stress-intensity factor, K_{Ic}, was obtained by the slope of σ_c versus $1/Y\sqrt{a}$ plots (σ_c is stress at the onset of crack propagation, Y is a geometrical parameter, and a is crack length). G_{Ic} was obtained by the relationship

$$G_{Ic} \frac{(1 - \nu^2)\, K_{Ic}^2}{E} \tag{1}$$

where ν is Poisson ratio and E is elastic modulus. Dumbell specimens were used for tensile tests. Scanning electron microscopy (SEM) was carried out (ISI DS180 and JEOL T–300 microscopes) on samples broken at room temperature. Surfaces were gold coated. Shear moduli were measured with a torsion tester according to ASTM (American Society for Testing and Materials) D 1043 61T.

Discussion

Effect of Cure Conditions. The development of good mechanical properties requires the attainment of a sufficiently high degree of conversion, although the highest fracture resistance develops when the conversion is somewhat lower than the maximum achievable (14). The cross-link process is delayed by an increase in viscosity, and it can be brought close to the topological limit only by opposing the reduction of molecular mobility. Increased conversion is typically obtained by curing or postcuring at high temperatures, which does not necessarily result in better mechanical properties. Curing at different temperatures can in fact lead to diverse network topologies. The effect of cure temperature (t_{cure}) on the fracture properties of toughened epoxies has been studied by Levita et al. (6) and by Butta et al. (7).

Figure 2 shows that the fracture toughness of 828–PIP increases on increasing t_{cure}. Networks obtained at high temperature are apparently looser and more flexible, as suggested by the progressive reduction of the glass-transition temperature (T_g) and elastic modulus in the rubbery plateau (Figures 3 and 4). This condition could result from hardener volatilization (15,

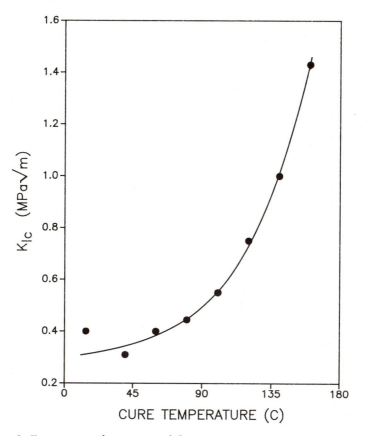

Figure 2. Fracture toughness, K_{Ic}, of the 828–PIP resin as a function of cure temperature.

16). Domenici et al. (17), however, did not observe any weight loss during the polymerization at 120 °C. Oxidation of piperidine, which can also occur, could diminish the cross-link efficiency of the hardener. It has frequently been observed (14, 18–20) that glasses obtained at low temperature have higher moduli, in spite of their lower cross-link density. The data in Figure 4 agree with such a finding. The effect has been attributed to the higher free-volume content in glasses cured at high temperature.

The addition of a small amount of reactive rubbers such as ATBNs can result in a substantial increase in fracture resistance. Unfortunately, the dependence of K_{Ic} on cure temperature becomes more pronounced, as shown in Figure 5 for two ATBNs. The presence of the rubber does not bring about toughening if the cure temperature is too low. In the examined formulations, fracture resistance develops only when t_{cure} is higher than 100 °C. The fact that K_{Ic} substantially increases when t_{cure} positively influences the properties of the matrix (Figure 2) supports the simple idea that matrix and composite ductilities are strongly related.

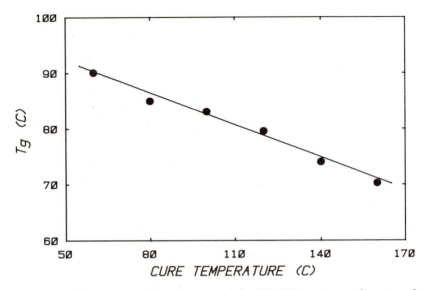

Figure 3. Glass-transition temperature of the 828–PIP resin as a function of cure temperature.

However, the presence of maxima in Figure 5 indicates that other parameters, such as the morphology and the internal structure of the soft phase, are to be considered. In order to ensure the formation of finely sized rubbery domains, the liquid rubber has to be soluble in the epoxy prepolymer. In addition, phase separation has to take place before gelation, when the viscosity is sufficiently high to prevent particle coalescence, but still low enough to favor nucleation and growth. Most phase separation takes place before gelation, although some localized change in composition occurs after gelation (21).

Figures 6 and 7 illustrate the morphological modifications brought about by curing ATBN blends at different temperatures. When t_{cure} is too low, phase separation is suppressed and the material is transparent, even after postcuring at elevated temperature. Increasing t_{cure} brings about an enlargement of particle size and a decrease of the number of particles. A comparison of data in Figures 5 and 8 suggests a strong dependence of K_{Ic} on particle dimensions. The effect of particle size on the deformation behavior of CTBN–DGEBA systems was studied by Sultan and McGarry (22), who showed that an enlargement of particles was paralleled by an increase in fracture energy. The dimensions reported in that paper were smaller than those observed in ATBN formulations, probably a consequence of the diverse AN content of the rubbers. The apparent volume fraction of the segregated rubber was found to be higher than the volume of the added rubber.

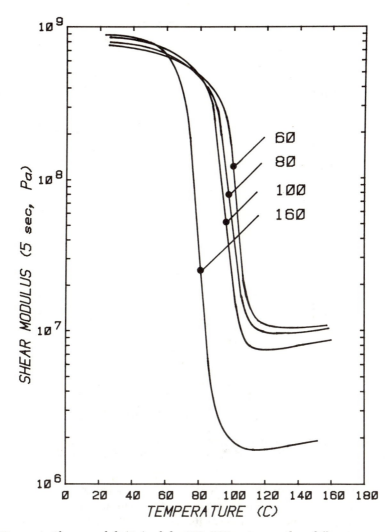

Figure 4. Shear moduli (5 s) of the 828–PIP resin cured at different temperatures as a function of temperature.

The relationship between the volume of added rubber and that of the separated phase is complex in toughened polymers. In high-impact polystyrene, for instance, 6% of polybutadiene can give rise to more than 50% of soft phase (23). An increase in the volume of the soft phase over that of the added rubber has also been observed in some toughened epoxies (8, 9, 24). There are two reasons for this observation. First, the particles can incorporate some matrix material as a rigid filler. The presence of extraneous material in the rubber particles has been reported repeatedly (8, 9, 25). Second,

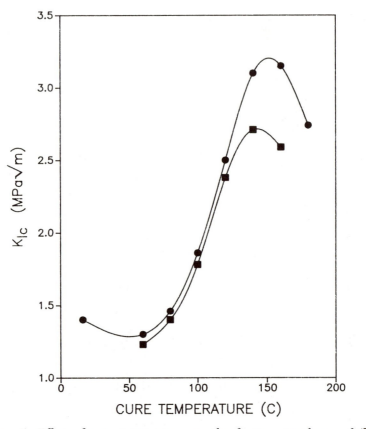

Figure 5. Effect of cure temperature on the fracture toughness of (■)
828–PIP–1300×21 and (●) 828–PIP–1300×16 formulations.

some resin can molecularly dissolve into the particles as a result of phase
equilibria (21, 26). This resin can react with the terminal groups of the rubber
and yield copolymers that increase its strength. Thus the deformation behavior of the particles is influenced by both the amount of dissolved resin
and the extent of rubber–resin reactions.

Sayre et al. (27) have shown that the fracture toughness of ATBN- and
CTBN-modified epoxies decreases by exposing the materials to radiation
that excessively increases the cross-link density of the rubber. Figures 9 and
10 reveal the different internal structures of particles formed at different
temperatures.

Changing the AN content of the rubber influences the solubility parameter and, consequently, the final morphology. Rowe et al. (28) showed that
the average particle size decreases on increasing the AN content of CTBN
rubbers. Data in Figure 8 indicate that the effect is similar with ATBNs.

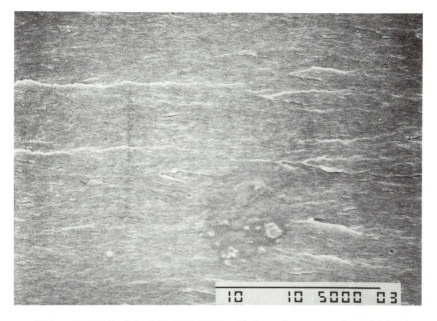

Figure 6. SEM of the 828–PIP–1300×21 formulation cured at 15 °C.

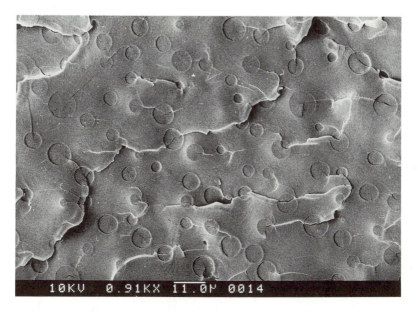

Figure 7. SEM of the 828–PIP–1300×21 formulation cured at 120 °C.

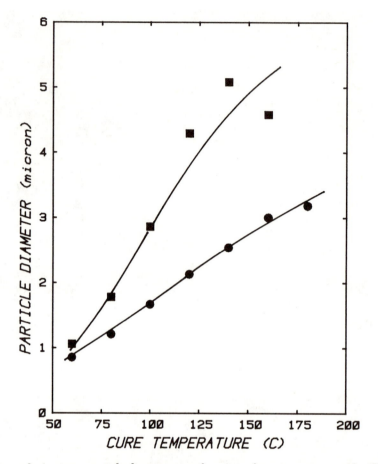

Figure 8. Average particle diameter as a function of cure temperature for (■) *828–PIP–1300×21 and* (●) *828–PIP–1300×16 formulations.*

ATBN 1300×21 gives rise to particles about twice the size of particles originated by ATBN 1300×16. This result contradicts the previous statement about the relationship between K_{Ic} and particle diameter. That conclusion cannot be generalized because it is not expected to be valid when a new parameter, the AN content, comes into play. The AN content apparently overrides the importance of particle size.

The morphology of CTBN–828 is similarly modified by the cure temperature and the AN level. Again, phase separation does not take place if the cure temperature is too low.

Chain Extension. The mechanical properties of linear polymers markedly depend on the average chain length (i.e., on molecular weight)

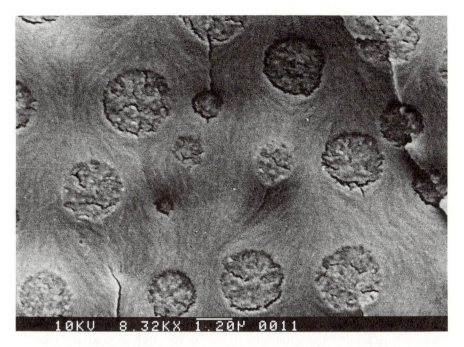

Figure 9. SEM of the 828–PIP–1300×21 formulation cured at 60 °C.

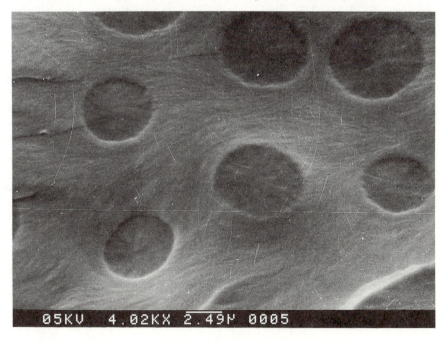

Figure 10. SEM of the 828–PIP–1300×21 formulation cured at 160 °C.

until the dependence almost vanishes above a certain level. In the case of cross-linked polymers the relevant parameter is the cross-link density, that is, the number of chemical junctions per unit volume or the average molecular weight of the segments among cross-links (M_c). The formation of junctions has the important effect of limiting molecular mobility, particularly long-range mobility, to such an extent that many cross-linked glasses do not even yield. As a consequence, the fracture energy dramatically drops. The mirror-like fracture surfaces of highly cross-linked polymers indicate that plastic processes are almost totally inhibited.

An easy way to moderate brittleness is therefore to increase M_c, which can be done easily by using high-molecular-weight flexible hardeners such as polyamines or polysulfides. Unfortunately, the increase of fracture energy is often not very important and does not compensate for the deterioration of other properties. A wiser approach consists in spacing out the junctions with stiff molecules. In this way modulus and distortion temperature are not unduly reduced. The diminished cross-linked density favors postyield deformability and constitutes an effective energy sink.

Bisphenol A (BPA) is a basic chemical for the production of DGEBA resins, whose molecular weight is determined by the ratio of BPA to epichlorohydrin (29). The general formula of DGEBAs can be written as shown in structure **I**, with n ranging from 0 to 20 in commercial grades. BPA is by far the most important modifier of DGEBA resins. High-molecular-weight resins can be obtained from low-molecular-weight products by reacting them with BPA in the presence of a basic catalyst (30). An immediate molecular weight increase can be achieved by mixing low-molecular-weight epoxies with BPA and curing with polyamines, as in the case of some adhesives that are curable at room temperature. Two competitive processes have been shown to take place: chain extension and cross-linking (31, 32). The ratio between the rate constants of such processes, k_1/k_2, depends on the nature of the catalyst used. The variation with time of epoxy group concentration is schematically shown in Figure 11. Products of rather high molecular weight (MW >4000) can form before cross-linking starts.

BPA is also used in rubber-modified formulations (8, 9, 11, 33–38) because it magnifies the effect of the rubber. Riew et al. (35) observed that the fracture energy of a 5% CTBN–DGEBA system increased by a factor of

I

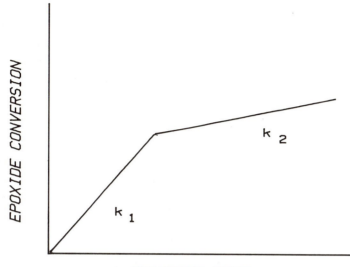

Figure 11. Extent of epoxide reaction vs. time. (Reproduced with permission from ref. 32. Copyright 1973 Wiley.)

4 when 27% BPA was added. Moreover, good results have been obtained with the use of nonreactive rubbers, either liquid or solid. For this reason BPA has been used frequently in toughened systems. Levita et al. (8) specifically investigated the role of BPA.

Figure 12 shows how the fracture toughness of the low-molecular-weight 828 resin increases on adding BPA. The effect is similar to that brought about by the cure temperature in Figure 2. The important difference is that T_g is not adversely affected by the addition of BPA (Figure 13). The effect on tensile behavior is even more important. The unmodified DGEBA is substantially brittle because of a lack of postyield deformability (Figure 14). On the contrary, at 30% BPA yielding is followed by a consistent softening and by the formation and propagation of a neck.

The deformation at break, ~0.08, by far underestimates the real deformability of the material, because it was obtained simply by specimen elongation to the original length ratio. A more realistic estimate was obtained by observing the behavior of tiny air bubbles dispersed within the test pieces. The unmodified DGEBA broke soon after the formation of a shear band. Bubbles located outside the band retained the original shape, whereas those lying within the band developed cracks at the equator but did not show any appreciable departure from the spherical shape (Figure 15).

On the contrary, bubbles located in the sheared region of the 30% BPA blend became ellipsoidal in shape, with axial ratios greater than 2 (Figure

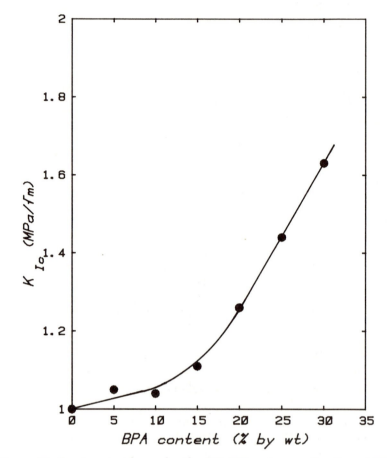

Figure 12. Fracture toughness for the 828–PIP resin as a function of BPA content.

16). The local deformation at the bubble equator was estimated to be higher than 40%. Such a high network deformability easily gives rise to crack blunting in rubber-modified formulations. A further indication that the increase of K_{Ic} in Figure 12 entirely relies on postyield processes is the fact that yield stress is independent of BPA content for both neat and rubber-modified resins (Figure 17).

Rubber-modified formulations greatly benefit from the greater ductility produced by BPA. Fracture toughness steadily increases when BPA is added (Figure 18). The higher toughness of ternary BPA–DGEBA–rubber blends over that of binary DGEBA–rubber blends was supposed to arise substantially from modifications of soft-phase geometric parameters.

Riew et al. (35) proposed that the effect of BPA was production of a

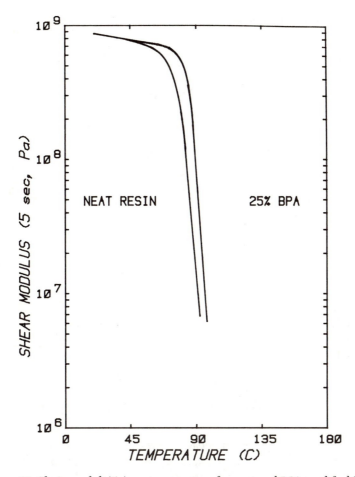

Figure 13. Shear moduli (5 s) vs. temperature for neat and BPA-modified (25%)
828–PIP resin.

bimodal particle-size distribution. Small particles (<0.1 μm) were believed
to be particularly effective in favoring shear yielding. Dilational processes,
on the contrary, were attributed to larger particles (≥1 μm). However, it
is not clear, on the basis of stress analysis, why particles with the same
properties (differing only in size) should activate different deformation processes. Later (9) it was pointed out that, because the molecular weight of the
resin increases when it reacts with BPA, the free energy of mixing decreases
rapidly and the separation of a larger amount of soft phase is favored.

   BPA acts as a toughener partly because the volume fraction of the separated phase tends to be higher than in binary blends; subinclusions are
easily visible in ternary blends. This effect probably depends on the kinetics

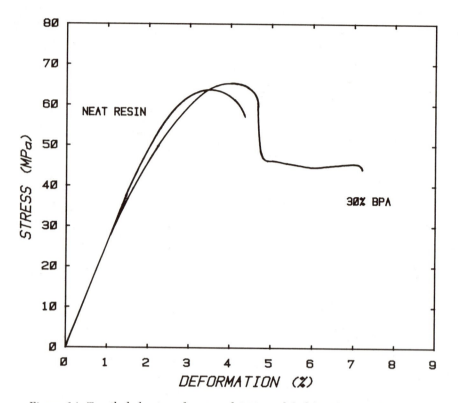

Figure 14. Tensile behavior of neat and BPA-modified (30%) 828–PIP resin.

of phase separation. Perhaps (8) because of the very high reaction rate (BPA is an effective catalyst), a certain amount of rubber remains undissolved before gelation and gives rise to complex morphologies. It is not explained, however, why nonreactive rubbers act as tougheners when BPA is used. The presence of reactive groups would otherwise be mandatory.

However, data in Figure 12 clearly indicate that the most important effect of BPA consists in modifying the properties of the matrix. When a crack propagates, most of the energy is absorbed to deform the matrix. Because in polymers the plastic processes are basically time-dependent, tough systems are more sensitive to deformation rate than brittle ones (11).

High-Molecular-Weight Components.   Commercial grades of DGEBA resins have molecular weights in the 400–6000 range. Although pure DGEBA is solid at room temperature, low-molecular-weight prepolymers are viscous liquids, mixtures of several components. For many technologies, the liquid form is by far the best. High-molecular-weight

Figure 15. Failure of spherical cavities located in a shear band of the 828–PIP resin.

Figure 16. Deformation and failure of spherical cavities located in the sheared region of the 828–PIP resin modified with 30% BPA.

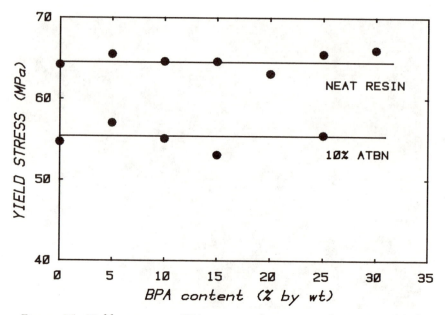

Figure 17. Yield stress vs. BPA content for neat and ATBN-modified
828–PIP–1300×21 resin.

grades are used mainly for coating. They are also useful for prepregs manufacturing, typically with the solvent process. Curing agents frequently used
are diaminophenylmethane and DDS, because they react slowly at room
temperature (long shelf life).

The mechanical properties of a series of epoxies cured with DDS (MW
380–2600) were studied by LeMay and Kelley (39) in both the glassy and
the rubbery states. They found that, in spite of the high cross-link density,
a simple relationship from the theory of rubber elasticity holds.

$$G = \frac{\rho RT}{M_c} \tag{2}$$

where G is the equilibrium shear modulus, R is 8.314 kJ/mol·K, ρ is the
specific gravity, and T is the absolute temperature. Cross-link density can
be obtained readily by measuring the equilibrium modulus above T_g. Similar
results were reported by Ilavsky et al. (40) and by Oleinik (41).

In the present study a different method was used to characterize the
networks obtained with DGEBA epoxies cured with DDS. Long ago Nielsen
(42) proposed that the increase of the glass-transition temperature brought
about by cross-links is inversely proportional to M_c:

$$T_g - T_{g_0} = \frac{3.9 \times 10^4}{M_c} \tag{3}$$

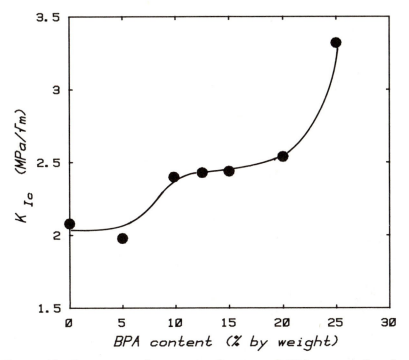

Figure 18. Fracture toughness as a function of BPA content for the 828–PIP–1300 × 21 formulation.

where T_{g_0} is the glass-transition temperature at the gel point. Figure 19 shows that this simple equation accurately fits the experimental data. It indicates that M_c substantially coincides with the average molecular weight of the resin. The T_g values were obtained by shear modulus versus T plots.

Fracture toughness linearly depends on molecular weight (Figure 20) and, as in the case of BPA-based formulations, it substantially relies on postyield deformability. Yield stress is almost independent of the molecular weight of the resin (Figure 21). The tensile behavior is very similar to that shown in Figure 14. This behavior is not surprising, because in both systems the network extensibility is obtained by chain extension.

Bucknall and Yoshii (9) added CTBN to high-molecular-weight epoxies and obtained materials with high fracture energy. More recently, Yee and Pearson (10) reported that the improvement in terms of G_{Ic} was proportional to the molecular weight of the resin. The present results indicate that fracture toughness tends to level off at high molecular weights. There are problems when using high-molecular-weight epoxies, because it is difficult to ensure that all reactive groups are consumed. DDS is a sluggish curing agent and the concentration of epoxy groups is low. However, the tendency of fracture parameters to level off seems reasonable because a similar trend is observed in thermoplastics.

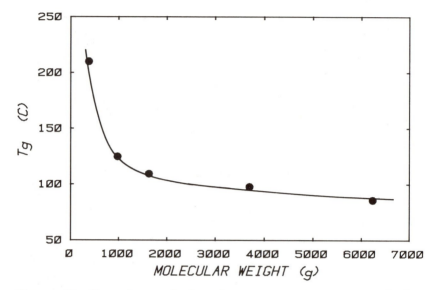

Figure 19. Dependence of the glass-transition temperature of the DGEBA–DDS resin on the molecular weight of the epoxy prepolymer.

Plastic processes that take place during crack propagation can be found on the fracture surface. The SEM micrographs in Figures 22–25 show surfaces produced at a low deformation rate. Some aspects are worth mentioning. When the fracture produces a rough surface, typically in case of stress whitening, the rubber particles appear at the bottom of roughly circular cavities. On the contrary, if the surface is a plane, the particles are cut in two pieces, and the fracture surfaces of the resin and the rubber are almost at the same level (see, for example, Figure 7). This result indicates that when the moving crack approaches a particle, the interface substantially holds and the particle is not detached from the matrix. For this reason particles never appear as hillocks. However, in the 828–DDS formulation the crack can either propagate through or detach the particles.

On the surface shown in Figure 22 both mechanisms were active. The mode of propagation changes as soon as the ductility of the resin increases. In Figure 23 the protuberances have disappeared and only shallow depressions are visible. The cavities get deeper as the ductility rises (Figures 24 and 25). This rise clearly indicates that most of the plastic work responsible for the rapid increase of K_{Ic} in Figure 18 comes from matrix deformation. When the fracture surface is heated above T_g, the deformed matrix retracts and the size of cavities is substantially reduced (Figure 26).

Conclusions

Epoxies are the only class of thermosets for which rubber toughening has proven to be efficacious. The technique has been applied to unsaturated

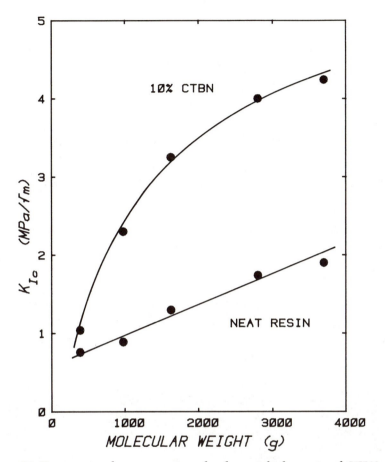

Figure 20. Fracture toughness vs. resin molecular weight for neat and CTBN-modified DGEBA–DDS–1300×13 resin.

polyesters, polyimides, and even phenolics with much less satisfactory results. On the other hand, epoxy resins have been toughened with a variety of substances, including reactive liquid acrylonitrile copolymers (the classic CTBN and ATBN), solid rubbers, fluoroelastomers, acrylic rubbers, polysiloxanes and ether–ester copolymers. Even with acrylonitrile–butadiene–styrene copolymers (ABS), some toughening has been achieved.

Such a list of modifiers indicates that the main reason why epoxy resins are "toughenable" is that they are intrinsically ductile. This ductility permits the occurrence of two basic processes: crack blunting and voiding. Crack blunting is the result of plastic deformations at the crack tip. Voiding occurs within or around the soft particles, in a region that extends above and below the crack trajectory. Both mechanisms lead to a drop in the local stress that is averaged over a larger area. Unfortunately, matrix ductility rapidly deteriorates when the cross-link density is too high (if the intrinsic rigidity of

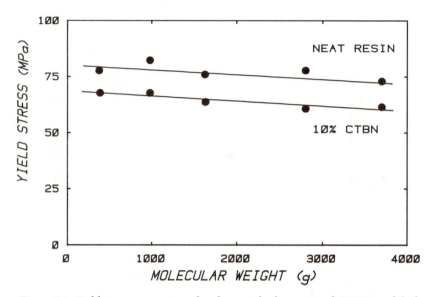

Figure 21. *Yield stress vs. resin molecular weight for neat and CTBN-modified DGEBA–DDS–1300×13 resin.*

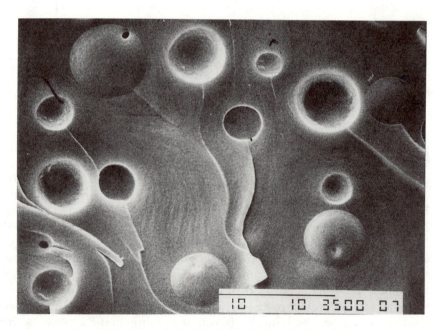

Figure 22. *SEM of the 828–DDS–1300×13 formulation.*

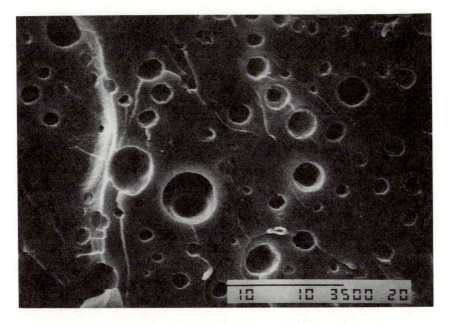

Figure 23. SEM of the 7161–DDS–1300 × 13 formulation.

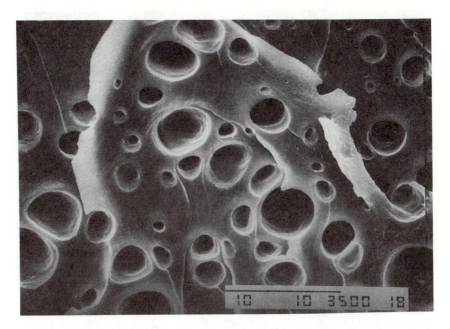

Figure 24. SEM of the 7170P–DDS–1300 × 13 formulation.

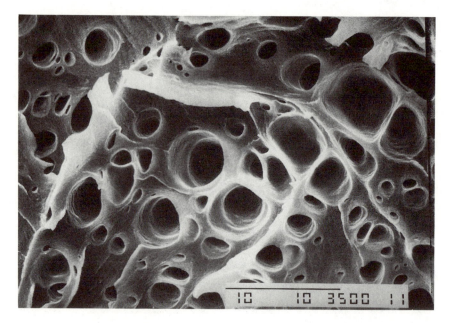

Figure 25. SEM of the 7180–DDS–1300 × 13 formulation.

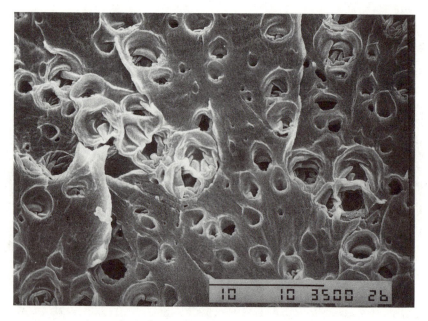

Figure 26. SEM of the 7180–DDS–1300 × 13 formulation after heating above T_g.

segments among cross-links is not enough to ensure a high T_g, it is necessary to diminish M_c) and the toughenability drops. This is the case with high-T_g resins such as tetraglycidyldiaminodiphenylmethane resin (TGDDM).

References

1. Cottrell, A. *Dislocations and Plastic Flow in Crystals;* Claredon: Oxford, 1953.
2. Oleinik, E. F. *Pure Appl. Chem.* 1981, *53*, 1567.
3. Mostovoy, S.; Ripling, E. J. *Pure Appl. Chem.* 1966, *10*, 1351.
4. Mostovoy, S.; Ripling, E. J. *Pure Appl. Chem.* 1970, *14*, 1901.
5. Robertson, R. E. *Appl. Polym. Symp.* 1968, 7, 291.
6. Levita, G.; Marchetti, A.; Butta, E. *Polymer* 1985, *26*, 1110.
7. Butta, E.; Levita, G.; Marchetti, A.; Lazzeri, L. *Polym. Eng. Sci.* 1986, *26*, 63.
8. Levita, G.; Marchetti, A.; Lazzeri, A.; Frosini, V. *Polym. Compos.* 1987, *8*, 141.
9. Bucknall, C. B.; Yoshii, T. *Br. Polym. J.* 1978, *10*, 53.
10. Yee, A. F.; Pearson, R. A. *Proceedings of the Second International Conference on the Toughening of Plastics;* London, 1985; p 2/1.
11. Bascom, W. D.; Ting, R. Y.; Moulton, R. J.; Riew, C. K.; Siebert, A. R. *J. Mater. Sci.* 1981, *16*, 2697.
12. Kinloch, A. J.; Shaw, S. J.; Tod, D. A. In *Rubber-Modified Thermoset Resins;* Riew, C. K.; Gillham, J. K., Eds.; Advances in Chemistry 208; American Chemical Society, Washington, DC, 1984; p 101.
13. Crosbie, G. A.; Phillips, M. G. *J. Mater. Sci.* 1985, *20*, 182.
14. Noordam, A.; Wintraecken, J. J. M. H.; Walton, G. In *Crosslinked Epoxies;* Sedlacek, B.; Kahovec, J., Eds.; de Gruyter: Berlin, 1987.
15. Manzione, L. T.; Gillham, J. K.; McPherson, C. A. *J. Appl. Polym. Sci.* 1981, *26*, 889.
16. Cuadrado, T. R.; Almaraz, A.; Williams, R. J. J. In *Crosslinked Epoxies;* Sedlacek, Ed.; de Gruyter: Berlin, 1987.
17. Domenici, C.; Levita, G.; Marchetti, A.; Frosini, V. *J. Appl. Polym. Sci.* 1987, *34*, 2285.
18. Findley, W. N.; Reed, R. M. *Polym. Eng. Sci.* 1977, *17*, 837.
19. Gillham, J. K. In *Developments in Polymer Characterization;* Dawkins, J. V., Ed.; Applied Science: London, 1982.
20. Enns, J. B.; Gillham, J. K. *J. Appl. Polym. Sci.* 1983, *28*, 2831.
21. Wang, T. T.; Zupko, H. M. *J. Appl. Polym. Sci.* 1981, *26*, 2391.
22. Sultan, J. N.; McGarry, F. *J. Polym. Eng. Sci.* 1973, *13*, 29.
23. Wagner, E. R.; Robertson, L. M. *Rubber Chem. Technol.* 1970, *43*, 1129.
24. Kunz, S. C.; Sayre, J. A.; Assink, R. A. *Polymer* 1982, *23*, 1897.
25. Manzione, L. T.; Gillham, J. K.; McPherson, C. A. *J. Appl. Polym. Sci.* 1981, *26*, 907.
26. Williams, R. J. J.; Borrajo, J.; Adabbo, H. E ; Rojas, A. J. In *Rubber-Modified Thermoset Resins;* Riew, C. K.; Gillham, J. K., Eds.; Advances in Chemistry 208; American Chemical Society, Washington, DC, 1984; p 195.
27. Sayre, J. A.; Kunz, S. C.; Assink, R. A. In *Rubber-Modified Thermoset Resins;* Riew, C. K.; Gillham, J. K., Eds.; Advances in Chemistry 208; American Chemical Society, Washington, DC, 1984; p 215.
28. Rowe, E. H.; Siebert, A. R.; Drake, R. S. *Mod. Plast.* 1970, *47*, 110.
29. Lee, H.; Neville, K. *Handbook of Epoxy Resins;* McGraw-Hill: New York, 1987.
30. Hata, N.; Kumanotani, J. *J. Appl. Polym. Sci.* 1973, *17*, 3545.
31. Alvey, F. B. *J. Appl. Polym. Sci.* 1969, *13*, 1473.

32. Son, P. N.; Weber, C. D. *J. Appl. Polym. Sci.* **1973**, *17*, **2415**.
33. Sayre, J. A.; Assink, R. A.; Lagasse, R. R. *Polymer* **1981**, **22**, **87**.
34. Siebert, A. R.; Rowe, E. R.; Riew, C. K.; Lipiec, J. M. *28th Annu. Tech. Conf. SPI*, 1973.
35. Riew, C. K.; Rowe E, H.; Siebert, A. R. In *Toughness and Brittleness of Plastics*; Deanin, R. D.; Crugnola, A. M., Eds.; Advances in Chemistry 154; American Chemical Society: Washington, DC, 1976; p 326.
36. Rowe, E. H.; Riew, C. K. *Plast. Eng.* **1975**, *31*, **45**.
37. Ting, R. Y.; Cottington, R. L. *J. Appl. Polym. Sci.* **1980**, *25*, **1815**.
38. Yee, A. F.; Pearson, R. A. *J. Mater. Sci.* **1986**, *21*, **2462**.
39. LeMay, J. D.; Kelley, F. N. *Adv. Polym. Sci.* **1986**, *78*, **115**.
40. Ilavsky, M.; Bogdanova, L. M.; Dusek, K. *J. Polym. Sci., Polym. Phys. Ed.* **1984, 22, 265**.
41. Oleinik, E. F. *Adv. Polym. Sci.* **1986**, *80*, **49**.
42. Nielsen, L. E. *J. Macromol. Sci. Rev. Macromol. Chem.* **1969**, *C3*, **69**.

Received for review February 11, 1988. Accepted revised manuscript September 12, 1988.

Fracture-Toughness Testing
of Tough Materials

Donald D. Huang

Polymer Products Department, E. I. du Pont de Nemours and Company,
Wilmington, DE 19898

*The J-integral method for measuring fracture toughness of tough
materials was applied to three toughened nylons. The J tests were
conducted from room temperature to –40 °C at displacement rates
as high as 26 mm/s. According to ASTM size recommendations, valid
plane-strain conditions were met in two of the three cases. Despite
falling short of the size criterion, additional J tests on grooved specimens of rubber-toughened semicrystalline nylon 6/6 indicate that
plane-strain conditions were also met in this case. Some comments
are given about the experimental aspects of applying the J method
to polymers.*

T HE *J*-INTEGRAL TECHNIQUE HAS BEEN USED TO CHARACTERIZE a variety
of polymers, including polyethylene (*1, 2*), polypropylene–ethylene block
copolymers (*2*), natural rubber (*3*), and a number of toughened blends (*4,
5*). The *J*-integral technique, proposed by Rice (*6*), is especially appealing
because a plane-strain toughness value can be obtained with specimens that
are significantly smaller than would be necessary for linear elastic fracture
mechanics (LEFM) testing. This ability is important for materials like toughened polymers because it is often impossible to make valid LEFM specimens
with the proper morphology.

J tests are not limited to specimens with small-scale yielding; they can
also be applied to large-scale plasticity cases (*7*). In the general elastic-plastic
case, *J* can be considered as the potential energy difference between two
loaded identical bodies having slightly different-sized cracks. It can also be
considered a measure of the characteristic crack-tip elastic-plastic field.

0065–2393/89/0222–0119$06.00/0

When plastic deformation does not occur, J takes on additional interpretations. For nonlinear elastic bodies, J represents the energy available for crack extension. For linear elastic bodies, J is equal to G, the strain-energy release rate (or the crack-driving force).

Begley and Landes demonstrated that the J-integral approach can provide a plane-strain ductile fracture toughness, J_{cl} (8). This material parameter represents the energy required to initiate a crack. It does not describe the propagation process. They demonstrated that J_{cl} values obtained in two different steels agreed very well with G_{cl} values obtained from specimens with much greater thicknesses.

To obtain valid fracture toughness values, J specimens are not required to be in elastic plane strain, but still must be in plastic plane strain. Some constraint of the crack-tip region is required to achieve this, but not to the degree needed in LEFM specimens. Consequently, the minimum thickness requirement is significantly smaller for J testing (as much as 0.2–0.3 of the size of the LEFM specimens). Earlier studies (e.g., *1*, *4*) have been conducted at slow rates (0.033 mm/s). Low-temperature (down to −80 °C) conditions were also employed to decrease the required minimum thickness. In the present study, we achieved the same goal by testing at high rates (up to 26 mm/s) and low temperatures (down to −40 °C) by using side-grooved specimens where necessary.

Multiple-Specimen J-Integral Method

The J-integral method that is under investigation is a multiple-specimen technique, similar to ASTM E 813–81. It was originally proposed by Landes and Begley (9). The first specimen is completely fractured to determine the ultimate displacement. Subsequent specimens are loaded to different subcritical displacements to obtain different levels of crack growth. A value of J is calculated from the area under the loading curve of each test. Crack growth (Δa) is marked and measured on the fracture surface. Resistance (J − Δa) curves are then constructed. J_{cl} is found graphically by finding the intersection of the resistance curve and the blunting line, $J = 2\,\sigma_y \Delta a$, where σ_y is the yield stress, as described later.

In the current investigation, ASTM E 813–81 recommendations were not strictly followed. The following modifications were made:

Modification 1. Crack growth (Δa) was measured at the center of the fracture surface. Because the crack fronts were thumbnail-shaped, this corresponded to the maximum crack growth. This is the original proposal by Landes and Begley (9). ASTM E 813–81 recommends making nine equally spaced measurements and averaging them in a prescribed manner. This recommendation was made because the average value would more accurately map the crack front when significant bowing had occurred. It had also been shown that an average value correlated well to other crack-extension measurement techniques such as unloading compliance and electrical potential. The single-point technique that was used here is more convenient and leads to a more conservative J_{cl} value (about 15% less than the value obtained by the averaged crack-growth method for the materials tested here).

Modification 2.   The span-to-depth (S/W) ratio was 3.5. ASTM E 813–81 recommends a ratio of 4. The span-to-depth ratio is important in the calculation of J. J can be expressed as (*10*)

$$J = J_e + J_p = \frac{\eta_e U_e + \eta_p U_p}{Bb} \tag{1}$$

where J_e and J_p are the elastic and plastic contributions to J; η_e and η_p are the elastic and plastic work factors, respectively; U_e and U_p are the elastic and plastic components of the total energy, U_T, respectively; B is the thickness; and b is the ligament. For bend specimens, the η_p coefficient is independent of S/W ratio. However, η_e has an S/W dependence. When the S/W ratio is 4 and the specimen is deeply notched (greater than 0.4), η_e and η_p are both 2. Therefore, for J calculations, the total energy does not have to be partitioned into its elastic and plastic portions. For this geometry,

$$J = 2 \frac{U_T}{Bb} \tag{2}$$

When the S/W ratio is 3.5, $\eta_e = 2.2$ and $\eta_p = 2$. This condition leads to a maximum error of 10% when the load-deflection curve is completely elastic. The error decreases as the ratio of elastic to plastic energy decreases.

Modification 3.   The resistance curve was fitted by using data points where crack growth was between two offset lines drawn parallel to the blunting line. The minimum offset was 0.6% of the ligament and the maximum offset was 6% of the ligament. ASTM E 813–81 recommends using parallel lines that are offset by 0.15 and 1.5 mm. Shih (*11*) has shown that the value of J is accurately predicted by these estimation procedures if the crack extension is less than 6% of the remaining ligament. This finding is a source of confusion in the ASTM standard. ASTM E 813–81 recommends that the test specimens should be a minimum size (i.e., $B,b > 25 (J_{cl}/\sigma_y)$). It also recommends fixed offsets for the maximum and minimum crack growths. Because the ligaments can be of any size, the fixed offsets will not always guarantee that the crack growths will be less than 6% of the ligament. However, the 6% criterion should not be considered definitive. Recent work (*12*) on polymers showed that linearity in the R-curve can exist at crack growths much greater than 6%.

Modification 4. The loading rates varied from 0.26 to 26 mm/s. This variation led to loading times as small as 0.05 s at the fastest speed. ASTM E 813–81 recommends that the loading time be greater than 6 s to avoid the confusion of measuring a dynamic fracture toughness.

Modification 5. The specimens were notched with single-toothed fly-cutters that were lapped to a radius between 5 and 12 μm. ASTM E 813–81 recommends a fatigue crack. In the current study, J_{cl} is independent of notch root radius in the 5–12-μm range. Earlier toughness studies of polymers have shown that notching with a sharp flycutter can be acceptable (4, 12).

Modification 6. The yield stress, σ_y, was the ultimate stress measured in tension at the appropriate rate. ASTM E 813–81 recommends using the 0.2% offset yield stress or an average between the ultimate stress and the 0.2% offset yield stress.

Experimental Details

Toughness testing was conducted on rubber-toughened crystalline nylon 6/6, rubber-toughened amorphous nylon, and a medium toughened crystalline nylon 6/6. These resins are commercially available and are listed in Table I. The resins were injection-molded into 100- × 254- × 12-mm plaques. Single-edged notched bend (SENB) specimens were cut from the plaques, notched to one-half of the depth (as described earlier), and tested in three-point bend. For materials A and B, further experiments were conducted using SENB specimens that were side-grooved with a blunt cutter (radius of curvature 250 μm). The total depth of the grooves varied from 10 to 50% of the thickness. ASTM recommends a total groove depth of 20%. The dimensions and notch directions are shown in Figure 1.

The crack growth was marked by freezing the test specimens in liquid nitrogen and then breaking them at 260 mm/s. An example of the crack-growth region is shown in Figure 2. The crack front is bowed and the *a* value is measured at the center of the specimen. The region next to the initial notch (between lines A and B) is the crack-growth region for this specimen. In *J* testing of polyethylene, it was shown that a stretch zone next to the initial notch formed first, followed by crack growth (1). This result was not observed in the materials in this investigation. This region AB continuously grew with increasing *J* levels. If it were the stretch zone due to blunting, it would increase with increasing *J* and then remain constant at *J*

Table I. Resins Used for Testing

Material	Description of Nylon 6/6	Trade Name
A	Rubber-toughened semicrystalline	"Zytel" ST801
B	Rubber-toughened amorphous	"Zytel" ST901
C	Toughened semicrystalline	"Zytel" 408

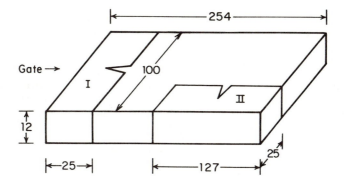

All dimensions are in mm.

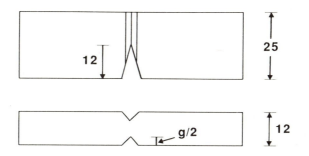

g = Total groove depth

Figure 1. Specimen geometries.

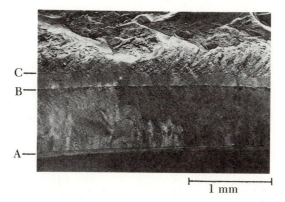

Figure 2. Material B, $J_{cl} = 15$ kJ/m².

values greater than J_{cl}. The reason for the second texture (BC) before the fast-fracture region is unclear. Its texture resembled that of the stretch zone found in the specimens that were loaded to sub-J_{cl} levels. In these specimens only blunting occurs, and only one texture was seen.

J tests were conducted on the toughened polymers at three different rates: 0.26, 2.6, and 26 mm/s at 23 °C and 50% relative humidity. Additional tests were conducted on materials A and B at 26 mm/s at –10 and –40 °C. Yield stress and modulus were measured at the same displacement rate. The resulting strain rates for these tests were within a factor of 2 of one another. Poisson's ratios were assumed to be rate- and temperature-independent and equal to 0.41.

Yield stresses were measured in tension by using injection-molded bars (ASTM D 638, Type I). The yield stress as a function of displacement rate is plotted for the three materials in Figure 3. The elastic moduli, E, for the room-temperature tests were measured in flexure by using injection-molded flex bars (3 mm thick). The moduli of the materials did not vary significantly as a function of rate (all were within 10% of the others). Because the moduli were essentially constant, an average value was calculated for each material. The results are listed in Table II. The elastic moduli

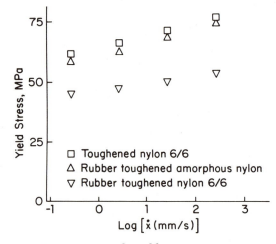

Figure 3. Tensile yield stress vs. rate.

Table II. Summary of Results

Material	Test Rate, mm/s	Yield Stress, MPa	J_Q kJ/m²	K_Q MPa$\sqrt{m}$
A	26.	50.1	27.5–32.5	8.05–8.75
	2.6	47.5	33.0–41.0	8.80–9.80
	0.26	45.1	39.3–44.5	9.60–10.20
B	26	68.7	16.0–19.0	6.22–6.77
	2.6	62.9	19.0	6.77
	0.26	58.6	18.0	5.69
C	0.26	61.9	18.5	7.41

NOTE: E_A = 1.96 GPa; E_B = 2.01 GPa; E_C = 2.47 GPa. Poisson's ratio = 0.41. All specimens were 12 mm thick.

of the materials at low temperature were calculated from the stiffnesses of the individual J tests by using the method of Haggag and Underwood (*13*).

A computer-controlled servohydraulic system was used for all mechanical testing. Software was developed to run the machine, acquire data in the form of load-displacement curves, and numerically integrate the curves to calculate J values. Crack growth was measured from the fracture surfaces by using a traveling microscope.

Results and Discussion

Test Variables.   Figures 4 and 5 show a sampling of the results for materials A and B, respectively. Overall, the data behave very well. The general shape of the R-curves is similar to those obtained for other materials (*1–5, 7, 8*). At low J values, the R-curve follows the blunting line defined

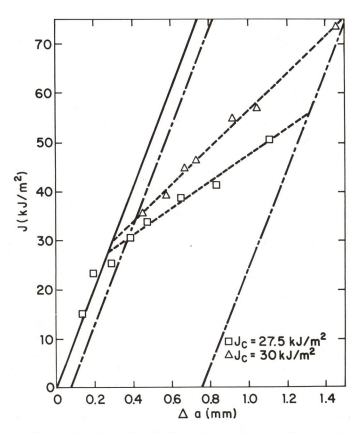

Figure 4. Material A, 26 mm/s, thickness 12 mm. Key: ---, R-curve that was fitted to the data from a set of specimens from one plaque; ——, blunting line calculated from appropriate yield stress; -·-·-, crack-growth window within which data must fall to qualify for linear regression.

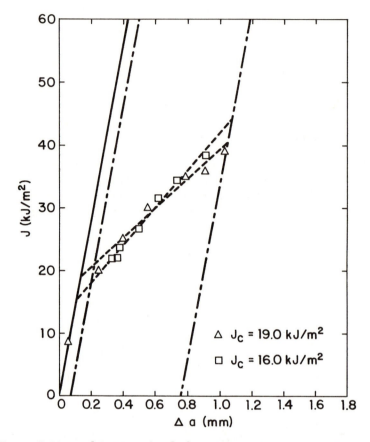

Figure 5. Material B, 26 mm/s, thickness 12 mm. Key: same as Figure 4.

by $J = 2\,\sigma_y\,\Delta a$. At higher values, the data fall off the blunting line. This part of the curve is essentially linear within the defined window. Although the J_{cl} values vary from plaque to plaque, the individual sets of data points do not show a great deal of scatter.

In the case of material A, the slopes of the fitted lines of the two J tests are very different, even though the J_{cl} values are similar. With material B, it appears that the scatter in J_{cl} values is a testing artifact; the data points for the two curves could very easily lie on the same line with minimal scatter. At this time, the significance of these differences is not clear. The J_Q values for all the materials (using specimen I) are listed in Table II.

The effect of varying notch root radius was studied because the initial notches were not made by fatiguing a prenotch, as recommended by ASTM. J tests were conducted with notch root radii of 5 and 12 μm. In this range, the J_{cl} values of both materials A and B are essentially independent of notch root radius. J_{cl} was 30 kJ/m² for material A and between 17 and 19 kJ/m² for material B. Some preliminary tests using larger radii (up to 1 mm) were also attempted. The tests were difficult to perform because the crack growth could not be measured. After the specimens were frozen, they did not necessarily break at the end of the crack-growth region. Instead, fracture occurred at unpredictable points around the blunt crack tip. Because the plane of the crack growth could not be seen, the amount of crack growth could not be measured. This problem occurred at radii down to 0.2 mm.

The spacing of the data in the crack-growth window is another important factor in obtaining consistent results. Figure 5 demonstrated the scatter that can be obtained in J_{cl} values when the data are properly spaced throughout the test window. Figure 6 shows an example of two tests with material A tested at 2.6 mm/s. Each set of data covers a different part of the window. Taken separately, the J_{cl} values vary from 33 to 41 kJ/m². Taken together, the data are still acceptable in terms of scatter, with J_{cl} equal to 35.5 kJ/m². The spacing of the data is more important when the R curve has a steep slope because minor data scatter can shift the blunting line intercept significantly.

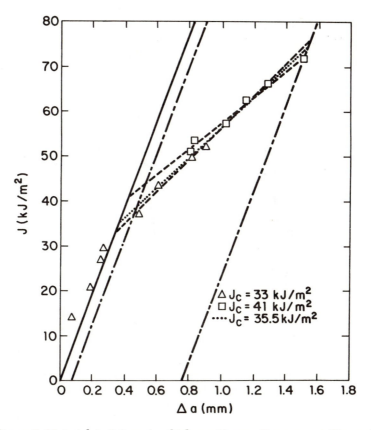

Figure 6. Material A, 2.6 mm/s, thickness 12 mm. Key: same as Figure 4.

To test isotropy of the injection-molded plaques, one set of specimens was prepared in the specimen II configuration (i.e., the specimens were notched normal to the flow direction instead of parallel to it). The results indicate no difference in toughness in these directions. The J_{cl} values were 31.5, 15.5, and 15.8 kJ/m^2 for materials A, B, and C, respectively. For materials A and B, the fracture toughnesses and R curves of the two directions were similar. However, for material C there was an unexpected amount of scatter (Figure 7) using specimen II. Experimentally, the test was difficult because fracture occurred unpredictably. Although the J_{cl} values are similar, the R curves are clearly different. This variation indicates that propagation properties may be different in the two directions.

Size Criteria. According to the ASTM size criterion for ungrooved SENB specimens, an experimental fracture toughness value, J_Q, may be considered a valid plane-strain value, J_{cl}, if the thickness, B, meets the

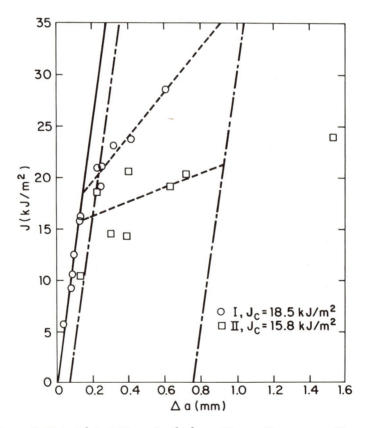

Figure 7. Material C, 0.26 mm/s, thickness 12 mm. Key: same as Figure 4.

requirement

$$B > 25 \frac{J_Q}{\sigma_y} \tag{3}$$

This empirical requirement was formulated on the basis of experimental evidence in *J* testing of metals. For the cases that have been studied, only material A does not meet this requirement. Because of the decrease in yield stress with decreasing test rate, slower rates place the specimen farther away from plane-strain conditions.

For material A, the J_Q values decrease as the test rate increases. The ASTM recommendation for size was never fully met, although the thickness was only 2–5 mm too thin at the fastest speed. Because the recommendation is empirical, the specimen may have been under plane-strain conditions at 26 mm/s. Thus, the value of 30 kJ/m^2 may be a valid plane-strain value. The drop in J_Q may have been caused by mixed mode conditions or rate sensitivity.

In contrast, the 12-mm thickness appears to be adequate for plane-strain conditions in materials B and C. For material B, the specimens are large enough to meet the ASTM recommendation at all speeds. The J_{cl} values are roughly constant, approximately 17.5 kJ/m^2. This result indicates rate insensitivity when the specimens are under plane-strain conditions. Material C was tested at only one speed, 0.26 mm/s, because the crack growth was not stable enough at higher speeds. The value of 18.5 kJ/m^2 should be a plane-strain value.

Grooved Specimens. Size independence must be demonstrated to determine whether the toughness values are actually plane-strain values. However, because of rubber particle agglomeration, it is not possible to obtain larger specimens through injection molding. Side-grooved specimens are an alternative method provided by ASTM E 813–81 to obtain plane-strain conditions. Side grooves place additional constraint on the crack tip because the plane-stress regions have, literally, been cut away. The crack front is much straighter than in an ungrooved specimen. According to ASTM E 813–81, the size recommendation for grooved SENB specimens is

$$B_{net} \geq 25 \frac{J_{cl}}{\sigma_y} \tag{4}$$

The results for the grooved specimens are presented in Figure 8.

For material B, the J_{cl} values for the grooved specimens range from 11 to 18 kJ/m^2, with the majority of the values around 15 kJ/m^2. Overall, these values compare very well with the range of 16–19 kJ/m^2 obtained by using

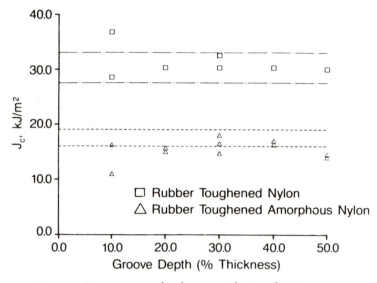

Figure 8. J_{cl} *vs. groove depth, materials A and B, 26 mm/s.*

ungrooved specimens, but they are on the low end of the range. The groove depth does not seem to grossly affect the results, although the 50% groove depth gives lower values and the 10% groove depth gives the most scatter. According to the ASTM size recommendation, all of the specimens were in plane strain, so it is encouraging that similar J_{cl} values were obtained.

For material A, the J_Q values are in the range of 28.5–36.8 kJ/m^2. This result compares favorably with the values of 27.5–32.5 kJ/m^2 obtained with ungrooved specimens. However, it is curious that in all cases, grooved or ungrooved, the ASTM recommendation was still not met. The groove depth did not seem to matter, although there seems to be more scatter in the 10% groove depth data.

One explanation for the similarity of the J_Q values of the grooved and ungrooved specimens could be that they are plane-strain values and that the ASTM size recommendation may be too conservative for this system. A second explanation may be that the recommendation is correct and that the J_{cl} value is artificially high. This error can occur because the energies used in the J calculations are obtained from the area under the load-displacement curve. In soft materials, such as these, there may be local deformation at the loading points. The energy required for this process would be included in the total energy that is measured by calculating the area under the loading curve. Thus, the measured energy is an overestimate of the energy required for crack initiation and growth. Further work needs to be done to determine whether the inclusion of the plastic deformation energy in these calculations will significantly alter the J_{cl} results as well as the resulting size recommendations.

The ductile fracture toughness obtained by the *J* method can be related to K_{cI}, the critical stress intensity factor, which is a measure of strength. The equation for specimens under plane strain is

$$K_{cI}^2 = \frac{E\,J_{cI}}{1 - v^2} \tag{5}$$

where *E* is elastic moduli and *v* is Poisson's ratio.

By calculating K_{cI} from J_{cI} and using the ASTM requirement for *K* testing $(B > 2.5(K_{cI}/\sigma_y)^2)$, a comparison of the recommended minimum thicknesses for each type of toughness test can be made. As shown in Table III, the *J*-integral method can decrease the minimum thickness requirement by a factor of 3–5.

Low-Temperature Tests

Low-temperature tests (23 to –40 °C) were conducted on ungrooved 12-mm-thick SENB specimens at 26 mm/s. The results are presented in Table IV. As expected, the yield stresses and moduli increased with decreasing temperature. In the case of material B, the specimen sizes met the ASTM recommendation at all temperatures. The plane-strain toughness values (J_{cI}) are on the low end of the room-temperature range of toughness values. However, taken as a whole, the toughness appears to be independent of temperature for this range. Calculation of K_{cI} from the J_{cI} values shows that the critical stress intensity factors are also temperature-independent for this range. Because the range of room-temperature values represents many tests

Table III. Comparison of Minimum Specimen
Thicknesses for *J* and *K* Testing

Material	Minimum Thickness for J, mm	Minimum Thickness for K, mm
A	15.0	70.3
B	5.8	20.3
C	7.5	35.8

Table IV. Low-Temperature Results for Rubber-
Toughened Amorphous Nylon

Temperature, °C	Yield Stress, MPa	E, GPa	J_{cI}, kJ/m²	K_{cI}, MPa√m
23	69	2.01	16.0–19.0	6.22–6.77
–10	86	2.04	14.8, 15.0	6.01, 6.06
–40	101	2.31	15	6.45

NOTE: Poisson's ratio = 0.41. All tests were conducted at 26 mm/s with 12-mm-thick SENB specimens.

and there are few low-temperature tests, more work needs to be done to verify these trends.

Similar low-temperature work is shown in Table V for material A. In this case, only the 23 °C tests were not conducted under plane-strain conditions according to ASTM. The −10 °C tests gave J_c and K_c values similar to those measured at room temperature. The decrease in plane-strain toughness values between −10 and −40 °C indicates that these values are temperature-dependent. This dependence implies that care must be taken when testing at low temperature to evaluate higher-temperature plane-strain fracture toughness or critical stress intensity factor. Although a modest decrease in temperature may give suitable values, further decreases in test temperature may give conservative values for this system.

In Table VI, J_{cl} and K_{cl} values are listed for the present study. In addition, literature values (4) of J_{cl} and K_{cl} for a variety of toughened polymers tested at 20 °C are listed. The minimum J_{cl} and K_{cl} values for material A are significantly better than those for the other materials. The toughness values of materials B and C, while lower, are still comparable to the best of the rest. While extensive J work has not been conducted on toughened polymers, material A ranks among the best in toughness and strength for polymers at any temperature. (Linear low-density polyethylene has a J_{cl} value of 18 kJ/m^2 and a K_{cl} of 7.2 MPa$\sqrt{m}$ at −100 °C (12).)

Table V. Low-Temperature Results for Rubber-Toughened Nylon 6/6

Temperature, °C	Yield Stress, MPa	E, GPa	J_{cl}, kJ/m^2	K_{cl}, MPa$\sqrt{m}$
23	50	1.96	27.5–32.5	8.05–8.75
−10	66	2.06	29.5, 31.3	8.54, 8.80
−40	92	2.19	21.0, 23.8	7.43, 7.91

NOTE: Poisson's ratio = 0.41. All tests were conducted at 26 mm/s with 12-mm-thick SENB specimens.

Table VI. Comparative Toughness for Other Toughened Polymers

Material	J_{cl}, kJ/m^2	K_{cl}, MPa$\sqrt{m}$
A	30.0	8.40
B	16.0	6.20
C	18.5	7.41
Polypropylene copolymer[a]	15.5	4.97
Rubber-toughened PVC[a]	3.0	3.09
ABS[a]	3.0	2.68

[a]From ref. 4.

Conclusions

The *J*-integral technique appears to be a useful method to measure plane-strain fracture toughness in tough materials with achievable specimen sizes. Test rates up to 26 mm/s were found to be satisfactory for *J* tests, provided that all material constants are measured at the same test rate. When applied to tough systems, ASTM E 813–81 need not be followed scrupulously. For example, the ASTM size recommendations were suitable for materials B and C and conservative for material A. *J* tests on side-grooved specimens gave results similar to those obtained from appropriately sized ungrooved specimens. The practice of low-temperature testing must be approached with caution because of the possibility of temperature-dependent toughnesses.

References

1. Chan, M. K. V.; Williams, J. G. *Int. J. Fract.* **1983,** *19,* 145.
2. Narisawa, I. *Polym. Eng. Sci.* **1987,** *27,* 41.
3. Lee, D. J.; Donovan, J. A. *Theor. Appl. Fract. Mech.* **1985,** *4,* 137.
4. Hashemi, S.; Williams, J. G. *Polym. Eng. Sci.* **1986,** *26,* 760.
5. So, P. K.; Broutman, L. J. *Polym. Eng. Sci.* **1986,** *26,* 1173.
6. Rice, J. R. *J. Appl. Mech.* **1968,** *35,* 379.
7. Broberg, K. B. *J. Mech. Phys. Solids* **1971,** *19,* 407.
8. Begley, J. A.; Landes, J. D. *The J-Integral as a Fracture Criterion;* in *Fracture Toughness;* ASTM STP 514, American Society for Testing and Materials: Philadelphia, 1972; pp 1–20.
9. Landes, J. D.; Begley, J. A. *Test Results from J-Integral Studies: An Attempt to Establish a J$_{IC}$ Testing Procedure;* in *Fracture Analysis;* ASTM STP 560, American Society for Testing and Materials: Philadelphia, 1974; pp 170–186.
10. Sumpter, J. D.; Turner, C. E. *Int. J. Fract.* **1973,** *9,* 320.
11. Shih, C. F.; Andrews, W. R.; de Lorenzi, H. G. et al. "Crack Initiation and Growth Under Fully Plastic Conditions: A Methodology for Plastic Fracture"; Report NP-701-SR; Electric Power Research Institute: Palo Alto, CA, Feb. 1978; pp 6.1–6.63.
12. Hashemi, S.; Williams, J. G. *Polymer* **1986,** *27,* 384.
13. Haggag, F. M.; Underwood, J. H. *Int. J. Fract.* **1984,** *26,* R63–R65.

Received for review February 11, 1988. Accepted revised manuscript December 16, 1988.

Fracture of Elastomer-Modified Epoxy Polymers

A Review

W. D. Bascom[1] and D. L. Hunston[2]

[1]Department of Materials Science and Engineering, University of Utah, Salt Lake City, UT 84112
[2]National Institute of Standards and Technology, Polymers Division, Gaithersburg, MD 20899

The fracture behavior of carboxyl-terminated butadiene–acrylonitrile elastomer-modified epoxy polymers in bulk, as adhesives, and as matrix polymers in composites is described. The dramatic increase in fracture energy over that of unmodified epoxies is discussed in terms of the elastomer being dispersed as small (1–5-μm) particles that increase the crack-front deformation zone size. Differences in fracture energies when these materials are used as adhesives are explained in terms of constraints on the deformation zone. The fracture energy of the modified epoxies is shown to exhibit a time–temperature dependence far greater than the unmodified epoxies, in bulk and as adhesives. In composites these toughened epoxies increase delamination resistance under quasi-static loading. However, the detailed mechanisms are not fully known, especially the effect of the rate dependence of these polymers on impact resistance.

THE FRACTURE BEHAVIOR OF EPOXY-BASED POLYMERS in bulk, as adhesives, and as matrix resins in fiber-reinforced composites has considerable importance because these materials are used for structural components in aircraft, automobiles, ships, and housing. In many instances the component is load-bearing so that its strength, stiffness, and toughness are critical to the function and reliability of the structure.

0065–2393/89/0222–0135$10.50/0

The base resins, from which essentially all structural adhesives and matrix polymers are formulated, are thermosetting polymers that, when cured, have minimal long-term creep and high tensile strength. Moreover, they begin as liquids or semisolids that can easily be applied to adherend surfaces. Unfortunately, cured thermosets are relatively brittle materials that (as adhesives for example) have fracture energies two or three orders of magnitude less than those of the metals and composites they are called upon to bond. Consequently, all structural adhesives are highly formulated resins; they are generally complex mixtures of the base resin with elastomers, plasticizers, and inorganic particulate fillers.

These systems generally have a common morphological feature, an elastomeric phase dispersed in and bonded to a matrix of the thermoset resin (*1*). In this morphology the elastomer imparts a high toughness to the material without serious losses in the inherent tensile strength, modulus, heat-distortion temperature, and similar properties of the base epoxy resin. Morphologically, these adhesives are related to the high-impact polystyrenes and other resins that have a low-modulus phase dispersed in a high-modulus matrix. In fact, there are several similarities in the fracture behavior of all these multiphase polymer systems, as has been demonstrated by McGarry and co-workers (*2, 3*) in their studies of rubber-modified polyester and epoxy polymers.

The general morphological requirements for two-phase toughened materials (i.e., elastomeric particles dispersed in a rigid matrix) are well documented. The effects of morphological details, however, are not clearly understood. Fracture energy increases with increasing concentration of the elastomer, up to the point where it becomes difficult to obtain the desired particle–matrix morphology. This point is often about 18% elastomer. Below that level a wide range of morphological features can be produced.

Generally, high-fracture-energy systems have a particle-size distribution in which many particles are in the $1-5$-μm range. The particles contain both epoxy and elastomer molecules, but the role (if any) of the particle composition in determining fracture energy is still a topic of discussion. The interfacial bond strength between the matrix and the particles is another factor whose role has not been clearly established. Finally, the matrix often contains some elastomer that is not phase-separated or is phase-separated in particles that are too small to see with the electron microscope. The effect of the matrix characteristics on fracture behavior is also under study by several groups. Yee and Pearson (*4*) have already shown that the cross-link density in the matrix can be very important.

Another factor that may be influential in determining fracture energy is the distribution of particle sizes. In this regard one type of morphology is of particular interest. When the cure chemistry is adjusted in the proper way, both large (>0.5-μm) and small (<0.2-μm) elastomeric particles are obtained. As a result, these formulations are often said to have a bimodal distribution of particle sizes. In reality the actual distribution may be quite complex. Nevertheless, this chapter will follow the general convention of using the term "bimodal" to identify these materials. The term "conventional distribution" will be used to identify materials in which the particle sizes are predominantly distributed around a central size value.

Although particle-size distribution provides the most obvious distinguishing feature of the bimodal systems, other differences are also present. To understand these differences one must consider how these materials are prepared. In most cases the elastomer is added to the epoxy before cure, and they are compatible. During cure the epoxy polymerizes with itself and with the elastomer. As the molecular weights increase, compatibility decreases; eventually phase separation occurs.

To change the particle-size distribution, one must change the chemistry. This change in chemistry can, of course, alter other features such as the degree of phase separation and the nature of the particle and matrix phases. Consequently, it is very dangerous to attribute differences in behavior obtained for these systems simply to particle morphology. A number of studies (5–7) have tried to address this problem by using preformed particles, but so far the results are limited. Pursuit of this approach, however, may provide a better basis for establishing structure–property relationships in the future.

What makes the bimodal systems of particular interest is that, for a given concentration of elastomer, they often show a significantly higher fracture energy than systems with the conventional morphology. However, as will be shown in this chapter, a comparison of fracture energies obtained by using a single set of experimental conditions is not sufficient to characterize the failure behavior of toughened systems and can, in fact, be very misleading.

Fracture Characterization

We will briefly describe the essential fracture mechanics pertinent to this review. Comprehensive discussions of polymer fracture can be found in the literature (8–10).

One approach to fracture mechanics is built on the Griffith failure criterion (11), which equates the strain energy density in a solid with the surface energy required to create new surface area at the instant of crack propagation. Thus the stress at crack initiation, σ_c, is given in general form by

$$\sigma_c = Y \left(\frac{E\gamma}{a} \right)^{1/2} \tag{1}$$

where Y is a dimensional parameter that depends only on specimen geometry, E is the tensile modulus, γ is the surface energy, and a is the crack length. Actually, crack initiation involves a number of energy-consuming processes within the immediate vicinity of the crack tip, including visco-elastic and plastic deformation. Irwin (12) demonstrated that the Griffith equation is formally correct if the surface-energy term, redefined by Irwin as the critical strain-energy release rate, G_c, is taken to include all these energy-dissipative processes, so that

$$\sigma_c = Y\left(\frac{EG_c}{2a}\right)^{1/2} \tag{2}$$

G_c is then an index of fracture toughness and, in the ideal brittle limit, $G_c = 2\gamma$. The deformation processes are limited to a small volume at the crack tip. The dimensions of this region (e.g., the radius of the deformation zone (r_c) or the crack-opening displacement at failure (δ_c), can also be taken as critical conditions for the onset of fracture (Figure 1). Analytical relationships exist between G_c and these dimensions, but their explicit forms depend upon the model assumed for failure and yielding in the deformation zone. For example, an elastic-plastic model gives

$$G_c = 6\,\pi\sigma_y\epsilon_y r_c(1 - \nu^2) \tag{3}$$

in plane-strain conditions (i.e., thick samples) where σ_y and ϵ_y are the yield stress and strain, respectively; ν is Poisson's ratio; and the loading is normal to the crack plane (indicated as mode I in Figure 2).

None of the failure parameters (δ_c, G_c, or r_c) is a true material constant for polymers. In general, they are time- and temperature-dependent. The fracture energy of isotropic solids is also a function of the stress state, so that it is necessary to define strain-energy release rates for each of the three principal modes of failure (Figure 2) (opening mode, G_{Ic}, in-plane shear, G_{IIc}, and torsional shear, G_{IIIc}) and develop a relationship to predict failure for mixed-mode loading. Once all of these effects are considered, however, the result is a completely general failure criterion for linear-elastic conditions. The strain-energy release rate is given by

$$G = \frac{P^2}{2b}\frac{dC}{da} \tag{4}$$

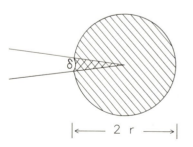

Figure 1. Schematic of the crack-tip deformation zone for an elastic-plastic material. The radius of the zone, r, and the crack opening displacement, δ, are shown.

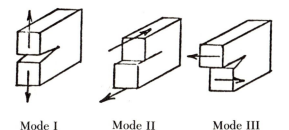

Mode I   Mode II   Mode III

Figure 2. The three basic modes of loading.

where P is the applied load, b is the material thickness in the vicinity of the crack, and dC/da is the change in the compliance, C, of the structure with crack length, a. The value of G at the failure load is the critical strain-energy release rate. Since dC/da is a "global" property of the specimen, equation 4 can be determined without knowing the crack-tip stress distribution. To develop equation 4 in this way, however, dC/da must be determined experimentally (*13*), and this is tedious and often inaccurate. Alternatively, there are analytical expressions for equation 4 in simple geometries. In recent years dC/da has been obtained by finite element analyses for many of the more complex specimens.

Another parameter that characterizes fracture resistance is the stress-intensity factor, K, a measure of the stress intensification at a crack tip relative to the far field stress. It is related to the fracture energy by

$$G_I = \frac{K_I^2(1 - v^2)}{E} \tag{5}$$

in plane-strain conditions.

There are a variety of fracture test specimen geometries (*10, 14*), many of which were developed for specific applications. For general use, a widely employed configuration is the compact-tension (CT) specimen shown in Figure 3. The failure can proceed by stable or unstable crack growth. In stable growth the crack propagation is driven by the cross-head motion of the test machine and stops (or rapidly slows down) when the cross-head is stopped. With unstable growth the crack rapidly propagates when a critical load is reached, and this causes the load to drop abruptly. Under some conditions the crack will arrest itself before the specimen is completely broken so that continued motion of the cross-head will reload the specimen and thus give a stick–slip crack motion.

The compact-tension specimen has two shortcomings. First, fracture usually occurs catastrophically for tough materials because, although the load falls as the crack extends by unstable growth, the sample dimensions are usually too small to permit crack arrest. The load-displacement diagram looks

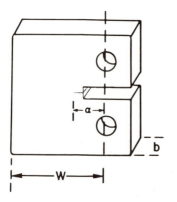

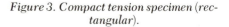

Figure 3. Compact tension specimen (rectangular).

like Figure 4A. Second, the calculation of G_{Ic} by equation 4 requires that the crack length be determined, and this determination can be difficult to do unambiguously.

These problems can be partially avoided by using a double cantilever beam (DCB) specimen that has been tapered for a constant change in compliance with crack length. Mostovoy et al. (*15–22*) used this specimen geometry in all of their work on mode I fracture studies on adhesives. The explicit form of equation 6 for a DCB is

$$G_{Ic} = \frac{4P_c^2}{b^2 E_b} \left(\frac{3a^2}{h^3} + \frac{1}{h} \right)^{1/2} \tag{6}$$

where h is the beam height measured normal to the crack plane at the crack tip and E_b is the bending modulus. If the beams are tapered as in Figure 5 so that the bracketed term is a constant (m), then G_{Ic} is independent of crack length and only the load at crack initiation need be measured. Two constant compliance gradient beams are illustrated in Figure 5: $m = 3$ cm^{-1} (A) and $m = 90$ cm^{-1} (B). It is important in all such tests to avoid inelastic bending of the beam arms, and for this reason the bulk polymer specimens (Figure 5A) are made quite stiff: $m = 3$ cm^{-1}. Because of the much higher modulus and yield stress of the metal adherends compared to the polymer, the adhesive specimen (Figure 5B) need not be as steeply tapered as the bulk specimen: $m = 90$ cm^{-1}. For neat specimens it is sometimes convenient to use side grooves to help control the crack path (*19*). For this geometry equation 6 is modified by replacing b^2 with $b b_n$ (*see* Figure 5).

Equation 6 has an interesting property in that for the tapered DCB specimen the load-displacement curve should show crack growth at constant load, as in Figure 4B. This will be the case if the fracture energy is independent of crack velocity. However, many polymers are strain-rate sensitive

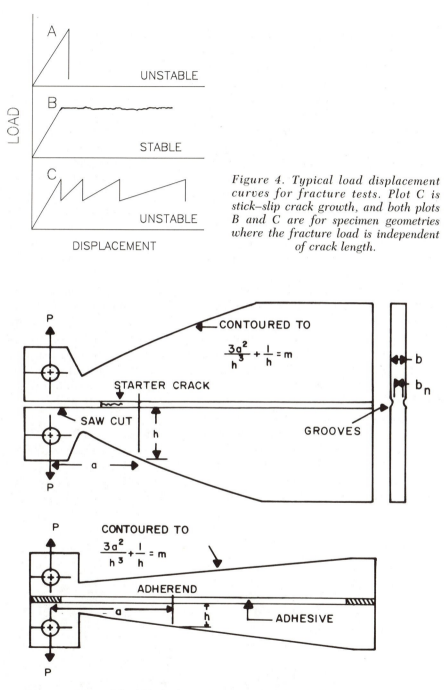

Figure 4. Typical load displacement curves for fracture tests. Plot C is stick–slip crack growth, and both plots B and C are for specimen geometries where the fracture load is independent of crack length.

Figure 5. Tapered double cantilever beam specimen for bulk (top) and adhesive (bottom) samples.

(i.e., G_{Ic} decreases with increasing crack-growth rate). In this case once the crack starts to grow it will propagate spontaneously, and the strain energy in the specimen must drop substantially before it is insufficient to force continued crack propagation. At that point the crack arrests. The result is a peaked load-displacement curve as shown in Figure 4C, with the peaks corresponding to an initiation energy and the valleys to an arrest energy.

The tapered DCB specimens for bulk polymer fracture are prepared by casting and curing the resin in silicone rubber molds or coated metal molds. Grooves are then machined along each side of the specimen to guide the crack through the center plane (Figure 5A). A saw cut is made at the loading end, and a starting crack is formed at the end of the saw cut. The adhesive fracture specimens are assembled as shown in Figure 6. Two metal halves are clamped together with polytetrafluoroethylene spacers at each end to establish the desired bond thickness. One side of the bond is closed with pressure-sensitive tape, and the liquid polymer is poured into the resulting cavity. After heat curing, the tape is removed and any excess resin is machined or abraded off.

Commercial structural adhesives are generally supplied as films of partially cured resin. To test these adhesives, the specimen is prepared by placing strips of the film between the tapered beams. These films usually contain a support or skrim cloth that establishes the bond thickness without any need for spacers. Details concerning the tapered DCB specimen and its use, as well as information on adhesive testing, can be found in refs. 15–28. In all of the work discussed here the adherends were aluminum alloys (5086 or 2024) that were given an acid–chromate etching pretreatment prior to bonding. The choice of alloy had no effect on the results.

It is important to test adhesive fracture under stress conditions other than mode I loading because most adhesive joints are designed to minimize peeling (mode I) loads in favor of shear loads. Two specimens designed for applying both shear and tensile forces are illustrated in Figure 7. In the first of these, the scarf joint (Figure 7A), the applied load is resolved in the bond

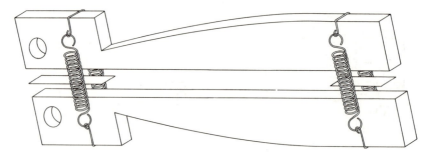

Figure 6. Assembly for making tapered double cantilever beam specimen for adhesive testing. Spacers for control of bond thickness and springs to hold specimen together during cure are shown.

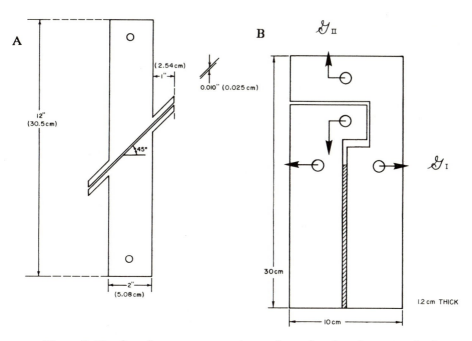

*Figure 7. Mixed-mode test specimens: A, scarf joint (bond angle at 45° to load);
B, ILMM specimen.*

line into mode I and mode II components, and their ratio changes with the
bond angle, φ. In the second specimen (Figure 7B) the shear and tensile
loads are applied independently, and this geometry is referred to as the
ILMM (independently loaded mixed-mode) specimen (27).

These samples are prepared and assembled in much the same way as
the tapered double cantilever beam specimens. The two side arms on the
scarf-joint specimen are clamped with spring clips to hold the two halves
together during sample preparation. The total fracture energy was computed
for the scarf-joint tests using a finite-element analysis developed by Trantina
(29–31). The ILMM test results were analyzed using equations for G_{Ic} and
G_{IIc} given by Mostovoy (15).

Unmodified Epoxies

Bulk and Adhesive Fracture.   Some perspective as to the magnitude
of polymer fracture energies relative to other materials is offered in Figure
8. G_{Ic} values of various polymers in bulk and as adhesives are scaled between
the very brittle inorganic glasses and the relatively tough metals. Epoxies
and other cross-linked thermosetting polymers are very brittle, whereas the
thermoplastics (e.g., polycarbonate and polysulfone) are much tougher.

Nevertheless, Figure 8 shows that the fracture energy of an epoxy can be greatly increased if it contains a dispersed phase of elastomer particles.

Table I compares the bulk fracture energies of a few epoxy resins based on a simple epoxy monomer, diglycidyl ether of bisphenol A (DGEBA), and cured with various curing agents. Values range between 80 and 160 J/m^2 with no obvious systematic dependence on the type of curing agent. The magnitudes of the fracture energies clearly indicate that inelastic dissipative processes are involved in the initiation of crack growth. The rupture of chemical bonds can account for no more than 10 J/m^2, less than 10% of the

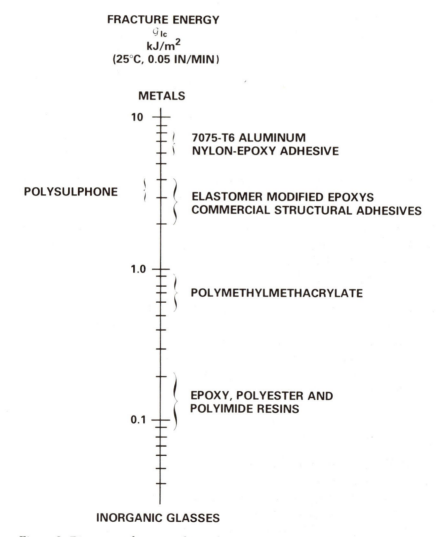

Figure 8. Diagram indicating relative fracture energies for different materials.

Table I. Bulk Fracture Energies of Unmodified Epoxy Resins

Curing Agent	G_{Ic} Bulk, J/m^2
Hexahydrophthalic anhydride[a]	136
Hexahydrophthalic anhydride[b]	153
"Nadic" methyl anhydride[c]	86
Piperidine[d]	154
Tetraethylpentamine[b]	88
m-Phenylenediamine[c]	148

NOTE: Diglycidyl ether of bisphenol A (DGEBA); 20 °C, 45% RH (relative humidity), cross-head speed of 0.13 cm/min.
[a]Ref. 25.
[b]Ref. 19.
[c]Unpublished data.
[d]Ref. 26.

measured energies (32). Even in these very brittle polymers, most of the fracture energy involves viscoelastic and plastic deformation of the resin at the crack tip. Post-failure scanning electron microscopic (SEM) examination of the fracture surfaces reveals clear evidence of plastic deformation in the region of crack-growth initiation (26, 33). Plastic yielding and flow of these highly cross-linked resins are intuitively difficult to accept in view of their apparently "brittle" behavior in tensile tests. Nonetheless, plastic flow does occur prior to failure in the small, highly stressed region of the crack tip. The stress condition in this region is triaxial, and therefore the process probably involves void formation followed by yielding.

The fracture energy of an adhesive bond made with an unmodified epoxy is generally equal to the fracture energy of a bulk sample of the same epoxy so long as the fracture is a cohesive, center-of-bond failure. Differences between bulk and adhesive fracture energies for a given epoxy have been reported (16, 34), but these differences may reflect the inherent uncertainties generally encountered in fracture measurements. Experimental scatter of ±10% or more is not uncommon. Also, even if the cure conditions are ostensibly the same for both bulk and adhesive specimens, they are not necessarily identical. For example, the rate of heating or cooling of the resin or the loss of a volatile component can be different for a resin in a thin bond line from that for a bulk casting.

The epoxies exhibit either stable or unstable crack propagation (*see* Figure 4), depending on temperature, strain rate, and composition. The specimen geometry also has an influence on crack-growth stability. Some of these factors have been addressed by Phillips (35), Gledhill et al. (34), and Kinloch and Williams (36). The data given in Figure 9 exemplify the observed variations in behavior. In this case behavior is measured as a function of temperature plotted against the corresponding yield stress. At cross-head speeds up to $\sim 10^{-5}$ m/s, fracture is unstable and there is a stress-intensity factor for both initiation and arrest (i.e., K_{Ici} and K_{Ica}, the peaks and the

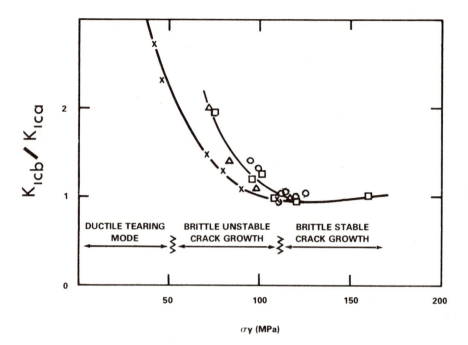

Figure 9. Normalized critical stress-intensity factors for various unmodified epoxies (open symbols) and modified epoxies (×) measured at different temperatures and then plotted against their yield stress values (see ref. 36).

troughs in the load-displacement diagram). Above cross-head speeds of 10^{-5} m/s, cracking is stable and there is only one K_{Ic} for each loading rate.

Gledhill et al. related this behavior to the critical crack-opening displacement, δ_c, illustrated in Figure 1. The δ_c for mode I loading, δ_{Ic}, is given for an elastic-plastic material in terms of the fracture energy by

$$\delta_{Ic} = \frac{G_{Ic}}{\sigma_y} (1 - \nu^2) \tag{7}$$

for plane-strain conditions. They were able to demonstrate that, for the bulk and adhesive fracture of tetraethylpentamine-cured DGEBA polymers, cracking was stable and at a constant G_{Ic} for $\delta < 1.0$ μm, but it became unstable with a systematic increase in G_{Ic} when $\delta > 1.0$ μm. Stated phenomenologically, when the yield stress is sufficiently low (or the time sufficiently long), the crack tip blunts and the material exhibits a high toughness. Gledhill et al. then argue that once a blunt crack begins to grow rapidly, it sharpens because the short time scale minimizes yielding. Consequently, a lower stress-intensity factor is needed to maintain crack growth, and thus unstable growth is produced.

Kinloch and Williams (36) developed a theoretical analysis of crack-tip blunting of epoxy polymers. They assumed that, in the absence of blunting, the critical stress-intensity factor for initiation is equal to the arrest value, K_{Ica}. If the crack-initiation stress-intensity factor measured in the experiments is designated K_{Icb}, then the ratio K_{Icb}/K_{Ica} is a measure of crack blunting. This ratio is unity for stable propagation. In Figure 9, adapted from ref. 36, K_{Icb}/K_{Ica} is plotted against σ_y for various unmodified and elastomer-modified epoxies tested in bulk and as adhesives. The influence of yield stress on blunting is clearly illustrated. Kinloch and Williams chose to draw a single curve through all the data points. A better fit of the data is obtained, however, if separate curves are drawn for the unmodified epoxies and for the elastomer-modified epoxies. This distinction points out the fact that the details of the crack-tip yielding are different for the two types of polymers, as will be discussed in the next sections.

The interdependence of fracture toughness and yield stress strongly suggests a time–temperature dependence of G_{Ic}. With the unmodified epoxies, the variation of G_{Ic} with the strain rate and temperature has been found to be small except near the glass-to-rubber transition (i.e., high temperatures and low strain rates) (37). In a study of the effect of temperatures and strain rate on G_{Ic} of a piperidine-cured DGEBA, Hunston and Kinloch (38) observed a very steep and sudden increase in toughness as the glass-to-rubber transition is approached by either increasing the temperature or decreasing the loading rate (Figure 10).

Composite and Mixed-Mode Adhesive Fracture.   Unmodified epoxies have also been studied extensively as matrix resins in fiber-reinforced composites and as adhesives in bonds subjected to mixed-mode loading. The resulting fracture energies vary from slightly higher to much higher than the corresponding mode I values for the neat resin. Such behavior is quite reasonable, but a detailed discussion of these results will be delayed until the section on elastomer-modified materials because comparisons of values for modified and unmodified materials are of most interest.

Elastomer-Modified Epoxies

Bulk Fracture.   The addition of a liquid carboxyl-terminated polybutadiene acrylonitrile (CTBN) elastomer to DGEBA epoxy resins can increase the fracture energy by factors of 30 or 40 over the unmodified resin. In Table II, G_{Ic} values are listed for a piperidine-cured DGEBA resin with increasing amounts of a liquid CTBN. The CTBN is initially soluble in the epoxy but phase separates during cure. In the concentration range up to about 18 wt %, the elastomer precipitates as spherical particles that, for the piperidine-catalyzed epoxy–CTBN system cured at 120 °C, have diameters of 1–3 μm (26, 28, 39–41). These particles can be seen in the fracture surface by using SEM. At concentrations above approximately 18% the elastomer

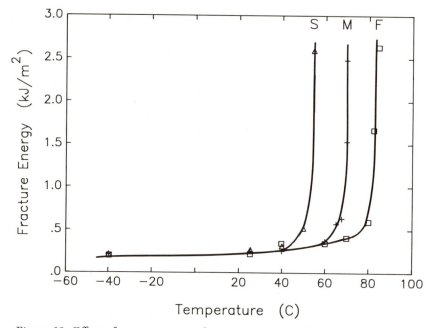

Figure 10. Effect of temperature and strain rate on the fracture energy of an unmodified epoxy near T_g. *Test machine cross-head speeds (mm/min): S, 0.05; M, 1.25; F, 50.*

Table II. Bulk Fracture Energy and Deformation Zone Size for CTBN-Modified Epoxy Resins

CTBN Concentration, wt %	Fracture Energy, G_{Ic}, J/m^2	Deformation Zone Size, $2r_c$, cm
0	121	0.0008
4.5	1050	0.0124
10	2720	0.021
15	3430	0.033
20	3590	0.070

SOURCE: Reprinted with permission from ref. 26. Copyright 1975 Wiley.

forms not a fine dispersion but an inhomogeneous blend, and there is a corresponding decline in the fracture energy.

One of the merits of elastomer–epoxy dispersions is that the large gain in toughness is accomplished with relatively little loss in the high tensile strength, modulus, thermal mechanical properties, etc., relative to those of the epoxy matrix resin. This condition is illustrated in Figure 11 for the piperidine-cured DGEBA system. Above 18% CTBN, where the resin composition passes from a dispersion to a blend, there is an abrupt decline in

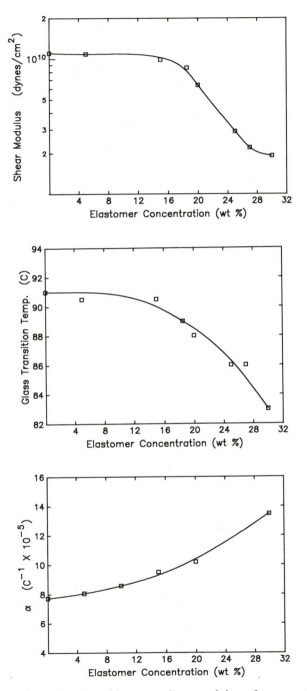

Figure 11. Effect of CTBN additive on shear modulus, glass transition temperature, and modulus, thermal expansion coefficient, $\propto$.

the properties. This retention of properties at the lower concentrations makes the dispersed elastomer morphology effective in formulating structural adhesives and composite matrix resins.

The mechanism of toughening in elastomer-modified epoxies has been thoroughly studied (2–4, 26, 42–44), and the following is a brief summary of the results. As with the unmodified epoxy resins, the CTBN–epoxy polymers exhibit deformation markings in the region of crack initiation. However, the extent of deformation is many times larger for the modified materials. In fact, a distinct stress whitening can be seen developing ahead of the tip of the precrack as the load on the test specimen is increased. SEM examination of the fracture surface in the region of stress whitening reveals closely spaced holes, as shown in Figure 12. These holes correspond to rubber particles that have cavitated. In the region of fast crack propagation beyond the stress-whitening, the elastomer particles appear as filled or partially filled holes, and the matrix exhibits much less deformation than in the stress-whitened region. The rapidly propagating crack appears to slice through the material and to allow little time for deformation and yielding.

The fractographic features of the stress-whitened region indicate a number of deformation processes. The hole formation suggests that there has been a dilational deformation because the holes are larger in diameter than the undeformed particles seen in the rapid-crack-growth region. This dilational deformation is consistent with the triaxial stress field that is known to be present at the crack tip. These stresses lead to rupture of the elastomer particles, which then relax back into the enlarged cavities and give the appearance of holes. Results from several different experiments support this idea, but the most convincing are those of Kinloch. He exposed the whitened region of the fracture surface to an organic fluid to swell the rubber. The SEM then revealed that the holes had become hills.

The stress concentrations associated with the particles and holes nucleate local shear yielding of the epoxy matrix, and this shear yielding is a significant part of the crack-tip deformation process. The holes may assist the yielding by relieving the triaxiality of the stresses. The tear markings in the stress-whitened zone (see Figure 12) are evidence of shear yielding. Strong supporting evidence for this mechanism is provided by the photographs of Yee and Pearson (4), which clearly show shear bands between cavitated rubber particles in modified epoxy materials.

The final stages of the deformation process involve the elongation and rupture of any elastomer particles that have not cavitated. Figure 13 is a SEM view into a crack showing "pillars" of elastomer stretched across the open crack tip. The insert in Figure 13 shows the specimen configuration; a thin strip of polymer was held in a U-shape and a notch was cut at the top of the U. The microscope was then focused into the notch. Although this process of elongation, bridging, and rupture of the elastomer particles may at times be a major contributor to the fracture energy (45), it is not considered an important factor in the highly toughened systems (46).

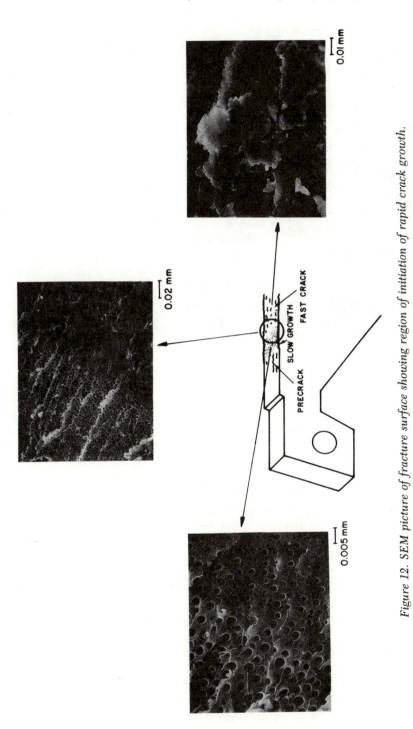

Figure 12. SEM picture of fracture surface showing region of initiation of rapid crack growth.

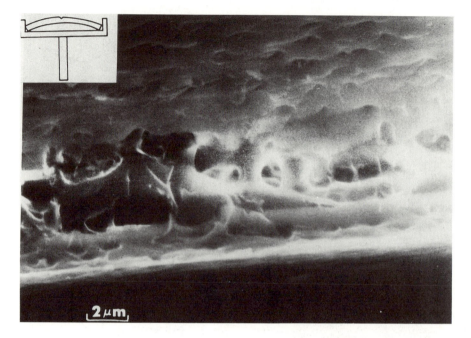

Figure 13. SEM view into a crack in a 15% CTBN-modified DGEBA cured with piperidine. Insert shows specimen.

The CTBN–epoxy system described in the previous paragraphs is a relatively simple composition compared to the commercial formulations used as adhesives, coatings, encapsulants, and matrix resins. Many of these commercial resins are developed to have the bimodal distribution of particle sizes mentioned earlier. This bimodal distribution can be obtained by "precooking" the epoxy (DGEBA) with bisphenol A (BPA) to form an adduct. When this adduct is diluted with CTBN dissolved in DGEBA and cured with piperidine, the dispersed elastomeric phase includes both small particles (<0.5 μm in diameter) and large particles (1–5 μm in diameter).

As mentioned earlier, bimodal distribution systems often have a higher toughness for a given rubber content when measured under the usual laboratory test conditions. Also, a larger proportion of the rubber is often precipitated in the bimodal systems, and thus less rubber remains dissolved in the epoxy matrix. Consequently, because there is a lower rubber content and less dissolved rubber to decrease the modulus and lower the glass-transition temperature (T_g), such systems are often considered superior to those with the conventional particle-size distribution. Some of the results from the work of Riew et al. (41) on bimodal CTBN–epoxy resins are given in Table III. The chemistry of these materials is outlined in ref. 41. At the concentration of BPA giving the highest toughness, the T_g is higher than for

**Table III. Some Properties of Epoxy–Bisphenol A–
CTBN Polymers**

Concentration of Bisphenol A, phr	G_{Ic}, kJ/m²	T_{hd},[a] °C
0	2.1	78
6	2.1	83
12	3.3	78
24	8.6	82
36	3.7	74
45	0.7	71

NOTE: Base material contains DGEBA (100 parts), CTBN (5 parts), and piperidine (5 parts), in addition to bisphenol A. It is cured at 120 °C for 16 h.
[a]Heat-distortion temperature.

the conventional particle-size distribution obtained when the BPA concentration is zero.

Detailed investigation of the failure mechanisms for CTBN–epoxy polymers with bimodal size distributions has not been undertaken until recently (42, 47, 48). One of these studies (47) involved a SEM examination of the fracture surfaces for specimens of a piperidine-cured CTBN–DGEBA–BPA epoxy. The photomicrographs in Figure 14 give views of the stress-whitened region. Figure 14A shows the boundary between the precrack (dark area) and the stress whitening. The massive yielding in the stress-whitened region is shown in closer detail in Figures 14B and 14C. As in the case of fracture surfaces for CTBN–epoxy materials with conventional particle-size distribution, holes in the stress-whitened region presumably result from dilation and rupture of the elastomer particles. Most of the larger holes, however, contained round inclusions (Figure 14D) that are believed to be hard (high epoxy content) regions formed in the particles. Both large and small particles were visible in the region of fast crack growth (beyond the stress whitening), but they have not been as severely deformed as in Figure 14C.

The massive shear yielding of the matrix resin evident in Figure 14 is difficult to explain if the epoxy has not been plasticized by the rubber. However, the fact that the T_g and the modulus are essentially the same as those for the unmodified epoxy is not indicative of plasticization. Possible explanations for this anomaly have been given by Kinloch et al. (46) and by Bascom and Hunston (49), based on the work of Sayre et al. (50) and Kunz et al. (51). However, the question is clearly in need of further study, especially because these formulations yield high fracture resistance with minimal trade-off in other mechanical properties.

The extensive yielding evident in the fractography of both the conventional and bimodal particle-size-distribution CTBN–epoxy materials suggests that their fracture behavior should be time- and temperature-dependent. This was shown to be the case by Hunston et al. (42, 48, 52), who determined

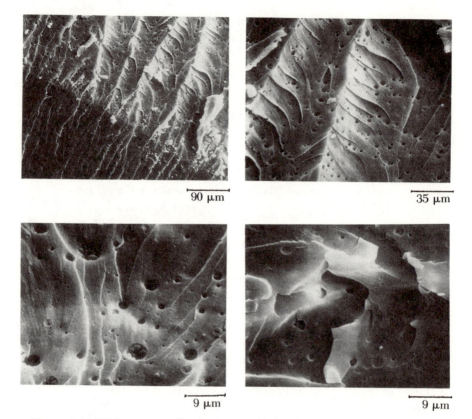

Figure 14. SEM pictures of a fracture surface of material with a bimodal distribution of particle sizes.

G_{Ic} by using CT specimens in experiments at cross-head speeds of 0.05–50 mm/min (0.002–2.0 in./min) and temperatures between –60 and +60 °C. They propose that the failure process is dominated by viscoelastic deformations, and thus there should be an inverse relationship between test temperature and loading rate. The experiments show that increasing the temperature produces effects similar to those obtained by decreasing the loading rate. If the behavior is viscoelastic and the relationship between these two parameters is simple, the data can be analyzed by using time–temperature superposition. Hunston et al. (48) selected time to failure, t_f, as the parameter to characterize the loading rate; t_f is the time from initial application of the load until the failure point in the constant cross-head speed experiment.

Some results from this study are given in Figure 15 for two piperidine-cured DGEBA–CTBN compositions and a piperidine-cured DEGBA–BTA–CTBN composition. Indeed, as shown in this figure, the data

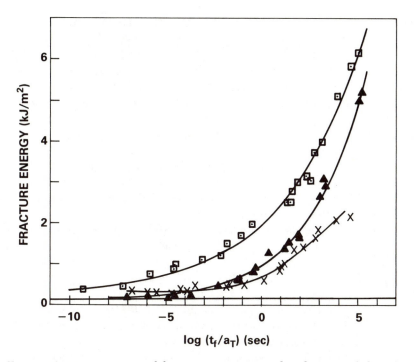

Figure 15. Master curves of fracture energy vs. reduced time to failure for three different formulations of elastomer-modified epoxy. Key: with no BPA, □, 15% CTBN, and ×, 5% CTBN; with BPA, ▲, 5% CTBN. Solid line at bottom of graph represents unmodified epoxy.

can be shifted successfully by using the principle of time–temperature superposition so that G_{Ic} is presented as a function of a reduced time to failure, t_f/a_T, where a_T is a function of temperature. Low values of t_f/a_T correspond to low temperature and high strain rates; high values of t_f/a_T correspond to high temperatures and low strain rates.

Figure 15 presents some interesting information. The unmodified epoxy exhibits very little rate or temperature dependence over the range of test conditions presented here; the data are shown as a line at the bottom of the graph. At higher temperatures and slower strain rates, G_{Ic} of the 5% elastomer-modified epoxy with the bimodal distribution of particle sizes greatly exceeds that for the 5% CTBN–epoxy with the conventional particle-size distribution; it is almost as high as that for the 15% composition. This observation has been made in the past and cited as an advantage of the bimodal systems. A reduction in the reduced time to failure of only three decades, however, brings the bimodal system to near equivalence with the conventional 5% CTBN composition. The curves come together at strain rates that correspond to low-temperature impact testing. In this region there appears to be no advantage to the bimodal system. Moreover, if designs are based on measurements made in the range where the bimodal system is superior, this higher rate sensitivity can lead to significant overestimates of toughness for the bimodal system. Further reductions in temperature (or increases in loading rate) bring the toughness of both modified-epoxy systems down to a value only slightly greater than that of the unmodified epoxy.

The most interesting point raised by the curves in Figure 15 concerns the mechanism of failure. Studies of the fracture surfaces suggest that quite similar deformation processes are involved for both the bimodal and conventional systems. Moreover, Hunston and Bullman (48) found the temperature dependences (i.e., the expressions for a_T) to be identical within experimental scatter for all of the systems shown in Figure 15. Then why are there differences in the rate dependences? At the moment there is no clear answer to this intriguing question.

In analyzing the data in Figure 15, Hunston et al. (48, 52–54) suggest that the results can be represented empirically by

$$G_{Ic} = Ae^{\Delta H/RT}(t_f)^m - G_{Icb} \tag{8}$$

This equation contains four constants, and although they are strictly empirical, they do have meaning in terms of the measured behavior. An examination of the equation shows that A provides a measure of magnitude of the toughening, m assesses the rate sensitivity, ΔH measures the temperature dependence, and G_{Icb} is the limiting toughness at low temperatures or high loading rates.

The data presently available are not sufficient to clearly establish relationships between these parameters and the morphology of the materials. Nevertheless, the following observations can be offered for the four materials systems that have been examined. The thermal parameter, ΔH, is quite similar for all four materials and is identical to that obtained by applying a similar time–temperature superposition process to yield data obtained for two of the materials (55). Consequently, there is a close connection between yielding and fracture, as might be expected in light of the mechanism just discussed. The parameter, m, is significantly different for the bimodal system from that for the three conventional distribution systems tested; the latter have identical values for m. The parameter A increases as the rubber concentration increases, but is also a function of the morphology.

The CTBN–epoxy resins discussed thus far contained only a liquid CTBN additive. Many commercial adhesive or matrix resin formulations use both a liquid CTBN and a solid rubber in an attempt to obtain a multiple particle-size distribution. Bascom et al. (56–58) studied a rubber-modified epoxy of this type and determined the effect of the amount of each type of elastomer additive on fracture energy as a function of strain rate. The composition of the base epoxy resin is given in Table IV. Different amounts of a liquid CTBN and solid CTBN rubbers were added to this base resin. The solid rubbers

**Table IV. Composition of the Base Epoxy Resin
of a Commercial Modified Epoxy**

Component	Approximate Weight %
Epoxides	73
(diglycidyl ether of bisphenol A)	
(epoxidized novolac, epox, equiv wt 165)	
Diphenols	20
(bisphenol A, tetrabromobisphenol A)	
Catalysts	7
(dicyandiamide, substituted urea)	

NOTE: Resin obtained from Hexcel Corp.
SOURCE: Reprinted with permission from ref. 57. Copyright 1980
Butterworth.

were precooked with the constituents listed in Table IV before the liquid
CTBN was added. The composition and numerical designation of the epoxy
materials are listed in Table V.

Fracture energies of the polymers listed in Table V are plotted as a
function of relative strain rate in Figure 16. Because the strain rate at the
crack tip cannot be determined, the inverse of the time to failure (time
between the initial application of the load and the fracture point) was used
as a relative measure of strain rate. The data include results obtained from
impact tests that used an instrumented Izod impact tester (data points at
$\sim 7.5 \times 10^2$ s^{-1}). The impact energies were converted to G_{Ic} values by using
the analysis of Plati and Williams (59). The results in Figure 16 are in many
ways similar to those discussed previously for the simpler rubber–epoxy
compositions. The unmodified epoxy (composition HX–205) has a low toughness, and G_{Ic} exhibits very little strain-rate dependence. All of the rubber-
modified compositions are many times tougher than the base epoxy, and the
G_{Ic} is strain-rate dependent. Composition F–185, which contains both liquid
and solid rubbers, exhibits relatively high toughness in low-strain-rate testing, but has a higher strain-rate sensitivity than the other compositions. This
result is similar to that for the CTBN–DEGA–BPA system discussed previously. In both cases the morphology involves both large and small particles.

Adhesive Fracture—Mode I.   There are some significant differences in the adhesive fracture of elastomer-modified epoxies, compared to
their bulk fracture behavior. All of these differences relate to the adhesive
joint geometry (i.e., bond thickness, bond angle, etc.). The simplest example, and the first to be recognized, was the effect of bond thickness on
adhesive G_{Ic}. In Figure 17 the fracture energies are given as a function of
bond thickness for a 15% CTBN, piperidine-cured, DGEBA epoxy tested
as an adhesive at 233–323 K (28). At each temperature G_{Ic} goes through a

Table V. Elastomer Content of Epoxy Resins

Sample[a]	Liquid CTBN,[b] wt %	Solid Elastomer,[c] wt %
HX 205[d]	0	0
HX 206	8.1	0
HX 207	8.1	1.0
HX 210	0	8.1
F 185	8.1	5.4

[a]Hexcel Corp.
[b]BF Goodrich, Hycar 1300×13, MW 3500.
[c]BF Goodrich, Hycar 1472, MW 260,000.
[d]Base epoxy system given in Table IV.

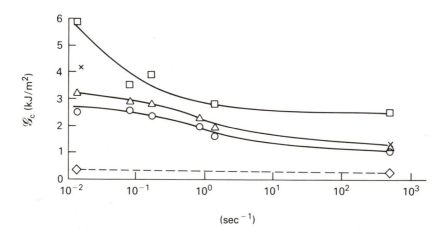

Figure 16. Fracture energy vs. the reciprocal of time to failure. Key: ◇,
HX–205; ○, *HX–206;* ×, *HX–207;* △, *HX–210; and* □, *F–185 (see Tables IV*
and V and ref. 54).

maximum (or approaches a maximum in the 233 K experiments). Failure
occurred by stable crack growth at bond thicknesses less than the maximum,
but cracking was unstable at bond thicknesses greater than the maximum.
The maximum in the data shifted to smaller bond thicknesses with increasing
test temperature.

The appearance of the stress-whitened zone as the bond line thickness
was reduced provides another important feature of the failure. In Figure
18, the stress-whitened zone is shown schematically. In specimens with bond
thickness above the maximum point, stress whitening is contained within
the bond line (Figure 18A) and the failure is center-of-bond. Below the
maximum the stress whitening fills the bond line and the locus of failure is
near the interface, with the crack jumping from one side of the bond to the
other (Figure 18B) as it propagates. It was established by SEM fractography
that, although failure occurred near the interface, crack growth was decidedly
in the resin (26).

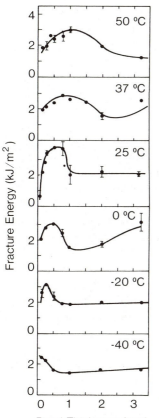

Figure 17. Fracture energy as a function of bond thickness at a series of different temperatures (15% CTBN-modified DGEBA cured with piperidine).

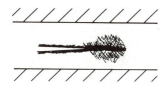

THICK BOND   **THIN BOND**

Figure 18. Schematic of deformation zone in adhesive bonds of different thickness.

These observations lead to the following proposal for the bond thickness effect (54, 60). As Figure 18B would suggest, the zone of stress whitening is increasingly constrained as the bond thickness is decreased. This limitation effectively reduces r_c, the damage zone radius, and thus, according to equation 3, decreases G_{Ic}. It was argued (26) that the maximum in adhesive G_{Ic} occurs when the bond thickness, h, is essentially equal to $2r_c$, the critical diameter of the damage zone measured in bulk specimens. This hypothesis was given some support when equation 3 was used to calculate r_c at the different test temperatures from measured values of bulk G_{Ic}, σ_y, and ϵ_y. Values of h_{max} and $2r_c$ are compared in Table VI, and the agreement is reasonably good, considering the assumptions involved. However, this explanation of the bond thickness effect does not account for the decrease in adhesive G_{Ic} at bond thicknesses greater than h_{max}. This point is addressed in more detail later.

This temperature dependence of the bond thickness effect on adhesive fracture energy implies that there should be a corresponding strain-rate dependence. This rate dependence was confirmed in work by Hunston et al. (42, 52, 54) and Kinloch and Shaw (60). However, in the more complete studies (54–58) the best agreement between predicted and experimentally determined bond thickness for the maximum was obtained by assuming plane-stress conditions at the crack tip. It is difficult to understand how plane-stress conditions could prevail in a highly constrained bond line. This problem has been discussed in ref. 49, where it is suggested that, among other factors, the presence of cavitated elastomer particles may relieve the triaxiality of the crack-tip stress field. In any event, a simple plane-stress model has had some success in predicting the optimum bond thickness, but more realistic models are needed before the thickness effect can be quantitatively understood.

In recent years considerable progress has been made in analyzing the mode I failure of the tapered double cantilever beam adhesive fracture specimen. This work provides a qualitative explanation for the maximum in G_{Ic} vs. bond thickness. It began with the work of McGarry et al. (61) on the stress distribution in the bond line of a tapered double cantilever beam (TDCB) adhesive specimen. They showed with a linear elastic finite element analysis (FEA) that the crack-tip stress distribution does not decrease as $r^{-1/2}$ ahead of the crack, but instead decays more slowly than expected along the bond line, as shown in Figure 19. This effect, which increases as the bond thickness decreases, is due to the difference in moduli between

Table VI. Comparison of Bond Thickness at Maximum Adhesive Fracture Energy with Crack-Tip Deformation Zone Diameter for Bulk Resin Fracture

Temp., K	h_{max}, cm	$2r_c$, cm
323	0.10	0.12
310	0.075	0.078
298	0.060	0.033
273	0.050	0.013
253	0.025	0.008
233	<0.01	0.004

SOURCE: Reproduced with permission from ref. 28. Copyright 1976 Gordon and Breach.

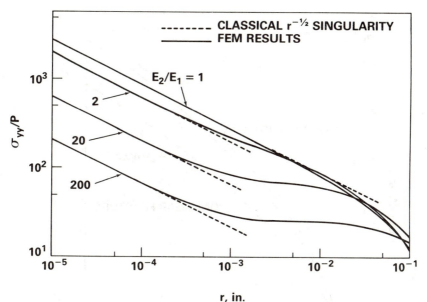

Figure 19. Local tensile stress, σ_{yy}, normalized with applied load, P, as a function of distance, r, ahead of crack tip. Results from ref. 59 for different ratios of moduli for adherend and adhesive (E_2/E_1).

the metal adherend (E_2) and the adhesive (E_1). For aluminum and epoxy, $E_2/E_1 = 20$. Subsequently, Wang (62) conducted a FEA analysis using a nonlinear constitutive equation for the adhesive based on data for commercial and model materials. The results are presented in terms of a *J*-integral that can be viewed in a simplistic way as a generalization of *G* for nonlinear systems.

Figure 20 presents a comparison of results obtained with simple beam theory, linear FEA, and nonlinear FEA for a TDCB adhesive specimen. Beam theory and linear FEA give essentially the same result. The curve for a more realistic constitutive equation, however, suggests that both may underestimate the toughness of the bond when nonlinear (high fracture energy) materials are involved. Nevertheless, the Wang analysis found that the trends, such as dependence of adhesive J_{Ic} on bond thickness, were unaltered.

Most recently, Wang (62), in cooperation with Hunston and Kinloch (55), reexamined the effects of bond thickness on adhesive fracture. The FEA of Wang suggests that two opposing effects are active in the bond line as the thickness is reduced. First, there is a progressive reduction in the volume in which the damage zone can develop because of constraint in the thickness direction. This is the same effect proposed earlier by Bascom et al. and studied by numerous authors. In addition to this effect, Wang's analysis also

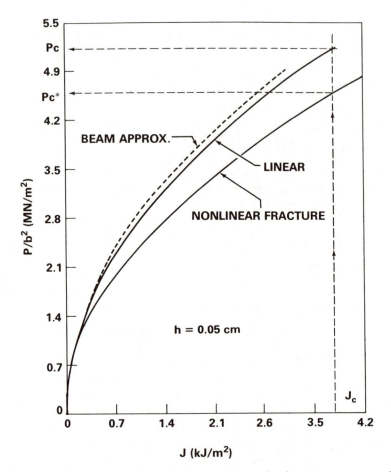

Figure 20. Comparison of results for load normalized with specimen width plotted against applied J. Bond thickness, h, is 0.05 cm.

finds that the zone extends farther down the bond line ahead of the crack tip as the thickness is decreased. This extension is caused by the high stresses farther than expected ahead of the crack tip, as shown in Figure 19.

The analysis shows that both effects are present and interact in the following way. At large bond thicknesses ($>>2r_c$), the adhesive and bulk samples show similar behavior. As the bond thickness is decreased, the extension of the deformation zone down the bond line causes the zone size to increase. This increase produces an increase in the J_{Ic} of the adhesive, although the exact relation between J_{Ic} and zone size is complex because the details of the stress–strain field in the zone change as the bond thickness is decreased. When the bond thickness is decreased further, the constraint imposed by the interaction between the adherends and the zone becomes important. This constraint eventually produces a decrease in the zone size (i.e., a reduction in height) and a related decrease in J_{Ic}. The net result is that J_{Ic} (fracture energy) first increases and then decreases as the bond thickness is reduced.

In an effort to experimentally confirm the conclusions from the Wang analysis, Hunston and Rushford (63) took high-speed movies of the stress-whitened region at the crack tip for CTBN–epoxy adhesive specimens. Their results confirm the analysis; as the bond thickness approaches the crack-tip diameter ($2r_c$), the deformation zone fills the bond line and extends ahead of the crack tip for a much longer distance than would be expected from observations on the same polymer in bulk.

These temperature, loading rate, and bond thickness effects shed new light on an aspect of adhesive behavior that has been known for many years but has not been clearly understood. It is customary to designate a temperature range in which an adhesive exhibits its maximum fracture behavior. For example, Mostovoy (64) determined the adhesive G_{Ic} as a function of temperature for a number of commercial structural adhesives, and in most cases there is a maximum in toughness. The dotted line in Figure 21 represents one of the commercial adhesives tested by Mostovoy. This is typical of the type of data often provided for adhesives in that only a single bond thickness is used (i.e., the thickness recommended by the manufacturer for applications). To understand this behavior, however, the influence of bond thickness and loading rate, as well as temperature, must be taken into account. For example, the data points and solid curve in Figure 21 are taken from Figure 17 at a bond thickness of 0.25 mm. Although a completely rigorous comparison would include the influence of loading rate as well, the curve for the model adhesive is quite similar to that for the commercial material.

Adhesive Fracture—Combined Mode.   Rarely, if ever, is a structural adhesive joint designed to experience pure mode I loading. In general, the bond line and the flaws within the bond line will see some combination of mode I, II, and possibly III loads (Figure 2). In the case of an isotropic specimen (e.g., bulk fracture), mode I has been determined (65, 66) to be the lowest energy fracture mode. Thus, the crack always propagates along a path normal to the direction of maximum resolved tension. In other words, a crack in an isotropic plate will propagate in a mode I fracture style, regardless of how the initial flaw is oriented with respect to the applied load. Such is not necessarily the case in adhesive fracture. Here, the crack propagation must be constrained to the resin layer. This requirement can force the crack to propagate in the bond direction at or near the interface, regardless of the orientation of the bond line. Consequently, attention must be given to adhesive fracture in combined stress conditions, because meas-

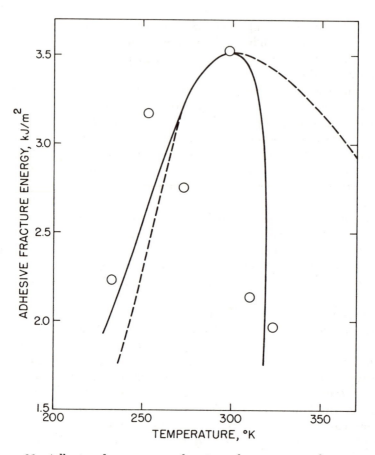

Figure 21. Adhesive fracture as a function of temperature for commercial material (dashed line) and model system from Figure 16 (points and solid line).

urements on bulk samples may be of little use in predicting adhesive behavior.

Ripling, Mostovoy, and Patrick (15) determined the mode II adhesive fracture of a simple unmodified epoxy resin (DGEBA–tetraethylpentamine [TEPA]) by using the specimen illustrated in Figure 7B. Trantina (29–31) determined the mixed-mode adhesive fracture behavior of the epoxy resin by using the scarf-joint specimen (Figure 7A). Bascom et al. (25) used the same specimen to study the mixed-mode adhesive fracture behavior of two other unmodified epoxy resins (DEGBA–hexahydrophthalic acid [HHPA] and DGEBA–piperidine).

The results of these tests are listed in Table VII with the notation G_{IIc} and $G_{(I,II)c}$ for mode II and mixed-mode fracture energies, respectively. The mode I values are also presented and are, in all cases, less than the $G_{(I,II)c}$ values. The rationale for the higher $G_{(I,II)c}$ values comes from observations

**Table VII. Effect of Fracture Mode on Adhesive
Fracture Energy of Epoxy Resins**

Resin	G_{Ic}	G_{IIc}	$G_{(I,II)c}$
TEPA[a]–DGEBA	70	1450	130[b]
HHPA[c]–DGEBA	116	–	140[b]
Piperidine–DGEBA	121	–	255[b]

NOTE: All results are given in Joules per square meter. Data from
refs. 19, 25, and 29.
[a]Tetraethylpentamine.
[b]Data from scarf joint with bond line at 45° angle to direction of
applied load.
[c]Hexahydrophthalic anhydride.

of crack growth. Trantina (*31*) and Bascom et al. (*25, 65*) established that the
precrack (AB in Figure 22) advances across the bond line (BC in Figure 22)
during loading (i.e., a mode I-type fracture) but arrests when it reaches the
adhesive–adherend interface. The specimen must then be loaded to a higher
level in order to force the crack to grow in an unfavorable direction (CD in
Figure 22).

In other work (*25, 65*) with the scarf-joint specimen, it was shown that
the roughness of the adherend surface can strongly influence the value of
$G_{(I,II)c}$. The reason for this roughness effect can be understood in terms of
the locus of crack growth. Because failure initiates near an interface (point
C in Figure 22), the surface roughness can affect the crack-tip deformation
process, the degree to which the actual crack area exceeds the nominal area,
and the local direction of crack growth. In general, the fracture energy is
greater for rough surfaces than for highly polished adherends. However,
surface asperities may be sufficiently sharp to act as local stress concentrations, which add to the crack-tip stress field and reduce the measured toughness.

In the scarf-joint test, the locus of failure is apparently interfacial, but
by using radioactive tracers in the adhesive, it was established (*25*) that the
fracture actually occurred in the resin, albeit deceptively close to the interface. As outlined, the nature of the mixed-mode stress field directs the crack
toward the interface, and this direction results in a locus of failure near, but
not at, the interface. Failure to recognize this common trait of mixed-mode
tests has lead to the erroneous conclusion that there was some weakness at
the adherend–adhesive interface.

Although the mixed-mode adhesive fracture behavior of brittle unmodified epoxies is qualitatively understood, the behavior of the CTBN–
elastomer epoxy compositions and commercial structural adhesives is a very
different situation. Relative to the mode I fracture energies, some of the
literature results report mixed-mode fracture energies that are higher (*64,
68*), others report lower values for mixed-mode fracture (*25, 27, 67, 69, 70*),
and still others see little difference, at least until the mode II component is
very high (*71, 72*). A complete explanation of this complex area will require

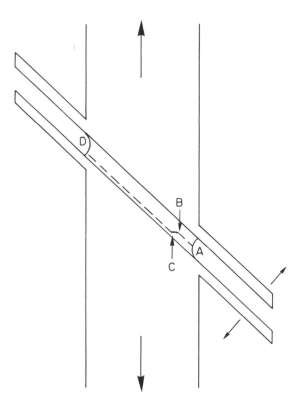

Figure 22. Diagram of crack growth path in scarf joint. Loading direction indicated by arrows at top and bottom.

much more detailed research. Nevertheless, it is useful to examine the results that are available.

The fracture mechanics literature (9, 10, 12) teaches that in an isotropic material crack propagation is normal to the maximum resolved tensile stress (i.e., mode I-type failure) because this is the direction of easiest crack propagation and lowest fracture energy. The question immediately arises as to whether the elastomer-modified epoxy materials behave differently. Work by Hunston (48) established that in bulk samples these polymers do not exhibit any unusual behavior in this respect. Consequently, any anomalies that may exist in mixed-mode fracture must be the result of effects that take place within the bond line.

Three important factors should be kept in mind when considering mixed-mode adhesive fracture of tough adhesives. First, the size of the deformation zone is of the same order as the bond thickness, so the actual direction of crack initiation is uncertain. This situation differs from the mixed-mode fracture of unmodified epoxies, where the initiating direction is known (i.e., point C in Figure 22). Second, the proximity of the metal adherend surfaces may influence the shape of the deformation zone, and thus affect the zone size and the direction of crack initiation. Bascom has speculated (25) that the high shear stresses at the boundary between the deformation zone and the adherend may initiate instability before the deformation zone can achieve the size characteristic of the material tested in bulk. This situation is analogous to mode I adhesive failure when the bond thickness is $<<r_c$. Finally, some of the mixed-mode experiments were done with commercial film adhesives that contain skrim cloths. This situation compounds the uncertainties of the stress state in the bond line under combined loading.

One fact does emerge from these studies; mode I testing is not always a sufficient failure criterion for adhesive fracture. This fact was clearly shown by Bascom and Timmons (27) for model adhesive systems and very recently by Sancaktar et al. (73) for a commercial film adhesive with and without a skrim cloth. This evidence is enough to alert design engineers that fail-safe designs of adhesively bonded structures cannot be based on mode I testing alone.

Interlaminar Fracture of Composites.   Another area of application for toughened epoxies is as matrix resins in fibrous composites. By increasing the fracture energy of the matrix, the delamination resistance and impact resistance of composites can be significantly increased. This fact has spurred interest in thermoplastic matrix polymers because of their inherent higher fracture energy compared to conventional thermosetting polymers. However, the thermoplastic matrix materials have deficiencies that are not readily overcome; creep and low solvent resistance are the most common of these. Thermoplastics that have been designed to overcome these problems have proved difficult to process (74). In addition, there have been difficulties in obtaining good adhesion to the reinforcement, notably carbon fibers (75).

Consequently, there has been continuing interest in elastomer-modified epoxies as matrix polymers. Work by Bascom et al. (57) demonstrated that composites with CTBN–epoxy matrix polymers had significantly higher mode I interlaminar fracture resistance than the composites with the corresponding unmodified epoxy matrix. This early work and subsequent studies demonstrate that the fracture processes that occur in delamination are far more complex than in adhesive fracture or bulk fracture of these materials. For example, Hunston (76–78) found a discontinuous relationship between matrix fracture energy and delamination resistance. As shown in Figure 23, composites with the lower mode I fracture energy polymers have mode I interlaminar fracture energies about twice that of the bulk polymer.

For the tougher matrices, there is a less advantageous trade-off. Increasing the resin fracture energy by 3 J/m^2 produced only about a 1-J/m^2 improvement in interlaminar delamination resistance. These results can be rationalized by noting that for the low-toughness polymers the deformation zone size is small relative to the fiber–fiber spacing between plies. As the

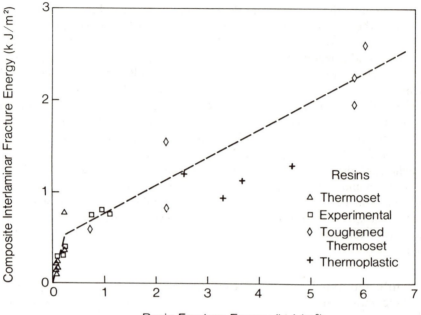

Figure 23. Comparison of resin fracture energies and composite interlaminar fracture energies for many different resin systems. Lines indicate general trends, but are not meant to indicate that the trends are necessarily linear.

matrix fracture energy is increased, the deformation size becomes larger and soon interacts with the fibers. This interaction results in a less complete transfer of resin toughness to the composites. The fibers do not totally inhibit the crack-tip deformation zone, however, and deformation of the polymer around the fiber is an important energy-dissipating mechanism that enhances interlaminar fracture energy. The fibers also contribute to the resistance to fracture by diverting the crack and bridging the opening behind the crack tip. This fiber effect probably explains why the interlaminar fracture energy is greater than the polymer bulk fracture energy for brittle resins. Indeed, the role of the fiber in delamination is reflected in the fact that the fiber diameter can affect the interlaminar G_{Ic} (79).

Impact damage and damage tolerance are more complex than the interlaminar fracture of unidirectional laminates. Damage during the impact event may involve shear (mode II) deformation, especially if there are large bending deflections. Subsequent propagation of interlaminar cracks resulting from impact probably occurs by mode I delamination, especially under compressive loads. However, delamination testing of unidirectional laminates does not properly imitate impact or even postimpact delamination. In structural laminates adjacent plies are often at different angles, so there is little opportunity for the nesting of fibers between plies. Moreover, there can be a discernible resin-rich layer between plies. This layering is not the case for unidirectional laminates, where the interply separation is often quite small. These differences help explain why impact damage tolerance does not correlate exactly with matrix fracture energy or interlaminar fracture energy (*80*). Nevertheless, the use of tough matrices, such as elastomer-modified epoxies, has a clear benefit.

Currently, there are three approaches to introducing elastomer toughening into epoxy matrix composites.

1. in situ phase separation (*81*), similar in principle to the elastomer-modified epoxy adhesives
2. interlayers (*82*) where a film of elastomer is introduced between the plies of epoxy and fiber
3. thin elastomeric films applied directly to the fibers (*83*)

The first two approaches are in early stages of commercial development, and the latter is in the research and development phase. Important questions remain on the toughening mechanisms in these systems, and this is clearly a fertile area for study.

Conclusions

The elastomer-modified epoxies represent an exciting class of materials that exhibit many interesting properties. The most important and intriguing aspect of their behavior, however, is the dramatic increase that can be obtained in fracture energy with a minimum sacrifice in other properties. Extensive studies have now elucidated the mechanism of toughening in these materials, and empirical models to describe the effects of temperature and loading rate on their fracture properties have been developed. An approach has also been outlined to establish relationships between the morphology of these two-phase systems and their failure behavior.

The application of modified epoxies as adhesives has been studied in mode I and mixed-mode loading. The understanding of mode I behavior has been highly developed and includes many aspects of the rate, temperature, and bond-thickness effects. The failure in mixed-mode loading has also been investigated, but much more remains to be done in this area.

The use of toughened materials as matrix resins in composites has also shown considerable promise. Tougher polymers have been found to give composites with improved interlaminar fracture behavior. The relationship between the two properties has been established for mode I loading under simple laboratory conditions. Work to model this behavior and extend these studies to more complex loading and test conditions is now underway.

Acknowledgments

This review is dedicated to the late Robert L. Cottington, whose attention to experimental detail revealed subtle but critically important aspects of adhesive fracture. We also thank our many co-workers who contributed so much to this work, especially, Tony Kinloch (Imperial College, London, UK), Al Siebert (BF Goodrich, Brecksville, OH), and Jim Rushford.

This chapter was adapted from "The Fracture of Epoxy and Elastomer-Modified Epoxy Polymers", W. D. Bascom and D. L. Hunston, in *Treatise on Adhesion and Adhesives*, R. L. Patrick, Ed., Dekker, NY, 1989.

Certain commercial materials are identified in this paper in order to specify the experimental procedure adequately. In no case does such identification imply recommendation or endorsement by the National Institute of Standards and Technology, nor does it imply that the materials are necessarily the best available for the purpose.

References

1. Bolger, J. C. In *Treatise on Adhesion and Adhesives;* Patrick, R. L., Ed.; Dekker: New York, 1973; Vol. 2, p 1.
2. Sultan, J. N.; Laible, R. C.; McGarry, F. J. *J. Appl. Polym. Sci.* 1971, 6, 127.
3. Sultan, J. N.; McGarry, F. J. *Polym. Eng. Sci.* 1973, 13, 29.
4. Yee, A. F.; Pearson, R. A. "Toughening Mechanism in Elastomer-Modified Epoxy Resins"; NASA Report CR-3718, 1983; NASA-3852, 1984.
5. Meeks, A. C. *Polymer* 1974, 15, 675.
6. Pocius, A. V.; Schultz, W. J.; Adam, R. E. *Proceedings of the Adhesion Society;* 1986; p 4.
7. Kelley, F. N., personal communication.
8. *Fracture, An Advanced Treatise;* Liebowitz, H., Ed.; Academic: New York, 1968; Vol. I–VII.
9. Tetelman, A. S.; McEvily, A. J. *Fracture of Structural Materials;* Wiley: New York, 1967; p 51.
10. Knott, J. F. *Fundamentals of Fracture Mechanics;* Butterworths: London, 1973.
11. Griffith, A. A. *Philos. Trans. R. Soc. London A* 1920, 221, 163.
12. Irwin, G. R. In *Handbuch der Physik;* Flugge, S., Ed.; Springer: Berlin, 1956; Vol. 6, p 551.
13. Gray, T. G. F. *Bull. Mech. Eng. Ed.* 1971, 10, 151.
14. Kinloch, A. J.; Young, R. J. *Fracture Behavior of Polymers;* Applied Science: London, 1983.
15. Ripling, E. J.; Mostovoy, S.; Patrick, R. L. *ASTM Spec. Tech. Publ.* 1963, 360, 5.
16. Mostovoy, S.; Ripling, E. J. *J. Appl. Polym. Sci.* 1966, 10, 1351.
17. Mostovoy, S.; Ripling, E. J. *J. Appl. Polym. Sci.* 1969, 13, 1083.
18. Patrick, R. L.; Brown, J. A.; Vehoeven, L. E.; Ripling, E. J.; Mostovoy, S. *J. Adhes.* 1969, 1, 136.
19. Mostovoy, S.; Ripling, E. J. *J. Appl. Polym. Sci.* 1971, 15, 641, 661.
20. Ripling, E. J.; Mostovoy, S.; Corten, H. T. *J. Adhes.* 1971, 3, 107.
21. Mostovoy, S.; Ripling, E. J.; Bersch, C. F. *J. Adhes.* 1971, 3, 125.
22. Ripling, E. J.; Mostovoy, S.; Bersch, C. F. *J. Adhes.* 1971, 3, 145.

23. Patrick, R. L.; Gehman, W. G.; Dunbar, L.; Brown, J. A. *J. Adhes.* **1971**, *3*, 165.
24. Mostovoy, S.; Ripling, E. J. "Fracture Characteristics of Adhesive Joints"; Final Report NAVAIR N00019-72-C00250; Materials Research Laboratory: Glenwood, IL, 1972.
25. Bascom, W. D.; Timmons, C. O.; Jones, R. L. *J. Mater. Sci.* **1975**, *10*, 1037.
26. Bascom, W. D.; Cottington, R. L.; Jones, R. L.; Peyser, P. *J. Appl. Polym. Sci.* **1975**, *19*, 2545.
27. Bascom, W. D.; Jones, R. L.; Timmons, C. O. In *Adhesion Science and Technology;* Lee, L.-H., Ed.; Plenum: New York, 1975; Vol. 9B, p 501.
28. Bascom, W. D.; Cottington, R. L. *J. Adhes.* **1976**, *7*, 333.
29. Trantina, G. G. *J. Compos. Mater.* **1972**, *6*, 191.
30. Trantina, G. G. *J. Compos. Mater.* **1972**, *6*, 371.
31. Trantina, G. G. "Combined Mode Crack Extension in Adhesive Joints"; T&AM Report No. 352; University of Illinois: Urbana, 1971.
32. Berry, J. P. In *Fracture Processes in Polymeric Solids;* Rosen, B., Ed.; Interscience: New York, 1964; p 257.
33. Patrick, R. L. In *Treatise on Adhesion and Adhesives;* Patrick, R. L., Ed.; Dekker: New York, 1973; Vol. 3, p 163.
34. Gledhill, R. A.; Kinloch, A. J.; Yamini, S.; Young, R. J. *Polymer* **1978**, *19*, 574.
35. Phillips, D. C.; Scott, J. M.; Jones, M. *J. Mater. Sci.* **1978**, *13*, 311.
36. Kinloch, A. J.; Williams, J. C. *J. Mater. Sci.* **1980**, *15*, 987.
37. Mostovoy, S.; Ripling, E. J. *Appl. Polym. Symp.* **1972**, *19*, 395.
38. Hunston, D. L.; Kinloch, A. J., unpublished data.
39. Rowe, E. H.; Siebert, A. R.; Drake, R. S. *Mod. Plast.* **1970**, *47*, 110.
40. Siebert, A. R.; Riew, C. K. *Org. Coat. Plast. Chem.* **1971**, *31*, 555.
41. Riew, C. K.; Rowe, E. H.; Siebert, A. R. In *Toughness and Brittleness of Plastics;* Deanin, R. D.; Crugnola, A. M., Eds.; Advances in Chemistry 154; American Chemical Society: Washington, DC, 1974; p 326.
42. Bitner, J. L.; Rushford, J. L.; Rose, W. S.; Hunston, D. L.; Riew, C. K. *J. Adhes.* **1981**, *13*, 3.
43. Kinloch, A. J.; Shaw, S. J.; Tod, D. A.; Hunston, D. L. *Polymer* **1983**, *24*, 1431.
44. Kinloch, A. J.; Shaw, S. J.; Hunston, D. L. *Polymer* **1983**, *24*, 1355.
45. Kunz-Douglas, S.; Beaumont, P. W. R.; Ashby, M. F. *J. Mater. Sci.* **1980**, *15*, 1109.
46. Kinloch, A. J.; Shaw, S. J.; Tod, D. A.; Hunston, D. L. *Polymer* **1983**, *24*, 1341, 1355.
47. Bascom, W. D.; Hunston, D. L.; Rushford, J., unpublished data.
48. Hunston, D. L.; Bullman, G. W. *Int. J. Adhes. Adhes.* **1985**, *5*, 69.
49. Bascom, W. D.; Hunston, D. L. In *Treatise on Adhesion and Adhesives;* DeVries, K. L.; Anderson, G., Eds.; Dekker: New York, 1988; Vol. 6, p 123.
50. Sayre, J. A.; Assink, R. A.; Lagasse, R. R. *Polymer* **1981**, *22*, 87.
51. Kunz, S. C.; Sayre, J. A.; Assink, R. A. *Polymer* **1982**, *23*, 1897.
52. Hunston, D. L.; Bitner, J. L.; Rushford, J. L.; Oroshnik, J.; Rose, W. S. *Elastomers Plast.* **1980**, *12*, 133.
53. Hunston, D. L.; Bascom, W. D. In *Rubber-Modified Thermoset Resins;* Riew, C. K.; Gillham, J. K., Eds.; Advances in Chemistry 208; American Chemical Society: Washington, DC, 1984; Chapter 7, p 83.
54. Hunston, D. L.; Kinloch, A. J.; Shaw, S. J.; Wang, S. S. In *Adhesive Joints;* Mittal, K., Ed.; Plenum: New York, 1984.
55. Hunston, D. L.; Kinloch, A. J., unpublished data.
56. Bascom, W. D.; Ting, R. Y.; Moulton, R. J.; Riew, C. K.; Siebert, A. R. *J. Mater. Sci.* **1981**, *16*, 2657.

57. Bascom, W. D.; Bitner, J. L.; Moulton, R. J.; Siebert, A. R. *Composites* **1980**, *11*, 9.
58. Bascom, W. D.; Moulton, R. J.; Rowe, E. J.; Siebert, A. R. *Org. Coat. Plast. Chem.* **1978**, *39*, 1964.
59. Plati, E.; Williams, J. G. *Polym. Eng. Sci.* **1975**, *15*, 470.
60. Kinloch, A. J.; Shaw, S. J. *J. Adhes.* **1981**, *12*, 59.
61. Wang, S. S.; Mandell, J. F.; McGarry, F. J. *Fracture of Adhesive Joints;* R76-1, Massachusetts Institute of Technology: Cambridge, MA, 1976.
62. Wang, S. S., personal communication.
63. Hunston, D. L.; Rushford, J. L., unpublished data.
64. Mostovoy, S.; Ripling, E. J. In *Adhesion Science and Technology;* Lee, L.-H., Ed.; Plenum: New York, 1975; Vol. 9B, p 513.
65. Irwin, G. R. *J. Appl. Mech.* **1957**, *24*, 361.
66. Erdogan, F.; Sih, G. C. *J. Basic Eng.* **1963**, *85D*, 519.
67. Bascom, W. D.; Oroshnik, J. *J. Mater. Sci.* **1978**, *13*, 1411.
68. Brussat, T. R.; Chiu, S. T.; Mostovoy, S. *Fracture Mechanics of Structural Adhesive Bonds;* LR 27614-5, Air Force Materials Laboratory, Wright-Patterson Air Force Base: Dayton, OH, 1976.
69. Bascom, W. D.; Hunston, D. L.; Timmons, C. O. *Org. Coat. Plast. Chem.* **1978**, *38*, 179.
70. Bascom, W. D.; Hunston D. L. In *Adhesion 6;* Allen, K. W., Ed.; Applied Science: London, 1982; p 185.
71. Johnson, W. S.; Mall, S. *ASTM Spec. Tech. Publ.* **1985**, *876*, 189.
72. Mall, S.; Johnson, W. S. *ASTM Spec. Tech. Publ.* **1985**, *893*, 322.
73. Sancaktar, E.; Jozavi, H.; Baldwin, J.; Tang, J. *J. Adhes.* **1987**, *23*, 233.
74. Gosnel, R. B. In *Engineered Materials Handbook;* Reinhart, T. J., Ed.; ASM: Metals Park, OH, 1987; Vol. 1, p 97.
75. Bascom, W. D.; Cordner, L. W.; Hinkley, J. L.; Johnston, N. J. *Proceedings of the American Society for Composites, 1st Technical Conference;* Technomic: Lancaster, PA, 1986; p 238.
76. Hunston, D. L. *Compos. Technol. Rev.* **1984**, *4*, 176.
77. Hunston, D. L. *Tough Composite Materials;* Noyes: Park Ridge, NJ, 1985; pp 1–15.
78. Hunston, D. L.; Moulton, R. J.; Johnston, N. J.; Bascom, W. D. *ASTM Spec. Tech. Publ.* **1985**, *937*, 74.
79. Bradley, W. L.; Cohen, R. N. In *Mechanical Behavior of Materials-IV: Proceedings of the 4th International Conference;* Carlsson, J.; Ohlson, N. G., Eds.; Pergamon: New York, 1984; pp 595–601.
80. Sinclair, T.; Johnston, N. J.; Baucim, R. M. NASA-TM 100518, 1988.
81. Gawin, I. *31st Int. SAMPE Symp.* **1986**, *31*, 1204.
82. Evans, R. E.; Masters, J. E. *ASTM Spec. Tech. Publ.* **1985**, *937*, 413.
83. Kawamoto, J.; McGarry, F. J.; Mandell, J. F. *Impact Resistance of Rubber Modified Carbon Fiber Composites;* PPSt-85-5, Massachusetts Institute of Technology: Cambridge, MA, 1985.

RECEIVED for review March 25, 1988. ACCEPTED revised manuscript January 23, 1989.

Thin Elastomer Films in Glassy Polymers

Frederick J. McGarry

Department of Materials Science and Engineering, Massachusetts Institute of Technology, Cambridge, MA 02139

Elastomer films 10^2–10^3 Å thick, surrounding occluded glassy domains, can toughen glassy thermoplastics by promoting crazing and shear banding. In glassy thermosets they promote shear and cavitation. When continuous or discontinuous reinforcing fibers, glass, and graphite are coated with such thin films, polymer matrix composite properties can be maintained while resistance to impact damage and impact strength are increased. The film reduces residual thermal stresses and stress concentrations around the fibers. It also promotes plastic flow in the adjacent matrix material.

IMPACT-RESISTANT POLYSTYRENE BECAME AVAILABLE commercially about 40 years ago (*1*). It was made by dissolving styrene–butadiene rubber in styrene monomer and then polymerizing the solution to form a cross-linked network. Granulation and mechanical milling were necessary to achieve a moldable material. In effect, this process broke up the rubber network and made subsequent melt forming possible (*2*).

Before long, however, it was discovered that stirring the solution caused the rubber to form discrete particles that persisted as the styrene polymerized. Then the final continuous polystyrene phase could be melted and molded without further processing. The solution-grafted rubber particles formed by this route contained substantial fractions of occluded polystyrene as glassy domains within the particles, and the effective volume fraction of the rubber was significantly increased. The insoluble gel content so formed could be several factors greater than the rubber actually added to the solution initially. At the time, rubber was the more expensive component of the

0065–2393/89/0222–0173$06.00/0

mixture. The fact that impact strength improved with increasing gel content spurred much work to maximize the effectiveness of the rubber via this mechanism (3).

An early alternative to producing impact polystyrene used melt-mixing of the rubber and the polystyrene. This product lacked graft bonding between the two phases, and it did not extend the rubber with the glass occlusions. As a result, its properties were comparatively inferior, and a higher volume of rubber was needed to achieve significant toughness. Because of this, its production declined and essentially stopped.

A similar experience took place with acrylonitrile–butadiene–styrene copolymers (ABS). Graft copolymerization of styrene–acrylonitrile and acrylonitrile–butadiene rubber displaced melt blending for the same reasons. Again, a gel content action was found to operate. In retrospect, it can be seen that almost from the start the morphology of the rubber in both systems involved thin films rather than simply solid particles.

For nearly the next 20 years, the role of the rubber was believed to be that of a crack stopper–deflector and a crack bridger, restraining the flanks of an advancing crack and thereby impeding its growth. In 1965, however, Bucknall and Smith observed crazes initiating at the particles, which subsequently deformed and constrained the craze from growing thicker and longer (4).

Bucknall and Smith used optical microscopy in their work. Soon others used electron microscopy to confirm that multiple crazes were generated at the particle–matrix interface when the material was under tensile stress. The elegant prior work on crazing by Kambour (5) provided a valuable context within which to study the subject.

The higher resolution of electron microscopy showed unmistakably that the rubber morphology was essentially filmlike, especially around the outer periphery of each particle. Rubber-wall thicknesses were in the range of a few hundred to a few thousand Angstroms. The entire particle was described as "a thin rubber bag full of hard marbles". For a while the cusps in the outer rubber film evoked much interest because they seemed to be the loci of craze generation.

Interest in using rubber to toughen glassy cross-linked polymers also was developing. Some of the first work of this type was reported in 1970 (6). Low-molecular-weight liquid copolymers of butadiene–acrylonitrile were reacted with liquid epoxy resins to form two-phase solid mixtures in which the presence of the particles caused significant increases in fracture toughness. The particles themselves contained two phases, rubber and glass, and their similarity to high-impact polystyrene in this respect was noted (6).

In the cross-linked systems, however, the mechanism of toughening appears to be different. Cavitation and elongation of the particles seemed to be caused by shear deformation in the surrounding glassy matrix, which is not capable of crazing (7, 8). A number of workers have contributed significantly to this technology, and a recent monograph conveniently summarizes the current state of the art (9).

Both for its own sake and because of its importance in rubber-toughened systems, crazing in glassy thermoplastics has received intensive study during the past decade. Kramer and various colleagues have made outstanding contributions to our understanding of the microstructure of crazes and of how entanglements in the molecular network control the local plastic deformation response of a glassy polymer. His use of constrained thin film samples and transmission electron microscopy has been wonderfully productive in this respect (10, 11). We now know that the chain contour length between entanglements determines whether the glass shear bands or crazes; that the maximum strain in a craze also depends upon the entanglement density; that craze material is highly porous, fibrillar in nature, and capable of large extensions; and that craze stability (or breakdown) depends upon polymer molecular weight and the presence of foreign inclusions.

Argon (12) developed a molecular model for plastic flow in glassy polymers that invokes no free volume change to describe the incremental alignment of chain molecules by thermally activated rotations of chain segments. Subsequently, he and Bessonov (13) greatly enhanced this theoretical treatment by demonstrating its validity when applied to a rather wide variety of aromatic polyimides. Because of the contributions of these two researchers, Kramer and Argon, our current understanding of flow mechanisms in glassy polymers is quite good.

Copolymerization of styrene and butadiene has been known for some time. About 20 years ago it was recognized that block versions, obtained from improved catalysis practice, could produce unusual morphologies in which the (discontinuous) rubber phase would exist as spheres, rods, or layered lamellae, the latter often only a few hundred Angstroms thick (14). Recently Schwier et al. (15) showed how the topology of the rubber phase (the geometrical shape of the rods, for example) in a polystyrene–polybutadiene diblock greatly influences the crazing behavior and macroscopic ductility of the blend.

In these systems the rubber particle, the rod or the lamella, often cavitates and induces crazing in the surrounding glassy matrix. The sequence of this crazing seems to differ from that in conventional high-impact polystyrene (HIPS), though the macroscopic result—increased toughness—is the same (16). By adding low-molecular-weight polybutadiene to certain of these polystyrene–polybutadiene (PS–PB) diblock blends, the same group has achieved solid composite particles. Concentric spherical shells, with alternating layers of glass and rubber, give control over the mechanical compliance and physical size of the particle that can be used for model study of the toughening process (17). Detailed analysis of the model has generated some unanswered questions, but the value of this analytical approach is attractive (18).

In a very valuable analysis published in 1978, Ricco and colleagues (19) showed that the stress and strain perturbations in the surrounding elastic matrix produced by a rubber-coated spherical hard particle closely approximated those produced by a solid rubber particle, until the coating became very thin compared to the total diameter. This theorized what had been determined from experience with HIPS—a large volume fraction of occluded glass could exist within the rubber particle without impairing its toughening effect.

Alternatively, thin films of rubber could induce plastic flow. If the diameter of the hard core is 4 μm, rubber films as thin as 400 Å on the surface should be effective. Tse (20) explored this idea experimentally by using butadiene–acrylonitrile liquid rubbers as the coating on inorganic glass microspheres. If good adhesion existed between the glass surface and the rubber film, and if the film was well bonded to the surrounding epoxy matrix, the rubber toughened as effectively as solid particles of the same elastomer (20). In fact, on a unit volume basis, the rubber film was more efficient than a similar solid particle.

The same idea was explored recently in the context of graphite fiber-reinforced epoxy laminates. Thin rubber coatings on the fibers measurably increased the resistance of the composite plate to damage from flatwise impact blows (21). The mechanism operating in the latter case is still being studied.

In retrospect, it becomes very clear that thin elastomer films can and do function effectively to increase the fracture toughness or the impact strength of glassy polymers. This is true whether they are supported on a rigid substrate or simply exist within the rigid polymer matrix. In either case, good bonding to the matrix is essential to their function. Bonding induces some form of plastic deformation in the surrounding glassy medium as a precursor to its cohesive fracture.

Coated Fibers

To study coated fibers, it was necessary to develop a method whereby each fiber in a tow of several thousand could be coated uniformly, continuously, economically, and to a controlled thickness of the deposited film. This process employed the apparatus shown in Figure 1. The tow was spread flat by passage through an air fan block, which contained opposing jets of nitrogen gas angled and sequenced to separate the fibers as they passed through a transition chamber from a circular inlet to a slit die outlet. At the block exit was a shaped anvil that kept the fibers apart as they entered the coating bath tray.

In the bath were four components: the elastomer, epoxy resin, curing agent, and the liquid carrier. The concentration of the coating formulation in the carrier was the means for thickness control. The range of solids content,

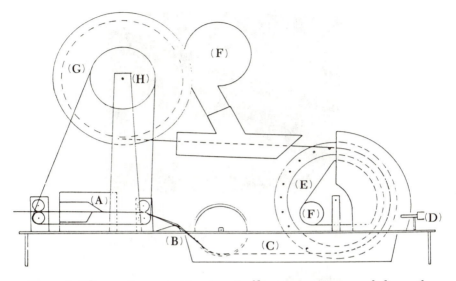

Figure 1. Fiber-coating apparatus. Key: A, fiber separator; B, parabolic anvil; C, coating solution bath; D, air nozzle; E, drying wheel; F, hot air dryer; G, take-up reel; H, disk clutch.

0.5–2.0% by weight, made possible a thickness range of 100–1200 Å on each individual fiber, depending upon fiber diameter and speed of passage through the bath. The tow exited the bath to a large drying wheel where gross drying was done by hot air jets, and then it traversed a long drying tube also serviced by counterflowing hot air. When it reached the windup spool it was essentially dry. Individual fibers were coated, and they did not stick together in the tow. Then the spool was put into a vacuum oven to remove any final traces of the carrier liquid and to react the elastomer with the epoxy, cross-linking it and adhering it to the fiber surface.

Scanning electron micrographs of uncoated and coated fibers, both glass and graphite, are shown in Figures 2 and 3, which are representative of the products from the process. The coatings are continuous, uniform, and stable. Stability can be confirmed by solvent immersion, which would remove unreacted elastomer or dislodge a coating that did not adhere well to the substrate.

The next step involved impregnating the coated fibers with a liquid epoxy solution, air drying, vacuum drying, and then oven curing to advance the matrix resin to the B stage. The dry prepreg is cut into suitable lengths and widths, laid up in the desired sequence and thickness, vacuum bagged, and press cured to produce the final composite plates. Many important processing details can be found in refs. 21 and 22. The laminates so produced are described in Tables I and II. These are high fiber volume fraction, continuous fiber reinforced epoxy laminates, in which the rubber content is

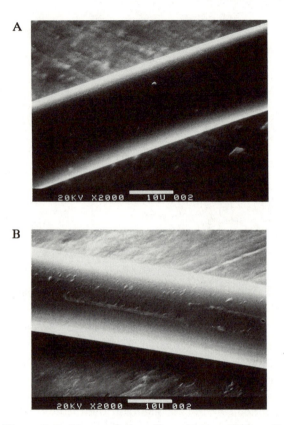

Figure 2. A, Uncoated glass fibers; B, coated glass fibers.

in the range of 0.5–2.5% by weight, reflecting film thicknesses on the fibers over the range of 100–1000 Å.

For thermoplastics the coated continuous fibers were chopped and blended with the plastics.

Materials

The elastomer used in this study is a low-molecular-weight copolymer of butadiene and acrylonitrile containing amine terminal groups (ATBN 1300×16, BF Goodrich). This copolymer is coreacted with a diglycidal ether of bisphenol A (DGEBA) epoxy (Celanese Epi-Rez 520–C) and catalyzed by an amine curing agent (CMD WJ60–8537; Celanese) in the conventional fashion. The proportions of rubber and epoxy can be varied over broad limits to produce a variety of stress–strain behaviors (Figure 4). This variability is one of the attractive features of the system, because it permits study of the effects of coating film properties on the final properties of the composite. The matrix epoxy resin is also a conventional DGEBA-type (Shell Epon 828), cured with a tertiary amine (Ancamine K–61–B). The fiberglass is 366 roving (a large bundle of fibers) from OCF with a silane coupling agent on its surface. The graphite fiber is Thornel 300, a PAN-based product from Amoco in a 3000-filament

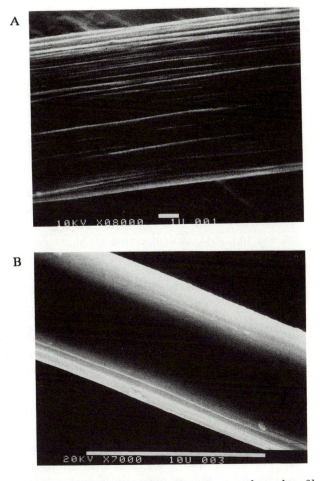

Figure 3. A, Uncoated graphite fiber; B, coated graphite fiber.

Table I. Glass–Epoxy Composites

Sample	Fiber Sample	Fiber Weight Fraction, %	V_f	V_m	V_r	V_v
			Volume Fraction, %			
CF0–I	A–1	86.0	71.4	25.8	0.0	1.97
CF0–II	A–1	86.5	72.2	25.0	0.0	2.80
CF1–I	A–2	85.3	68.3	25.4	0.89	5.45
CF1–II	A–2	84.3	68.2	27.6	2.30	4.53
CF2–I	A–3	86.3	70.3	22.9	2.30	4.53
CF2–II	A–3	85.1	69.3	25.2	2.26	3.32

NOTE: V_f, V_m, V_r, and V_v are volume fractions of fiber, matrix, rubber, and void, respectively.

Table II. Graphite–Epoxy Composites

Sample	Lamination Geometry	Fiber	Volume Fraction, %			
			V_f	V_m	V_r	V_v
RC00–I	0	Original	72.6	26.9	0.0	0.5
RC0–I	0	A–0	61.9	34.4	0.0	3.7
RC0–II	0	A–0	64.6	31.4	0.0	4.0
RC0–III	0:90	A–0	64.3	34.8	0.0	0.9
RC1–I	0	A–1	67.0	31.2	0.4	1.5
RC1–II	0	A–1	65.8	31.8	0.4	2.0
RC1–III	0:90	A–1	63.0	36.2	0.4	0.4
RC2–I	0	A–2	63.1	30.8	1.3	4.8
RC2–II	0	A–2	61.7	34.9	1.3	2.1
RC2–III	0:90	A–2	63.1	31.7	1.3	3.9
RC3–I	0	A–3	63.6	30.9	2.3	3.2
RC3–II	0	A–3	58.0	35.2	2.3	4.5
RC3–III	0:90	A–3	56.9	40.8	2.3	0.0

NOTE: V_f, V_m, V_r, and V_v are volume fractions of fiber, matrix, rubber, and void, respectively.

tow with a proprietary finish that was removed with an organic solvent prior to use in this research.

One very important set of experiments involved discontinuous chopped glass fibers coated with elastomer. These fibers were mixed with nylon 6 by passing through an extruder. Then the extrudate, in which the fibers were quite well aligned, was cut into lengths, laid up in a mold, and compression molded into plates for subsequent testing of mechanical properties. The fiber content was approximately 20% by weight.

Mechanical Tests

The effects of the rubber coating on several properties of the composites were of principal interest. These properties include interlaminar shear strength, impact strength, resistance to impact damage, cyclic tensile fatigue resistance, and transverse tensile strength. We sought an improvement in impact capacity without a serious loss in other important mechanical properties, such as strength and stiffness.

This balance of effects explains why it is so critical to keep the film thickness a very small fraction of the fiber diameter. The compliance and ultimate ductility of the elastomer offer the potential for deformation energy absorption. However, if the fiber is too decoupled from the surrounding matrix, its stiffening and strengthening actions are impaired. The coating also affects two other important factors. It reduces elastic stress concentrations in the matrix in the vicinity of the fibers, which often initiate fracture, and it can reduce the residual stresses caused by differential thermal contraction as the composite cools from melt-processing temperatures. All three of these roles—deformation, stress concentration reduction, and shrinkage

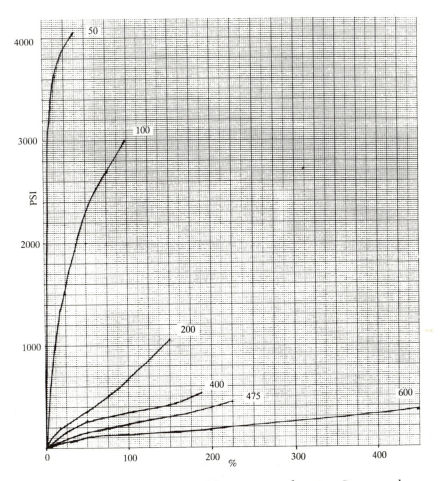

*Figure 4. Stress–strain curves for rubber–epoxy combination. Curve numbers
represent parts of rubber per 100 parts of epoxy.*

stress moderation—are treated in a finite element model analysis now being
developed.

The interlaminar shear strength (ILSS) was measured by the short-beam
bending test following the method of ASTM D 2344. The results are presented in Figures 5 and 6. With the glass fiber reinforcement there is a
modest increase in shear strength, followed by a leveling off as the rubber
film thickness increases. With the carbon fibers, the property is essentially
unaffected.

In both cases there is no loss in shear strength, and this was confirmed
by other types of tests measuring the same characteristic. Scanning electron

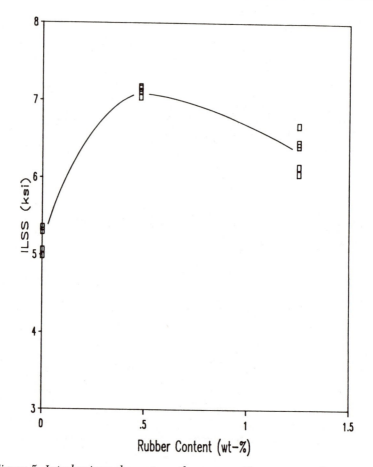

Figure 5. Interlaminar shear strength versus rubber content; glass–epoxy.

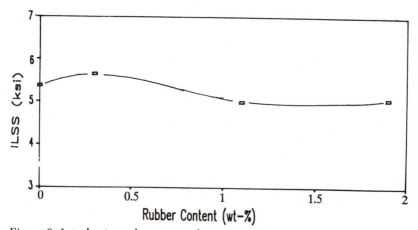

Figure 6. Interlaminar shear strength versus rubber content; graphite–epoxy.

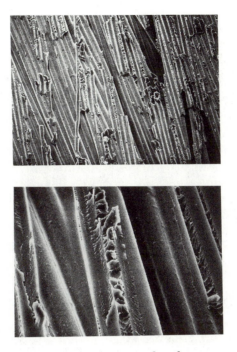

Figure 7. ILSS fractographs; glass–epoxy.

microscopic (SEM) fractographs, shown in Figures 7 and 8, illustrate the good adhesion that is obtained in the fiber–film–matrix connections in both systems.

We expected that the transverse tensile strength of unidirectionally fiber reinforced laminates would decrease as the rubber-film thickness increased because the tensile strength of the rubber is so much less than the strengths of the other two components. Such proved to be the case (*see* Figures 9 and 10), though the reduction with the glass fibers was not too great. Again, fractographs showed that good adhesion was operating, with the cohesive failures occurring primarily in the rubber layers. In both types of laminates, the decreases in transverse tensile strength seemed to be tolerable in terms of practical use characteristics.

With high volume continuous fiber reinforced composites, impact strength usually depends almost entirely on fiber properties and fiber–matrix debonding behavior, unless the matrix possesses great inherent fracture toughness itself (*23*). Such is the situation with epoxy-based laminates; the fiber characteristics and the debonding dominate the impact strength. Thus, it was gratifying to note a finite improvement in impact strength from the use of the rubber coating.

In Figure 11 several different kinds of test data are shown for the graphite fiber–epoxy laminates. In Figures 11A and 11B, short-beam Izod and Gard-

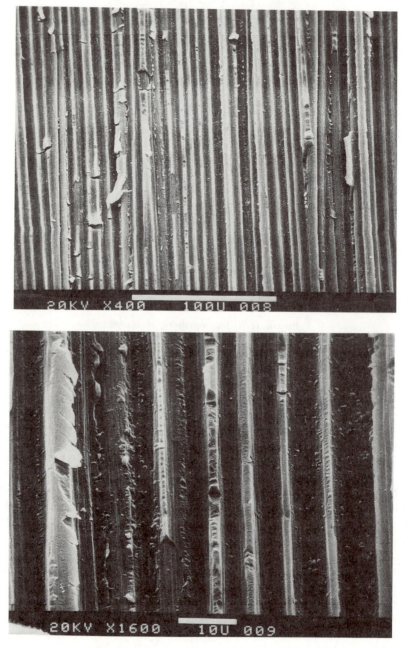

Figure 8. ILSS fractographs; graphite–epoxy.

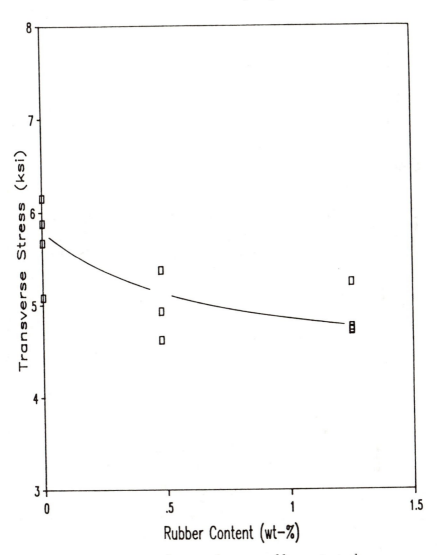

Figure 9. Transverse tensile strength versus rubber content; glass–epoxy.

ner results for unidirectionally reinforced laminates show significant increases with coating thickness. The magnitudes of the increases are beyond
experimental scatter, so the effect can be considered real. In Figure 11C, a
0/90 graphite–epoxy laminate, the instrumented driven dart impact energy
also increases in the midrange of coating thickness before decreasing at the
higher level of rubber content. This same kind of thickness sensitivity was
shown in measurements of resistance to impact damage. The residual interlaminar shear strength retention of nondestructively impacted beam spec-

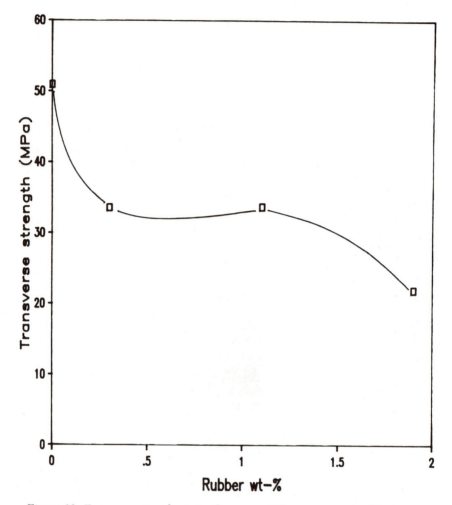

Figure 10. Transverse tensile strength versus rubber content; graphite–epoxy.

imens was greatest for fiber coatings in the intermediate thickness range. With no coating or the thickest coating, shear strength retention was less than for the intermediate range.

If the reinforcing fibers are discontinuous, either parallel aligned or randomly oriented, and if the matrix material has inherent fracture toughness, the impact behavior of the composite becomes more complicated. Several factors can contribute to the overall effect: gross matrix deformation, fiber pullout, fiber fracture, and fiber–matrix debonding are the principal ones. Thus, it would be expected that alterations in the zone of material between each fiber and its surrounding matrix could strongly affect the macroscopic impact properties of such a composite.

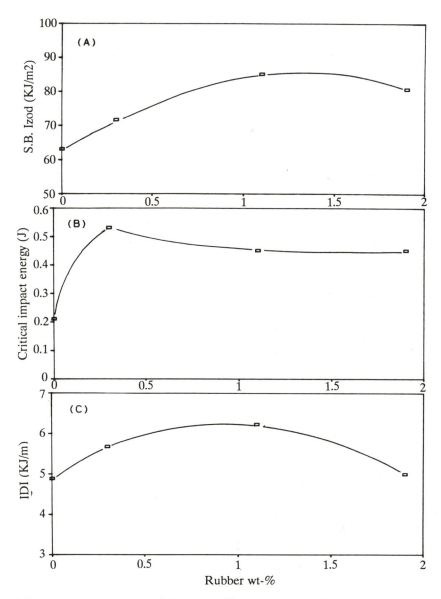

Figure 11. Impact strengths versus rubber content; graphite–epoxy. Key: A, Izod; B, Gardner; C, driven dart.

This effect was observed when rubber-coated short glass fibers were added to nylon 6 (20% weight fraction of the composite) in a parallel-aligned orientation. The notched Izod impact strength was measured in both the aligned and transverse directions. These data are shown in Figure 12, where significant monotonic increases with coating thickness are evident. The effect is large and unambiguous.

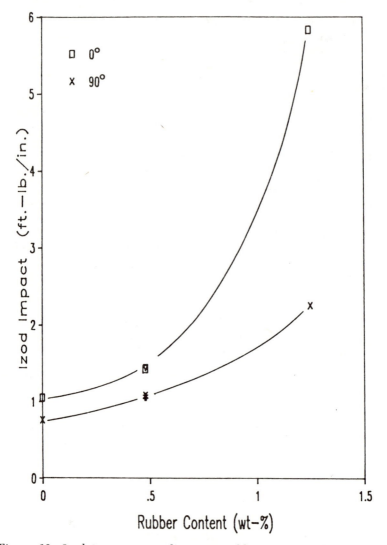

Figure 12. Izod impact strength versus rubber content; chopped glass–nylon 6.

Extensive SEM fractographic analysis of the broken specimens revealed no unexpected findings. The degree of fiber pullout and the volume of distorted matrix polymer around each fiber appeared to increase with coating thickness, as the impact energy increased. The same effect was evident in driven dart results (Figure 13). The energy absorbed both to maximum load and to specimen failure increased with coating thickness, probably for the same reasons. Again, the changes are quite large and may be enough to have technological value.

Previous research on the tensile fatigue behavior of glass fiber reinforced

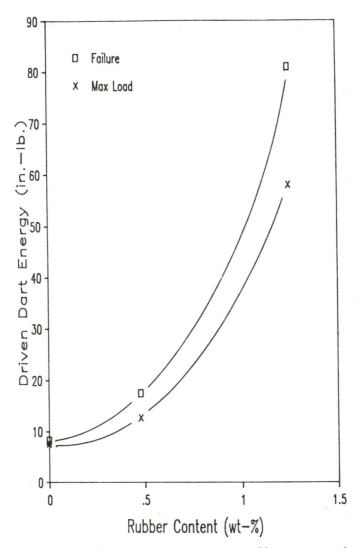

Figure 13. Driven dart impact energy versus rubber content; chopped glass–nylon 6.

polymer composites established that a dominant mechanism of degradation is abrasive damage on the surface of the fibers (24). This damage arises from very small relative motions between fibers (or between fibers and other components in the composite) that occur when the material is cyclically deformed in tension. The damage greatly accelerates the strength loss of the fibers. This strength loss is reflected macroscopically in a steeper slope of the S–N fatigue curve (stress vs. log number of cycles) for the composite than would occur if the abrasion were absent.

It was postulated that a thin elastomer film coating on each fiber might protect against this damage and thereby improve the composite fatigue resistance. Preliminary data from flexural samples that have been subjected to cyclic bending show such an improvement to a significant degree. More extensive tensile testing of similar composite samples reinforced with rubber-coated glass fibers is being done.

Conclusions

Thin films of elastomer on particulate cores, or as coatings on glassy occlusions, effectively toughen glassy polymers. They induce and promote plastic flow in the matrix, shear banding, crazing, and cavitation. If properly crosslinked and bonded to the matrix, they also deform and absorb energy before they fail cohesively. If the elastomer is used as a thin coating on reinforcing fibers, the composite can exhibit improved impact strength and resistance to impact damage. When glass is the reinforcing fiber, improvements in cyclic tensile fatigue resistance also may be realized.

References

1. Keskkula, H.; Platt, A. E.; Boyer, R. F. *Kirk-Othmer Encycl. Chem. Technol. 2nd Ed.* **1969,** *19,* 85.
2. Amos, J. J. *Polym. Eng. Sci.* **1974,** *14,* 7.
3. Boyer, R. F.; Keskkula, H.; Platt, A. E. *Encycl. Polym. Sci. Technol.* **1970,** *13,* 128–447.
4. Bucknall, C. B.; Smith, R. R. *Polymer* **1965,** *6,* 437.
5. Kambour, R. P. *Nature (London)* **1962,** *195,* 1299.
6. McGarry, F. J. *Proc. R. Soc. London A* **1970,** *319,* 59–68.
7. Henkee, C.; Kramer, E. J. *J. Polym. Sci., Polym. Phys. Ed.* **1984,** *22,* 721–737.
8. Glad, M. D. Ph.D. Thesis, Cornell University, 1986.
9. *Rubber-Modified Thermoset Resins;* Riew, C. K.; Gillham, J. K. Eds.; Advances in Chemistry 208; American Chemical Society: Washington, DC, 1984.
10. Yang, A. C.-M.; Kramer, E. J.; Kuo, C. C.; Phoenix, S. L. *Macromolecules* **1986,** *19,* 2010.
11. Donald, A. E.; Kramer, E. J. *Polymer* **1982,** *23,* 1183–1188.
12. Argon, A. S. *Philos. Mag.* **1973,** *28,* 839.
13. Argon, A. S.; Bessonov, M. J. *Philos. Mag.* **1977,** *35,* 917.
14. Vanzo, E. *J. Polym. Sci. Part A-1* **1966,** *4,* 1727.
15. Schwier, C. E.; Argon, A. S.; Cohen, R. E. *Polymer* **1985,** *26,* 1985.
16. Argon, A. S. et al. *J. Polym. Sci., Polym. Phys. Ed.* **1981,** *19,* 253.
17. Geblizlioglu, O. S.; Argon, A. S.; Cohen, R. E. *Polymer* **1985,** *26,* 529.
18. Boyce, M. E.; Argon, A. S.; Parks, D. M. *Polymer* **1987,** *28,* 1680.
19. Ricco, T.; Pavan, A.; Danusso, F. *Polym. Eng. Sci.* **1978,** *18,* 774.
20. Tse, M. K. Ph.D. Thesis, Massachusetts Institute of Technology, 1982.
21. Kawamoto, J. S.M. Thesis, Massachusetts Institute of Technology, 1984.
22. Rogg, C. A. S.M. Thesis, Massachusetts Institute of Technology, 1988.
23. McGarry, F. J.; Mandell, J. F.; Wang, S. S. *Polym. Eng. Sci.* **1976,** *16,* 609–615.
24. McGarry, F. J.; Mandell, J. F.; Hsieh, A. J.; Li, C. G. *Polym. Compos.* **1985,** *6,* 168.

Received for review October 7, 1988. Accepted revised manuscript February 23, 1989.

MORPHOLOGY AND MECHANICAL PROPERTIES

Controlling Factors in the Rubber-Toughening of Unfilled Epoxy Networks

Application to Filled Systems

Serge Montarnal, Jean-Pierre Pascault, and Henry Sautereau

Laboratoire des Matériaux Macromoléculaires, UA–CNRS 507, Institut National des Sciences Appliquées de Lyon, Bât. 403, 20 Avenue Albert Einstein, 69621 Villeurbanne Cédex, France

Bisphenol-A-type epoxy prepolymers cured with a sterically hindered diamine and toughened with carboxyl-terminated butadiene–acrylonitrile (CTBN) copolymer are used as a model system. This diamine's low reactivity yielded different size distributions of the rubber-rich domains. Various morphologies result from competition between the phase-separation and gelation (or vitrification) processes. The viscosity of the system at the cloud point determines the size distribution of the particles. When the cure temperature increases from 27 to 100 °C, the mean diameter of a 15% CTBN composition (18% acrylonitrile) increases from 0.5 to 1.1 μm, with a nearly constant volume fraction. The mechanical properties are strongly dependent on separated-phase morphology. An isothermal cure for 7 h at 75 °C strongly enhances the toughness at a constant-volume fraction. We also studied the influence of CTBN content on morphology and mechanical properties. This system was used in a coating process to reinforce composite materials.

THERMOSETTING RESINS IN THE GLASSY STATE are undesirably brittle. One way to improve the toughness of epoxy networks is to introduce small amounts of a carboxyl-terminated reactive liquid rubber copolymer of acrylonitrile–butadiene (CTBN), which forms a second phase during the cure (*1–3*). A general description of the phase separation was first provided by

Visconti and Marchessault (4). Phase separation produces rubber-rich domains, generally in the order of magnitude of 1 μm, that can promote toughening by various mechanisms determined by the amount of phase-separated and dissolved rubber. Manzione et al. (5, 6) showed that the rubber domain size could be controlled by varying either the temperature of gelation or the acrylonitrile content of the rubber additive. Williams et al. (7–9) developed a model to predict the fraction, composition, and average radius of the dispersed phase segregated during a thermoset polymerization. This model was based on a thermodynamic description through a Flory–Huggins equation and constitutive equations for nucleation, coalescence, and growth rates.

The bisphenol-A-type epoxy prepolymer used in this study was cured with a sterically hindered amine, 1,8-diamino-p-menthane (MNDA), and toughened with CTBN. This diamine's low reactivity allowed us to control the morphology development by changing the cure schedule. Different size distributions of the rubber-rich domains were obtained without influencing the chemical structure of the network. This independence was not expected with hardeners such as piperidine and dicyanodiamide (5, 6), which could lead to etherification and epoxy polymerization, depending on the cure schedule.

This chapter further defines the main factors that control final morphology and mechanical properties in the curing of epoxy polymers. Various ways to enhance the fracture toughness of reinforced thermosets have been investigated. The most widely used methods attempt to improve the network itself or to increase filler matrix adhesion. A mixed approach is the encapsulation of fillers with a thin layer of low-modulus elastomer, which can produce a tougher composite with a modulus in the same order of magnitude. This method has been suggested with rigid spherical inclusions (10), and further calculations have been done on particulate or fibrous composites (11, 12). A few experiments carried out on glass beads (13), glass fibers (14), and carbon fibers (15) have confirmed the conclusions of the theoretical assessments. In the second part of this chapter, we describe a coating process that uses glass beads and carbon fibers.

Experimental Details

Materials. All materials were purchased from commercial sources and used without purification. The epoxy prepolymer is a diglycidyl ether of bisphenol A (DGEBA), where the number average degree of polymerization, $\bar{n}$, is 0.15 (Lopox 200). Commercial MNDA (Aldrich Chemical Company) contains about 20% unknown impurities which involve an excess of epoxy groups for a calculated stoichiometric ratio $r = 1$.

The reactivity of this aliphatic and cycloaliphatic amine has been studied previously (16, 17). The secondary amine functions react only at high temperature; below 50 °C, vitrification occurs before gelation. Because of degradation reactions (homolytic cleavage of the macromolecular skeleton, favored by quaternary carbon atoms), curing cycles have been optimized at an isothermal temperature of 190 °C.

The CTBN (hycar 1300 × 8) with 18% acrylonitrile content was provided by the B. F. Goodrich Company. We used a CTBN–epoxy adduct prepared, as described previously (*18*), at 85 °C with mechanical stirring under a nitrogen flow, using triphenylphosphine, 0.15% by weight, as activator. The carboxyl-to-epoxy ratio (c/e) was set at 0.065 to ensure an excess of epoxy. To prepare plates for mechanical tests, DGEBA, CTBN adduct, and MNDA were mixed by mechanical stirring under vacuum at room temperature. Then the mixture was cast into a polytetrafluoroethylene-coated (Teflon) aluminum mold (170 × 150 × 4 mm) and heated in an oven. After it was cured, the mold was taken out and cooled slowly in air at room temperature. The samples for various analyses were cut from the plates as prepared and stored in a dry atmosphere.

Analysis. Soluble fractions were characterized by size exclusion chromatography (SEC) (Waters), with tetrahydrofuran (THF) as elution solvent. Eluted macromolecules were observed simultaneously by refractive index (RI) and by ultraviolet absorption (λ = 254 nm). We used four separating columns: 1000-, 500-, 100-, and 100-Å pore size.

Glass-transition temperature (T_g) and heats of reaction of the unreacted and partially cured samples were determined with a differential scanning calorimeter (Mettler TA 3000). The analyses were performed under argon atmosphere, with a heating rate of 10 K/min, in the temperature range from –100 to 250 °C.

A rheometer (Contraves) was used to monitor the viscosity of the system during the different isothermal cures. The cloud-point measurements were recorded in the range of 20 to 250 °C with a homemade light transmission setup that allowed us to determine turbidity changes with either linear heating rates (q) or isothermal cures.

Scanning electron micrographs (SEM) of fracture surfaces were obtained with a microscope (Jeol T200) after the surfaces were coated with a vacuum gold sputterer. The micrographs of fracture surfaces were enlarged to calculate the diameters and surface fraction of rubber domains. About 100 particles were plotted with a graphic tablet (HP 9111 A) associated with a microcomputer (HP 85) area–distance measurements program. Because of optical considerations, the diameter-detection level is >0.1 μm. Further details can be found in the "Cure Process, Transitions, and Morphology" section.

Tensile and compressive tests were performed at room temperature with a tensile machine (DY 25, Adamel–Lhomargy). For tensile tests, strain measurements were performed by using an extensometer (EX–10) at a strain rate of 3.3×10^{-4} s^{-1}, using ISO–60 specimens. Cylindrical rods with a height-to-diameter ratio of 2 were deformed in a compression cage between polished steel plates. The nominal strain was determined by using the average information of two linear variable differential transformer (LVDT) transducers. The strain rate was 8.3×10^{-4} s^{-1}.

A dynamic mechanical thermal analyzer (DMTA, Polymer Laboratories) was used to study the dynamic mechanical properties of epoxy matrix and composite materials. Samples were tested in a double cantilever, bending with a fixed displacement ($\approx$16 μm). Unnotched instrumented Charpy impact characterization was carried out as described in ref. 19.

Results and Discussion

Gelation, Vitrification, and Phase Separation. Our experimental mixture consisted of low-molecular-weight bisphenol-A-type epoxy resin

DGEBA ($\bar{n}$ = 0.15), diamine hardener MNDA (stoichiometric ratio r = 1), and CTBN (15% by weight), introduced as a CTBN–DGEBA adduct. The mixture was homogeneous at the start of the curing process at room temperature.

At a certain extent, P_{cl}, in the reaction, rubber-rich domains began to separate from the epoxy matrix. Visual observations of a cloud point during the cure indicated phase separation. For the 15% CTBN system, we determined the cloud-point time (t_{cl}) from the onset of opacity at different isothermal curing temperatures (Table I) and the duration of the phenomenon (Δt_{cl}). Both t_{cl} and Δt_{cl} decreased as the curing temperature T_c increased.

The isothermal cure of thermosetting resins is usually characterized by two important macroscopic phenomena: gelation and vitrification. Gelation corresponds to the incipient formation of an infinite cross-linked polymer molecule. Vitrification involves transformation from a liquid or rubbery state to a glassy state because of an increase in either molecular weight before gelation or cross-link density after gelation. These two phenomena are connected to a large increase in viscosity, but the system remains processable and fusible as long as gelation does not occur.

Viscosity measurements have been determined at different isothermal curing temperatures. The viscosity depends on both the extent of reaction and the temperature. For a given isothermal cure temperature T_c, viscosity increases slowly at the beginning ($1–10^3$ Pa·s) and then rapidly ($10^3–\infty$ Pa·s) near gelation or vitrification. We determined the cloud-point viscosity η_{cl} and the time t_η required for reaching a high viscosity (η = 50,000 Pa·s). Phase separation in the 15% CTBN system occurred well before gelation or vitrification, as shown by the values of t_{cl} and t_η (Table I).

Gelation occurs at a constant extent of reaction if the chemical mechanism remains the same, but the gelation times depend on the temperature and generally fit well with an Arrhenius plot. Figure 1 shows ln t_η versus

Table I. Viscosity and Cloud-Point Measurements at Different Isothermal Curing Temperatures for DGEBA–MNDA–CTBN

	t_η[b]		Cloud Points (15% CTBN)		
T_c[a]	0% CTBN	15% CTBN	t_{cl}[c]	Δt_{cl}[d]	η_{cl}[e]
27	1170 (s)[f]	1500 (s)	1100	78	3500
50	450 (s)	590 (s)	234	12.5	92
75	210 (is)[g]	275 (is)	53	4.5	7
100	75 (is)	92 (is)	22	1.8	2
120	–	42 (is)	10	<1	–

NOTE: DGEBA, $\bar{n}$ = 0.15; CTBN, 15% by weight.
[a]Curing temperature, °C.
[b]Time (min) when viscosity tends to infinity ($\eta \simeq$ 50,000 Pa·s). [c]Onset (min) of opacity.
[d]Time (min) during which opacity is visible.
[e]Viscosity at cloud point (Pa·s).
[f]Soluble in THF.
[g]Insoluble in THF.

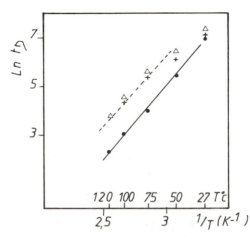

Figure 1. Arrhenius plot for t_η *and* t_{cl}. *Key:* +, t_η *with 0% CTBN;* △, t_η *with 15% CTBN; and* ●, t_{cl} *with 15% CTBN.*

$1/T$ for neat or rubber-modified systems. The viscosity, $\eta = 50,000$ Pa·s at $T_c = 27$ and 50 °C, was achieved experimentally sooner than predicted by the theoretical plot.

At these two temperatures all of the polymers are soluble in THF, even with a viscosity equal to 50,000 Pa·s, so vitrification comes before gelation. On the contrary, the gel points determined from the onset of insolubility correspond well to t_η at 120, 100, and 75 °C.

Figure 1 shows the systems with and without CTBN; t_η for the 15% CTBN system is higher than for the neat resin. Beyond $T_c = 75$ °C, the Arrhenius plot gives an activation energy of 46 kJ/mol. This value differs from those obtained with an aromatic diamine hardener by Chan et al. (20).

The time from the onset of the phase separation, t_{cl}, seems also to have an Arrhenius behavior (Figure 1) with the same activation energy as t_η. This result is unexpected because the extent of reaction at phase separation is not fixed, but should increase with the isothermal curing temperature. Theoretical models (8) show very small variations of cloud point (p_{cl}) with T_c, and this could explain the Arrhenius behavior of t_{cl}.

During the course of the reaction, samples are prepared and analyzed by differential scanning calorimetry (DSC). On each sample, $T_g(t)$ is determined at various curing times (Figure 2a). Vitrification occurs when the glass-transition temperature reaches the cure temperature. Vitrification comes after gelation at $T_c = 75$ °C, but not at 27 °C. The temperature at which the gelation time equals the vitrification time, $_{gel}T_g$, is 50 °C. As the extent of reaction at p_{cl} increases, $T_{g,cl}$ also increases from –15 °C at $T_c = 27$ °C to +5 °C at $T_c = 75$ °C.

Figure 2b shows a drastic increase in T_g with curing time and after the cloud point at 27 °C. Before t_{cl} it is a single-phase system with a low T_g as

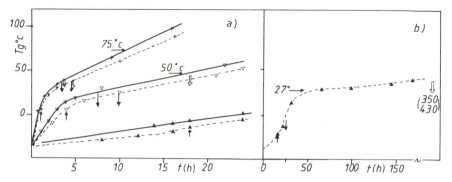

Figure 2. Variation of T_g versus reaction time at different curing temperatures
T_c. Key: +, 75 °C; ▽, 50 °C; ▲, 27 °C; ——, without CTBN; - - - -, with 15%
CTBN; ↑, t_{cl}; ↓, t_η; →, vitrification; and ⇩, gelation.

a result of the plasticization of DGEBA–MNDA copolymer by CTBN (T_g
= –60 °C). At phase separation we expect two T_g values, one for the hard
phase rich in epoxy–MNDA copolymer and another for the rubber phase.
Unfortunately, the T_g of the rubber phase is not discernible by DSC analysis.

At this curing temperature (T_c = 27 °C), vitrification occurs before
gelation (Figure 2b). The kinetics after vitrification are affected by local
viscosity and controlled by diffusion. With CTBN, gelation occurs after
350–430 h (about 18 days). Without CTBN, $T_g(t)$ is higher and no gelation
occurs after 60 days.

To complete the DSC analysis, the samples were dissolved in THF and
SEC analyses were performed. Using the height of the peak of the DGEBA
monomer, $n = 0$, we can compute the extent of reaction of DGEBA molecules ($1 - h_t/h_0$) (and not the epoxy conversion). The height of the peak h_t
decreases with time at different curing temperatures, and higher-molecular-
weight products are formed with time. In a plot of the weight-average molecular weight $\overline{M}_w$ versus ($1 - h_t/h_0$) (Figure 3), all the experimental points
belong to master curves for the neat system and the rubber-modified system.
$\overline{M}_w$ increases and tends to infinity at gelation, around 90% conversion of
DGEBA molecules.

All the results are summarized in a time–temperature–transformation
(TTT) isothermal cure diagram (5) (Figure 4).

- The vitrification curve is generally S-shaped, with T_{g_∞} and T_{g_0}
  the upper and lower boundaries, respectively (T_{g_∞} = 152 °C
  and T_{g_0} = –42 °C for the neat system) (17). Other competing
  reactions such as degradation can occur at higher temperatures;
  T_{g_∞} decreases after 18 h at 190 °C (17).

- Gelation is delayed with 15% CTBN (by weight) at isothermal
  curing temperature $T_c > 50$ °C (not visible on Figure 4, but
  found in Table I). The viscosity measurements have been con-

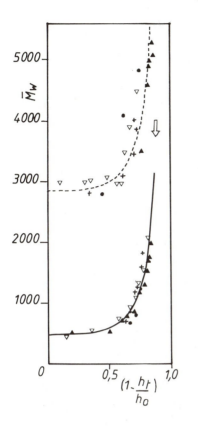

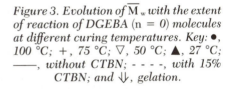

Figure 3. Evolution of $\overline{M}_w$ with the extent of reaction of DGEBA (n = 0) molecules at different curing temperatures. Key: ●, 100 °C; +, 75 °C; ▽, 50 °C; ▲, 27 °C; ——, without CTBN; - - - -, with 15% CTBN; and ⇓, gelation.

firmed by SEC. The extent of reaction of DGEBA molecules is lower for the same reaction time, with 15% CTBN at each curing temperature. Wang and Zupko (*21*) found that the viscosity increase comes sooner with 15% CTBN, but they used unreacted CTBN. In this case the carboxyl end groups act as an accelerator in the epoxy–amine reaction.

- Beyond 50 °C, the time to reach η = 50,000 Pa·s is a good approximation of the gel time. This approximation does not apply when $T_c < 50\,°C$, because then vitrification comes before gelation.

- At 75, 100, and 120 °C, phase separation occurs at low viscosity, markedly before gelation.

- At 27 °C phase separation occurs in a high-viscosity medium. As a glassy state forms, there is a two-phase ungelled thermoplastic system of low molecular weight.

Because of the low reactivity of MNDA, morphologies of partially cured specimens may be examined with optical microscopy or scanning electron

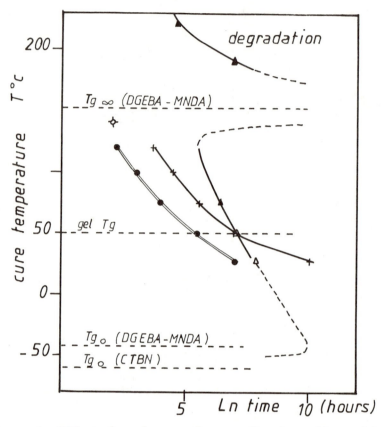

Figure 4. TTT isothermal cure diagram for the rubber-modified DGEBA–MNDA (15% CTBN). Key: ●, *cloud point;* +, *gelation;* △, *vitrification;* ⟡, *cloud point for the system cured from room temperature with a heating rate q = 5 K/min; and* ▲, *degradation.*

microscopy (SEM). Similar morphologies are obtained after 60 min (t_{cl} + 7 min) and after 180 min (t_{cl} + 127 min) at T_c = 75 °C, or after 6 days (after vitrification) and after 60 days (after gelation) at T_c = 27 °C (Figure 5). Within the limits of our accuracy, most of the phase separation occurred well before gelation (T_c > 50 °C) or vitrification (T_c < 50 °C), although other authors (4, 20) have reported some changes in phase composition after gelation.

Generally, polymers are not strictly processed under isothermal curing conditions, and phase separation occurs during the initial heating step. To clarify how processing conditions affect the properties of the final products, cloud points were determined with various linear heating rates q (Table II).

As q increases, the temperature T_{cl} of the onset of the phase separation increases. For a heating rate q = 5 K/min, T_{cl} = 140 °C with 15% CTBN

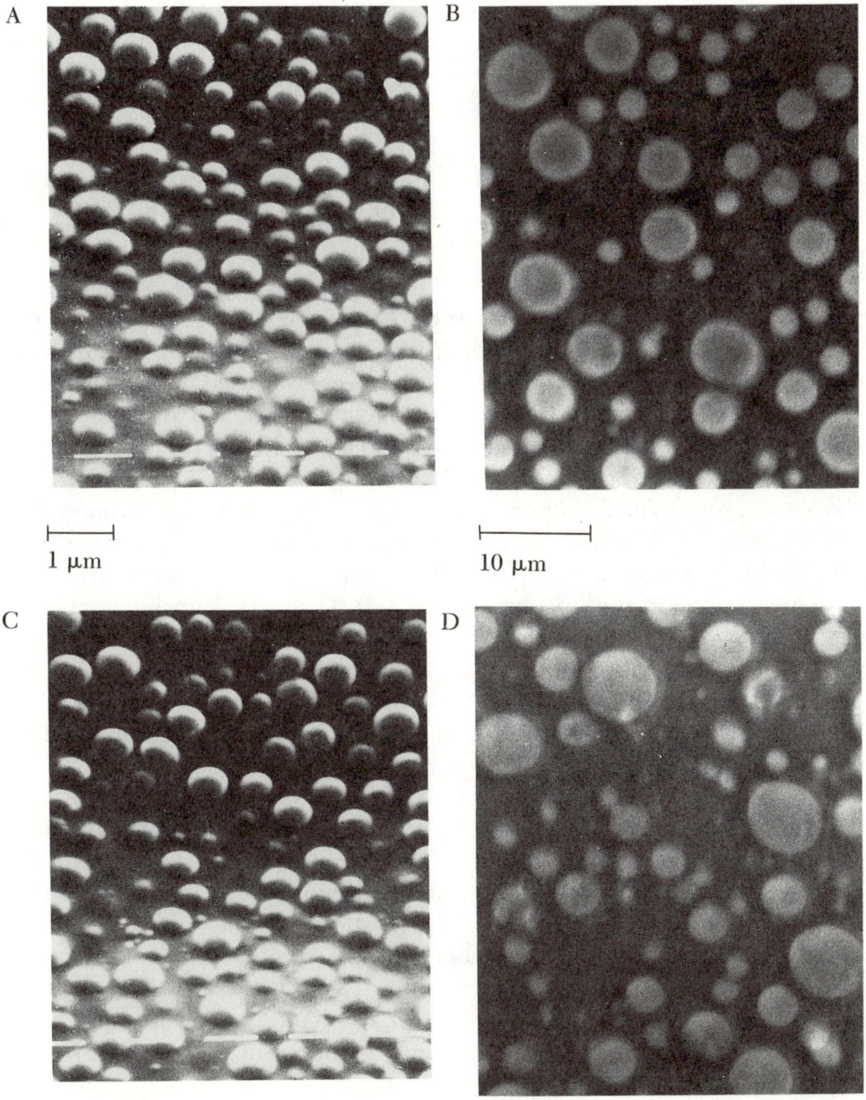

Figure 5. Morphologies of the precured systems observed with scanning electron microscopy for the precured samples at 27 °C or with optical microscopy for the precured samples at 75 °C. Key: A, 27 °C, 6 days; B, 75 °C, 1 h; C, 27 °C, 60 days; D, 27 °C, 3 h.

Table II. Cloud-Point Measurements at Various Heating Rates

% CTBN (wt)	q (K/min)	T_{cl} (°C)	t_{cl} (min)	ΔT_{cl} (°C)	Δt_{cl} (min)
15	2.2	105	37	2	2
	5.0	140	23	4	0.8
	6.7	145	18	3	0.5
10	5.0	123	20	9	2

and 123 °C with only 10% CTBN. As T_{cl} decreases, the extent of reaction p_{cl} at the cloud point decreases; p_{cl} is lower with 10% than with 15% CTBN. This result is in good agreement with our phase diagram determination (22).

Cure Process, Transitions, and Morphology. One purpose of this work was to develop widely different morphologies from the same initial content of CTBN (15%). A two-step cure process was used to develop a fully cured but distinct cure-dependent morphology. First the system was iso-thermally cured at different temperatures (T_c = 27, 50, 75, and 100 °C). Different schedules (t_c = 6 days, 24 h, 7 h, and 2 h, respectively) were chosen to ensure vitrification or gelation (Figure 4). Next the cured system was postcured by heating above the maximum glass-transition temperature ($_ET_{g_\infty}$) of the system to complete the reaction of the epoxy matrix. The postcure treatment was set for 14 h at 190 °C to avoid degradation reac-tions (17).

To compare and to understand how processing conditions affect the properties of the final product in a mold, samples were cured (q = 5 K/min) from room temperature to 190 °C and then postcured for 14 h at 190 °C.

After the postcure treatment (14 h at 190 °C), the glass-transition tem-perature of the neat resin was 152 °C. Values of $_ET_{g_\infty}$ for the rubber-modified samples were lower than for the neat system by 10–18 °C (Table III). The decrease in $_ET_{g_\infty}$ for cured rubber-modified epoxy systems arises from in-complete phase separation caused by a plasticization phenomenon that has been noted in varied rubber-modified epoxy formulations (5, 6, 18, 20). Unfortunately, the rubber–glass transition $_RT_g$ is not discernible by DSC. With the Fox equation (23), the fraction of dissolved rubber can be predicted from $_ET_{g_\infty}$ of the rubber-modified system, T_{g_1} of the neat epoxy matrix, and T_{g_2} of the rubber:

$$\frac{1}{_ET_{g_\infty}} = \frac{W_1}{T_{g_1}} + \frac{W_2}{T_{g_2}} \tag{1}$$

where W_1 and W_2 are the weight fractions of epoxy and rubber, respectively, in the rubber-modified epoxy matrix. The weight fraction $_RW$ of the phase-

Table III. Influence of the Curing Process on the Morphology of the Cured Rubber-Modified Epoxy

Property	6 days, 27 °C + 14 h, 190 °C	24 h, 50 °C + 14 h, 190 °C	7 h, 75 °C + 14 h, 190 °C	2 h, 100 °C + 14 h, 190 °C	q = 5 K/min + 14 h, 190 °C
$_E T_g$, °C	142	137	137	137	135
$_R V$	15	14	14	14	13
$\overline{D}$,[a] μm	0.53	0.79	0.90	1.09	1.05
(σ)[b]	(0.16)	(0.17)	(0.21)	(0.19)	(0.28)
$\overline{N}$[c], μm²	1.09	0.53	0.52	0.28	0.16
$\overline{d}_c$,[d] μm	0.81	1.15	1.34	1.57	1.96
(σ)	(0.19)	(0.40)	(0.30)	(0.40)	(0.50)
$\overline{d}_{p2}$,[e] μm	0.28	0.36	0.44	0.48	0.91
$V(\overline{d})$,[f] %	21	24	22	25	11.4
$V(\overline{D}, \overline{N})$, %	16	18	22	17	10

NOTE: DGEBA $\overline{n}$ = 0.15, MNDA, 15% CTBN by weight.
[a]Mean particle diameter.
[b]Standard deviation.
[c]Mean number of dispersed-phase particles.
[d]Mean distance between two particles, center to center.
[e]Mean distance between two particles, surface to surface.
[f]Volume fraction of dispersed phase, predicted from different measurements, from $\overline{d}$, $\overline{D}$, and $\overline{N}$.

separated rubber was obtained from the mass balance of W_2 calculated from equation 1 and the initial rubber used in the formulation: $_R W = 0.15 - W_2$.

The weight fraction can easily be converted to volume fraction $_R V$ by using the densities of the rubber (0.95 g/cm³) and of the epoxy network (1.22 g/cm³). The results are listed in Table III. The smaller decrease of $_E T_{g_x}$ is obtained with a precure at 27 °C, and the larger one with the heating rate $q = 5$ K/min. As expected from this method, the volume fraction of phase-separated rubber is less than the amount of rubber added, $V_0 = 18.5\%$. These results will be compared to corresponding data determined from SEM micrographs.

Figure 6 shows a nodular structure with good adhesion between the two phases. Quantitative microscopic analysis was conducted by using SEM micrographs.

Butta et al. (24) demonstrated that examination of fracture surfaces by using SEM reveals more rubber particles than a randomly cross sectional area (Figure 7). This finding indicates that stress concentration at equatorial planes can lead to a determination of the true diameters.

The average domain size (diameter $\overline{D}$) and number of domains per surface unit ($\overline{N}$) were determined. We also calculated interparticle distances, center-to-center ($\overline{d}_c$) and surface-to-surface ($\overline{d}_p$), as described by Wu (25).

Figure 8 gives the particle-diameter distribution curves for the various systems. All the isothermally precured specimens displayed a narrow uni-

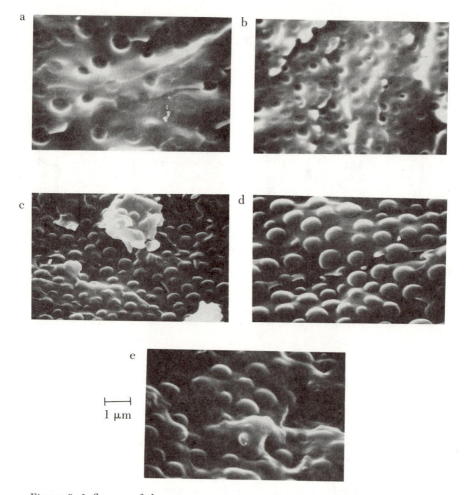

Figure 6. Influence of the precure reaction on fracture surfaces (SEM). (a) q = 5 K/min to 190 °C + 14 h at 190 °C; (b) 6 days at 27 °C + 14 h at 190 °C; (c) 24 h at 50 °C + 14 h at 190 °C; (d) 7 h at 75 °C + 14 h at 190 °C; (e) 2 h at 100 °C + 14 h at 190 °C.

modal distribution of particle sizes with a nearly constant width. Smaller particles (0.1–0.2 μm) are present, but are not discernible by this technique. The mean particle diameter $(\overline{D})$ increases with the isothermal precure temperature from 0.5 μm at 27 °C to 1.1 μm at 100 °C. These values are smaller than those obtained in other rubber-modified thermosets (7, 8, 18, 24). The number of rubber particles per surface unit decreases as the distances $\overline{d}_p$ and $\overline{d}_c$ increase (Table III) with the mean diameter $\overline{D}$.

For the sample precured with a heating rate of 5 K/min from room temperature to 190 °C in our molding conditions, the cloud point has been

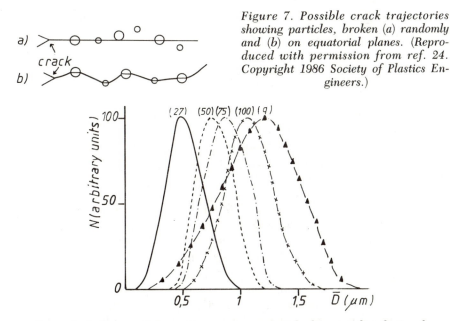

Figure 7. Possible crack trajectories showing particles, broken (a) randomly and (b) on equatorial planes. (Reproduced with permission from ref. 24. Copyright 1986 Society of Plastics Engineers.)

Figure 8. Influence of the curing process on particle-diameter distribution for the system DGEBA ($\overline{n}$ = 0.15)–MNDA and 15% CTBN. Key: ——, 6 days at 27 °C + 14 h at 190 °C; - - - -, 24 h at 50 °C + 14 h at 190 °C; -------, 7 h at 75 °C + 14 h at 190 °C; -×-×-×-, 2 h at 100 °C + 14 h at 190 °C; and -▲-▲-▲-, curing rate 5 K/min to 190 °C and 14 h at 190 °C.

measured at T_{cl} = 140 °C (Table II). This system shows a broad unsymmetrical distribution with predominantly large particles ($\overline{D}_{max} \simeq 1.2$ μm) and a tail indicating few small diameters. In this nonisothermal case, the temperature increases after the cloud point, and the fast gelation rate results in insufficient time for some rubber domains to grow.

The volume fraction V of the dispersed phase was determined for the five specimens by two methods. The first method was used by Butta et al. (24) in amino-terminated butadiene–acrylonitrile (ATBN)–epoxy blends. In this simple model, the spherical rubber particles are packed in a cubic lattice. The SEM micrographs show that this assumption may be justified. The interparticle distance $\overline{d}_c$ is related to the mean diameter $\overline{D}$ and the volume fraction of the rubbery phase through the following relationship:

$$V(\overline{d}) = \left(\frac{0.905\,\overline{D}}{\overline{d}_c}\right)^3 \qquad (2)$$

The values of $V(\overline{d})$ are listed in Table III.

For the second method we assume that all of the particles revealed by crack propagation are contained in a thin layer of thickness $\overline{D}$. We do not

assume any theoretical configuration, but rather we experimentally determine $\overline{N}$ and $\overline{D}$ thus:

$$V(\overline{D}, \overline{N}) = \frac{\overline{N}}{\overline{D}} \cdot \frac{4}{3} \pi \left(\frac{\overline{D}}{2}\right)^3 \tag{3}$$

These two methods are oversimplifications of the reality because rubber
domains are not spherical and have a size distribution. Moreover, the SEM
reference surfaces are not strictly planar, and we probably overestimate the
true rubber volume fraction. The values given in Table III for the two
methods are in the same order of magnitude, with the higher values given
by the cubic arrangement.

The initial volume fraction of CTBN in the system before curing is V_0
= 0.185. The high $V(\overline{d})$ or $V(\overline{D}, \overline{N})$ of the dispersed phase for the four
isothermally precured specimens, compared to the lower value of $_R V$ cal-
culated from $_E T_{g_\infty}$, indicates that a large amount of epoxy–MNDA copolymer
is present in the dispersed phase. Because of a lack of precision in the results,
the difference in $V(\overline{d})$ among the four isothermally precured samples is
difficult to estimate. Therefore, we consider $V(\overline{d})$ as constant and equal to
0.23 ± 0.02. However, we have a lower value of $V(\overline{d})$ (quite similar to $_R V$)
for the sample precured with a heating rate q = 5 K/min.

Butta et al. (24) obtained similar morphological results with ATBN–epoxy
blends. When T_c increases, the mean particle diameter increases with a
rather constant volume fraction. In this system, the ATBN also initiates epoxy
homopolymerization. Williams et al. (7–9) showed that phase separation
proceeds through a classic nucleation–growth mechanism, rather than
through spinodal separation.

The variation of morphology with cure conditions presumably results
from the effect of temperature on the rates of nucleation and growth of the
dispersed phase (5–9). At low T_c we obtained a system with smaller but
more numerous particles; this system can be produced by a high nucleation
rate and a low growth rate. On the other hand, at high T_c, low nucleation
rate and high growth rate will produce a system with larger but less numerous
particles.

Williams et al. (7–9) proposed a model in which the nucleation rate is
proportional to the temperature, to the surface tension between the phases,
and to the diffusion coefficient. As with the crystallization of linear polymers,
the free energy change ΔG^* required to form critical nuclei is proportional
to the temperature (T_{cl}). In our case, because we always use the same
CTBN–epoxy system, we can assume that the surface tension at the cloud
point is nearly constant. On the other hand, the diffusion coefficient (Dif)
can be estimated by the Stokes–Einstein equation:

$$\text{Dif} = (\text{Dif})_0 \, T/\eta \, (p, T) \tag{4}$$

where η is the viscosity of the continuous phase, which is a function of both extent of reaction and temperature. The growth rate, which involves mass transfer, is proportional to the diffusion coefficient. In a first approximation, it is inversely proportional to the viscosity.

Nuclei are formed and phase separation occurs at the cloud point. Therefore a plot of ln η_{cl}, the viscosity at the cloud point, versus the different parameters of the morphology (Figure 9) generates a linear relationship. Morphology is mainly controlled by viscosity during phase separation and at the cloud point, independent of the polymerization rate, because of the low reactivity of MNDA, even at 100 °C. It depends only on the competition between nucleation and growth rates. This result agrees with cloud-point measurements t_{cl} and Δt_{cl} (Table I).

For the system precured with a heating rate $q = 5$ K/min, the broad distribution with a tail toward low diameters results from the influence of the polymerization rate (kinetic restriction). The rubber fraction drops abruptly because of the high polymerization rate.

The morphology of the 15% CTBN system ($\overline{D}$, $\overline{N}$, or $V(\overline{d})$) is strongly affected by the use of MNDA with isothermal curing temperatures. The interparticle distance $\overline{d}_c$ may be chosen as a representative parameter of the system evolution.

This interparticle distance is mainly determined by the cloud-point viscosity. At high viscosity and low precuring temperature T_c, a high nucleation rate and low growth rate produce a system with smaller but more numerous particles. At low viscosity and high T_c, a low nucleation rate and high growth rate produce a system with larger and less numerous particles.

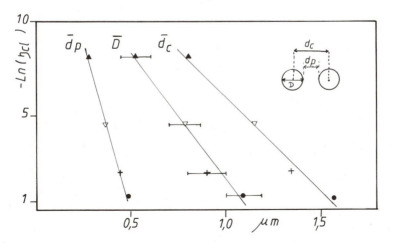

Figure 9. Influence of viscosity at the cloud point on different parameters of the morphology for the 15% CTBN–DGEBA–MNDA system. Key: ▲, 27 °C; ▽, 50 °C; +, 70 °C; and •, 100 °C.

For a nonisothermal cured system, the competition between nucleation and growth rates and polymerization rate has a broadening effect on the particle-diameter distribution and decreases the volume fraction of the dispersed phase.

Morphology and Mechanical Properties. The main deformation process in rubber-modified epoxy networks is shear yielding (20, 26, 27) associated with cavitation (28), which depends strongly on the morphology of the rubbery phase. In order to establish relationships between the morphology of these systems and the induced deformation mechanism, we have performed several mechanical characterizations.

In a first step we neglected the linear elastic fracture mechanics (LEFM). LEFM gives much information on crack propagation and the toughness of the material, but little concerning the deformation process and the initiation phase. We developed a conventional room-temperature approach with tensile tests to determine elastic moduli and fracture properties. Because of the brittleness of the samples, tensile–yield stress was not determined; compressive tests were performed instead.

Unnotched impact characterization was also carried out. With an instrumented impact tester (Charpy), total energy can be separated into two parts: energy of initiation W_i (work done to reach the greatest applied stress) and energy of propagation W_p (29). The balance between these two parts provides information about the mechanism of deformation and the absorption of energy during the impact test.

Influence of an Isothermal Precure Schedule on Mechanical Properties. We have detailed the morphologies of 15% CTBN–epoxy systems induced by a precure schedule that froze the nodular structure through vitrification (27 and 50 °C) or gelation (75 and 100 °C). The overall mechanical properties are summarized in Table IV for the 15% CTBN system cured in the standard conditions (q = 5 K/min up to 190 °C and 14 h at 190 °C). The precure schedule has no significant influence on the mechanical properties of the neat system (Table V).

The fact that the particle-size distribution changes without notable modification in the rubber volume fraction (V = 0.23 ± 0.02) allows discussion of the observed toughening effect with a single morphological parameter: the particle size $(\overline{D})$ or the interparticle distance $(\overline{d}_c)$.

The instrumented impact tests exhibit a clear maximum for the 75 °C precured samples with a surface resilience R_s = 46 kJ/m^2, which is an unusually high value for rubber-modified thermosets. This toughening effect compared to the standard network (R_s = 33 kJ/m^2) is accompanied by a low standard deviation. The highest value for R_s is obtained with the highest initiation energy ratio W_i (Table IV). This fact clearly shows the importance of rubber inclusion size during the initiation step; it is less important in the

Table IV. Mechanical Properties Dependent on Cure Schedule

Cure Schedule	Tensile Test			Compressive Test		Charpy Impact Test	
	E (GPa)	σ_R (MPa)	ϵ_R (%)	σ_y (MPa)	ϵ_y (%)	R_s (kJ/m²) (σ)	W_i (%)
6 days, 27 °C + 14 h, 190 °C	2.12	54	4	85	8.9	23 (3)	64
24 h, 50 °C + 14 h, 190 °C	2.11	60	8.2	76	7.6	32 (7)	73
7 h, 75 °C + 14 h, 190 °C	2.09	57	8.5	79	8.5	46 (4)	80
2 h, 100 °C + 14 h, 190 °C	2.07	61	8.1	84	8.2	28 (4)	73
q = 5 K/min + 14 h, 190 °C	2.15	60	6.6	76	7.1	33 (3)	74

NOTE: Room temperature, 15% CTBN samples.

Table V. Influence of Cure Process on Room-Temperature Mechanical Properties of DGEBA–MNDA Epoxy System

Cure Schedule	Tensile Test			Compressive Test		Charpy Impact Test	
	E (GPa)	σ_R (MPa)	ϵ_R (%)	σ_y (MPa)	ϵ_y (%)	R_s (kJ/m²) (σ)	W_i (%)
6 days, 27 °C + 14 h, 190 °C	2.12	54	4	85	8.9	23 (3)	64
q = 5 K/min + 14 h, 190 °C	2.15	60	6.6	76	7.1	33 (3)	74

crack propagation process. Moreover, the 27 °C precured samples are less impact-resistant than those cured with the standard schedule. If the mean diameter is too low (0.5 μm), it is insufficient to enhance the matrix toughening.

The Young modulus is nearly the same (Table IV), so this impact improvement cannot be imputed to a lowering of stiffness. Similar values for the Young modulus agree with the constant value of the rubber volume fraction. This agreement was predicted, for instance, by the Kerner model, which does not take rubber particle size into account. The tensile stress–strain curves (Figure 10 and Table IV), at low strain rate, lead to the same conclusions. Changes in the morphology parameters with V constant induce an important modification of the deformation process.

A very brittle behavior is evidenced for the 27 °C precured samples; those precured at 50, 75, and 100 °C are more ductile. An upper yield stress is even exhibited for T_c = 75 °C with the highest ultimate strain (ϵ_r = 8.5%). Shear yielding, the usual predominant deformation mechanism (6, 18, 26–31) in elastomer-modified epoxies, is thus reinforced by this mor-

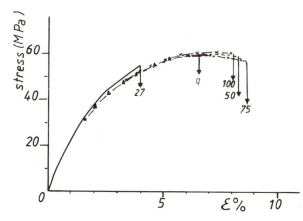

Figure 10. *Influence of different curing processes on the engineering tensile stress–strain curves of rubber-modified epoxies (15% CTBN).*

phology ($\overline{V}$ = 0.23 and $\overline{D}$ = 0.9 μm). Cavitation in or around rubber particles is another deformation process for such systems (7, 8, 24, 26). Tensile dilatometry experiments conducted on toughened epoxies by Yee and Pearson (28) show that this phenomenon is only in the order of magnitude of $\Delta V/V_0 \simeq 1\%$.

The influence of morphology on the deformation mechanism is less clear with compressive tests (Table IV) because the yield stresses are close together. The samples precured at T_c = 27 or 100 °C have the highest stress at yield point (σ_y) and strain at yield point (ϵ_y). Even those cured at 50 and 75 °C and under standard conditions are in the same domain with the lowest values. Thus, the shear yielding mechanism is inhibited with the morphology of 27 and 100 °C samples.

The influence of strain rate on the yield stress, established by using an Eyring model, may soon help us to understand the viscoelastic nature of yield stress and the influence of morphology on activation energy (27, 28).

The mechanical properties of the 15% CTBN system, at both low and high rates, exhibit an optimum toughness at T_c = 75 °C. The Charpy impact test increases the difference between the samples. Thus, a precure schedule may have a negative (27 °C) or positive (75 °C) effect, compared to the standard curing conditions.

We developed smaller rubber particles than those generally reported in the literature for the same (15%) initial elastomer content. Others have observed particles in the range of 1.2 μm (2), 1.6 μm (27), and up to 3.5 μm (24) or more (18).

The system containing the smallest particles (T_c = 27 °C, $\overline{D}$ = 0.5 μm) is only slightly tougher than the neat system. Therefore, even if phase separation seems necessary, it sometimes appears insufficient for toughening.

SEM observation of fracture surfaces gives us much information. For $T_c = 27$ °C (Figure 6b), voiding and cavitation occur only around a few rubber particles. The energy-dissipation process, with this morphology, seems to be limited because the numerous cut rubber domains do not contribute to the initiation of microshear bands between the particles.

Similar conclusions may be drawn from Figure 6a for the precured samples with a heating rate $q = 5$ K/min. However, in this case the less numerous particles induce plastic deformation in the matrix.

Figure 6c and 6d ($T_c = 50$ and 75 °C, respectively) show most of the spherical inclusions in the equatorial plane, acting as stress concentrators. No microcavitation is visible, but numerous microshear bands are observed between rubber domains. With many particles this process tends to keep the yielding localized (27). This localization can explain the appearance of a tensile yield stress and the observed improvement for this particle-size distribution.

Another strong difference between samples precured at 75 and 27 °C is the proportion of dissolved rubber. As shown in Table III, $_E T_g$ decreases as T_c increases. This shift in $_E T_g$ leads to a larger proportion of dissolved rubber and favors plastic deformation, which is a more effective toughening mechanism than stress-whitening (6). LEFM is being used to study the influence of morphology on the crack propagation step.

It appears possible to optimize mechanical properties at a constant volume fraction of the dispersed phase. This maximum is connected, for the 15% CTBN system, with an optimum in phase-separated morphology having a mean diameter of 0.9 μm for a narrow, unimodal distribution. These results show the importance of a precure schedule with hindered diamine–epoxy systems.

Influence of CTBN Content.   The rubber-modified epoxy networks were achieved as described in the experimental section. Various amounts of CTBN 1300 × 8 adduct were introduced, up to 25% by weight. The samples were precured with a heating rate $q = 5$ K/min from room temperature to 190 °C and then postcured for 14 h at 190 °C.

The $_E T_g$ values and morphology characteristics are listed in Table VI. Unfortunately, the glass-transition temperatures of the rubbery phase cannot be detected with the DSC measurements.

This slight decrease in the $_E T_g$ is due to an increase in the dissolved rubber in the epoxy–amine matrix. The $_E T_g$ dramatically falls to 70 °C for the 25% CTBN sample, a result indicating that the inversion-phase phenomenon (30) occurs close to 26%, as calculated by Williams et al. (7, 8). Mechanical properties will confirm this interpretation.

The morphology was characterized by using the same methods and parameters as before. Results are summarized in Figure 11 and Table VI.

Table VI. Influence of CTBN Content (wt %) on Morphology

Morphological Characteristics	5	10	15	20	25
$_E T_g$, °C	142	137	135	127	70
$V_0{}^a$, %	6.3	12.4	18.5	24.2	29.8
$_R V\ (_E T_g)$, %	3	8	13	19	–
$\overline{D}{}^b$, μm	0.77	0.86	1.05	1.22	1.63
(σ)	(0.22)	(0.22)	(0.27)	(0.35)	(0.33)
$\overline{N}{}^c$, μm²	0.12	0.34	0.16	0.19	0.21
$\overline{d}_c{}^d$, μm	2.6	1.76	2.09	1.92	2.19
(σ)	(0.9)	(0.48)	(0.48)	(0.62)	(0.49)
$\overline{d}_p{}^e$, μm	1.83	0.9	1.04	0.7	0.56
$V(\overline{d})^f$, %	2	8.6	11.4	19	30
$V(\overline{D},\overline{N})^f$, %	4	13.5	9	15	2

NOTE: Specimen was cured for 14 h at 190 °C. The values in the column heads are % CTBN (weight).
^{a}Initial calculated volume fraction of rubber phase.
^{b}Mean particle diameter.
^{c}Mean number of dispersed-phase particles.
^{d}Mean distance between two particles, center to center.
^{e}Mean distance between two particles, surface to surface.
^{f}Volume fraction of dispersed phase predicted from different measurements, from $\overline{d}$, $\overline{D}$, and $\overline{N}$.

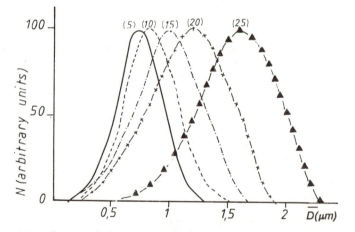

Figure 11. Influence of the amount of CTBN on particle-size distribution for the system DGEBA ($\overline{n}$ = 0.15)–MNDA cured 14 h at 190 °C.

All specimens displayed a unimodal distribution of particle sizes, with a regular increase in the mean diameter (0.8 μm at 5% to 1.2 μm at 20%). We noticed a larger increase of $\overline{D}$ with 25% CTBN (by weight) because of phase inversion. An increase in the CTBN content seems to increase the mean diameter and volume fraction of the rubbery phase, according to many observations (18, 26, 31).

We observed an increase in the width of the distribution curve (Figure 11) and the development of a tail toward low-diameter domains as the CTBN content increased. To our knowledge, this finding has not been mentioned in the literature. The cloud-point temperature T_{cl} increases with the percent of CTBN; T_{cl} = 123 °C for 10% and 140 °C for 15% (Table II). We used a precure with a heating rate q = 5 K/min, so the particle distributions may be connected with a higher T_{cl} and a lower viscosity η_{cl} at a high CTBN content. All of the mechanical results are listed in Table VII.

The Young's moduli are determined by using the tensile stress–strain curves shown in Figure 12. As expected, the elastic moduli decrease with the CTBN content and well fit with the generalized Kerner equation adapted by Manzione et al. (6) (Figure 13). Only the points corresponding to the phase inversion show an important discrepancy with the Kerner model. This relatively good agreement confirms the validity of our dispersed-phase volume fraction determinations by using SEM.

Brittle fractures occur with nearly all samples. However, the 25% CTBN samples exhibit a ductile deformation with the appearance of an upper-yield stress. A maximum toughening effect close to 15% CTBN content seems to be indicated (18, 32).

Compression tests at room temperature reveal similar upper-yield stresses related to the shear deformation process for all the samples. Figure 14 shows relative yield stress plotted versus rubber volume fraction, using the Ishai and Cohen equation (33):

$$\sigma_E/\sigma_{E_0} = 1 - 1.21\, V^{2/3} \qquad (5)$$

Table VII. Room-Temperature Mechanical Properties of Epoxy Systems with Different Percentages of CTBN

% CTBN (weight)	Tensile Test			Compressive Test		Charpy Impact Test R_s (kJ/m²) (σ)	W_i (%)
	E (GPa)	σ_R (MPa)	ϵ_R (%)	σ_y (MPa)	ϵ_y (%)		
0	2.88	40	1.8	130	10.6	17 (5)	65
5	2.78	71	3.9	109	9.5	21.0 (5)	63
10	2.40	70	4.7	95	7.9	16.0 (6)	64
15	2.15	60	6.6	76	7.1	33.0 (1)	74
20	1.70	46	5.1	65	7.5	36.0 (6)	82
25	1.05	14	7.0	23	4.9	33.0 (20)	63

NOTE: Heating rate q = 5 K/min and 14 h at 190 °C.

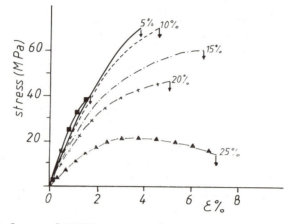

Figure 12. *Influence of CTBN content on the engineering tensile stress–strain curves of rubber-modified epoxies.*

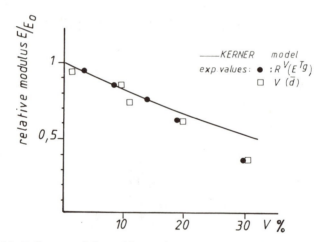

Figure 13. *Influence of the rubber volume fraction on the relative Young's modulus at 25 °C.*

σ_{E_0} characterizes the neat system. Relative yield stress decreases with an increase of dispersed-phase volume fraction.

Experimental agreement with this model is not really good for the highest rubber content levels because of the increased dissolved rubber in the matrix. As pointed out by Bucknall et al. (34), this model overestimates the yield stress because the rubbery particles are considered as voids. As a consequence, the lower experimental yield stress values most accurately predicted shear yielding.

The 25% CTBN specimen shows a particular behavior with a barreling deformation without fracture (in the limit of the tensile machine, 20,000 N).

The toughening effect attributed to the presence of rubbery particles is also well exhibited at high strain rates with the unnotched impact test (Figure 15 and Table VII). The main feature is the appearance of a surface resilience

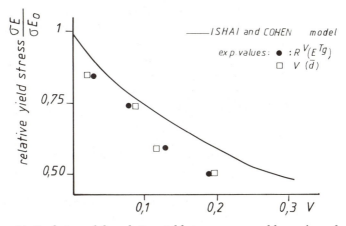

Figure 14. *Evolution of the relative yield stress versus rubber volume fraction; compressive test at 25 °C.*

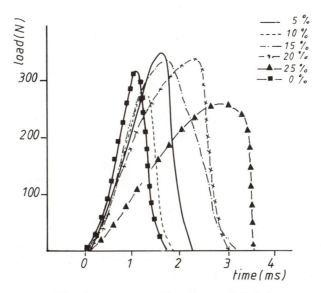

Figure 15. *Typical load–time curves of rubber-modified thermosets; instrumented impact Charpy, 25 °C.*

maximum (*18, 32*) between 15 and 20% CTBN in the range of 33–36 kJ/m^2 that is two times tougher than the neat system. This observation confirms, a posteriori, our choice to use 15% CTBN materials to study the influence of the cure schedule. This improvement results mainly from the the increase in the initiation energy W_i, which is rather constant at 63% below 15% CTBN and reaches 74–82% for 15–20% CTBN. Visual observation of the fracture surfaces evidenced stress-whitening and SEM revealed cavity enlargement, which is the result of particle debonding.

As shown by compressive tests, the shear yielding process is favored by an increasing rubber content. Two interactive parameters, volume fraction and mean diameter of dispersed phase, complicate the situation. Nevertheless, the interparticle distance d_c expresses these two terms. The lowest value obtained for the 20% sample induces a maximum overlapping of the stress fields of each particle, as suggested by Broutman and Panizza (*35*).

Application to Filled Systems.   When multiphase systems are used to make composites, the unfilled sample and the matrix of the composite may not have similar morphologies, even if they are prepared with the same curing process (*36*). On the other hand, theoretical studies (*4*) have shown that insertion of an interphase can improve impact strength without decreasing elastic mechanical properties of the composites. Broutman and Agarwal (*11*) showed that composite strain energy absorbed reached a maximum with a lower modulus interphase than that of the matrix. DiBenedetto and Nicolais (*37*) showed that, depending on the difference in the thermal expansion coefficients of the fillers and matrix, the introduction of an elastomeric interphase can relieve residual curing stresses. Electropolymerization has generally been used to deposit different copolymers at the surface of graphite fibers (*38*).

For these reasons we used the CTBN–DGEBA–MNDA system to develop a coating process involving glass beads and carbon fibers. Preliminary reports have already been published (*39–41*).

To prepare the rubbery coating, we used stoichiometric ratios that were slightly different from those used in the preceding sections. First the prepolymer DGEBA and CTBN were prepared with a carboxy-to-epoxy ratio $c/e = 0.5$ and reacted for 19 h at 85 °C. Then MNDA was added with an amine-to-epoxy ratio $a/e = 2.0$. This mixture was kept in bulk at 25 °C for 4 days before it was dissolved in methyl ethyl ketone (MEK). The concentration determined the thickness of the coating. Then the solvent was evaporated under vacuum at 120 °C for 12 h (glass beads, batch process) or at 85 °C for 15 s (carbon fibers, continuous process). During this operation some of the secondary amine hydrogen atoms reacted with the residual epoxy groups and the coating became insoluble in MEK or DGEBA. The treated fillers were then introduced in a DGEBA–dicyanodiamide (DDA, a/e =

0.6 and 1 parts per hundred resin (phr) of benzyldimethylamine, BDMA)
or DGEBA–methylenedianiline (MDA, a/e = 1.0) matrix.

Figure 16 describes the continuous-process setup used for the deposition
of our rubbery coating on carbon fibers (AS4, Hercules Carbon Fibers), their
impregnation by epoxy–MDA matrix, and filament winding for making a
composite (*41*). The amount of deposit and its thickness could be controlled
by thermogravimetric analysis (TGA) and SEM. We have been able to control
the sticking of the fibers up to a weight gain of 2% (coating thickness 0.1
μm). With a batch process the thickness of the coating on glass beads is
estimated at 3% of the bead radius.

This thin rubber coating cannot be detected by DSC measurements.
Evidence for an elastomeric interphase was displayed with viscoelastic measurements (*40*). In the composites with embedded glass beads beside the β-
relaxation peak of the DGEBA–DDA matrix (tan δ_{max} at –90 °C), a relaxation
peak (tan δ_{max} at –56 °C) corresponds to the CTBN glass transition (Figure
17). Discrimination of peaks is only possible at low frequencies (10^{-2} to 10^{-1}
Hz) because of the large difference in apparent activation energies between
the α_a relaxation of the CTBN adduct (220–230 kJ/mol) and the β_e relaxation
of the epoxy matrix (58 kJ/mol). Another relaxation α'_a in the range 50–90
°C is evidenced in the case of glass beads covered with the elastomeric

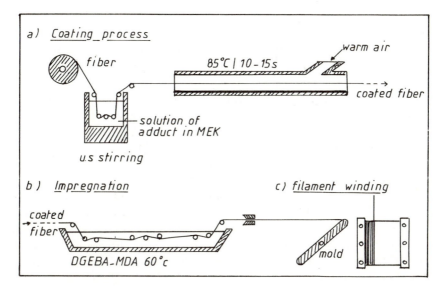

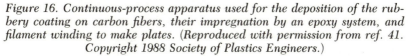

Figure 16. Continuous-process apparatus used for the deposition of the rub-
bery coating on carbon fibers, their impregnation by an epoxy system, and
filament winding to make plates. (Reproduced with permission from ref. 41.
Copyright 1988 Society of Plastics Engineers.)

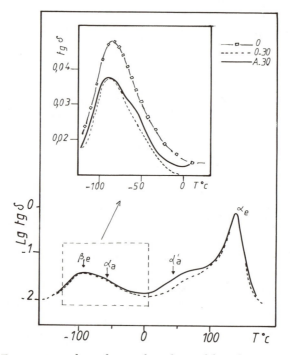

Figure 17. Temperature dependence of mechanical loss factor tan δ (torsion pendulum at 0.01 Hz) for different glass bead–filled epoxy–DDA systems. Key: 0, pure matrix; 0–30, 30% uncoated glass beads; A–30, 30% adduct-coated glass beads; e, epoxy matrix; and a, elastomeric adduct. (Reproduced with permission from ref. 40. Copyright 1987 Springer-Verlag.)

adduct. Pogany (42) and Arridge and Speak (43) interpreted this relaxation as arising from the motion of a loosely cross-linked network embedded in the cure epoxide network. In this case the structure of the interphase may be represented by a two-phase system, a CTBN elastomeric phase and a loosely cross-linked DGEBA–MNDA network.

For coating glass beads we used a batch process with a final 12-h treatment at 120 °C. Such a long treatment was not possible with continuous processing of carbon fibers; it was limited to 15 s at 85 °C. Because of this limitation in this case and with an epoxy–MDA matrix, we observed only one relaxation at −40 °C, instead of at −56 °C. This observation indicates that the structure of the interphase is strongly dependent on the cure treatment after the coating.

Typical load-versus-time curves of the impact tests for unidirectional composite materials are shown in Figure 18. Impact strength was improved with this elastomeric adduct on carbon fibers (AS4). We observed a maximum for the total resilience R_S and for the initiation energy ratio W_i for 0.75%

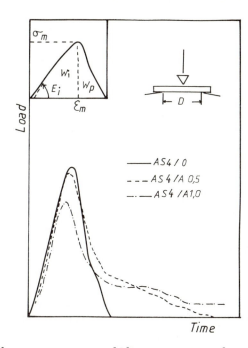

Figure 18. Load-versus-time curves of Charpy impact tests for composites made from untreated (AS4/0) and treated (AS4/A0.5 and AS4/A1.0) carbon fibers. Volume fraction was 60% with a DGEBA–MDA matrix (span-to-depth ratio = 20). (Reproduced with permission from ref. 41. Copyright 1988 Society of Plastics Engineers.)

by weight of elastomeric adduct (sample AS4/A0.5). With higher amounts of coating on carbon fibers (AS4/A1.0), the failure was not complete.

This toughening effect is also clearly demonstrated by both three-points bending and impact tests at room temperature for the glass-beads-filled epoxy thermoset. We used LEFM to compare our adduct treatment of glass beads with a classical silane treatment (γ-glycidoxytrimethylsilane A187) (Figure 19). The critical stress intensity factor K_{Ic} values are clearly improved by the adduct coating, which is better than a classical silane treatment. LEFM applied to the impact test led us to the same conclusion.

These conclusions were confirmed by SEM surface fracture observations. We had poor adhesion between untreated glass fillers and DGEBA–DDA matrix, and good adhesion with silane or adduct-treated glass beads. For adduct-treated glass beads we also observed a trace of plastic deformation of polymeric materials on the filler surface (Figure 20). Similar differences have been observed for untreated and treated carbon fiber composites.

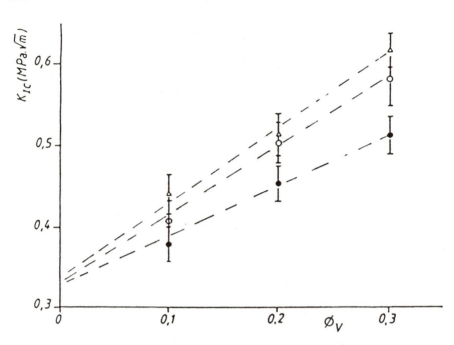

Figure 19. Influence of glass bead volume fraction on critical stress intensity factor K_{Ic} at 25 °C. Key: ●, without treatment; ○, silane treated; and ▲, adduct treated. (Reproduced with permission from ref. 39. Copyright 1985 Association des Matériaux Composites.)

Summary and Conclusions

The effect of a precure schedule on the transitions and morphology of a rubber-modified epoxy system was investigated by using light transmission microscopy, SEM, viscosity, DSC, and SEC. A TTT isothermal-cure diagram was plotted. During the cure, the rubbery phase separated well before gelation ($T_c > {}_{gel}T_g$) or vitrification ($T_c < {}_{gel}T_g$). The duration of opacity Δt_{cl} is one or two orders of magnitude lower than $(t_{gel} - t_{cl})$ or $(t_g - t_{cl})$. So, contrary to previous publications (3, 5, 6), we think that $(t_{gel} - t_{cl})$ is not the main parameter and that the morphology developed during the precure process depends on the viscosity of the system determined at the cloud point. Because Δt_{cl} is short, the variation of viscosity during the phase separation process is neglected in a first approximation. Thus it seems that the diffusion process is less important than the germination step, which depends strongly on interfacial tension and viscosity.

The decrease in ${}_E T_g$ for the rubber-modified systems was caused by dissolved rubber. The volume fraction V of the dispersed phase increased with the initial CTBN fraction V_0. For 15% CTBN, V is nearly constant at 0.23 ± 0.02, with various types of isothermal precuring and narrow particle-

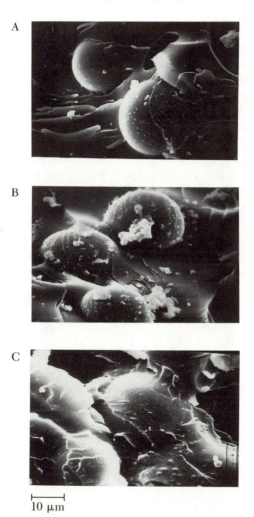

A

B

C

$\vdash\!\!\!\!-\!\!\!\!-\!\!\!\dashv$
10 μm

*Figure 20. Influence of the glass-bead treatment on fracture surfaces (SEM).
Key: A, without treatment; B, silane treated; C, adduct treated. (Reproduced
with permission from ref. 39. Copyright 1985 Association des Matériaux Composites.)*

size distribution. For the same CTBN content, V decreased with a heating
rate of 5 K/min from room temperature to 190 °C. This treatment produced
a broader particle-size distribution.

    With this hindered diamine MNDA and with 15% CTBN, we varied
the precure schedule to obtain different morphologies for the dispersed phase
at a nearly unexpected constant volume fraction. A strongly marked optimum
for all the mechanical properties appeared with a precure schedule of 7 h
at 75 °C (with mean diameter $\overline{D}$ = 0.9 μm). This improvement was mostly

attributable to the increase in shear yielding associated with an easier plastic deformation with the dissolved rubber.

The mechanical properties of the fully cured rubber-modified epoxy systems were investigated as a function of CTBN content and different precure temperatures T_c. With 25% CTBN the phase-inversion phenomenon occurred. Both volume fraction V and mean diameter $\overline{D}$ of the dispersed phase increased with CTBN content. The best toughening effect was achieved with 15–20% CTBN.

We used our knowledge about this CTBN–DGEBA–MNDA system to describe a coating process for glass beads and carbon fibers. Impact strength was improved with our elastomeric adduct. These interlayers have high potential for improving the property losses caused by the introduction of a brittle filler in a brittle matrix, especially when the thermal expansion coefficients are different.

Acknowledgments

The authors are grateful to ORCHEM for financial support and to J. F. Gerard for his contribution on coating carbon fibers.

References

1. Rowe, E. H.; Siebert, A. R.; Drake, R. S. *Mod. Plast.* **1970**, *47*, 110.
2. Sultan, J. N.; McGarry, F. J. *Polym. Eng. Sci.* **1973**, *13*, 29.
3. *Rubber-Modified Thermoset Resins;* Riew, C. K.; Gillham, J. K., Eds.; Advances in Chemistry 208; American Chemical Society: Washington, DC, 1984.
4. Visconti, S.; Marchessault, R. H. *Macromolecules* **1974**, *7*, 913.
5. Manzione, L. T.; Gillham, J. K.; McPherson, C. A. *J. Appl. Polym. Sci.* **1981**, *26*, 889.
6. Manzione, L. T.; Gillham, J. K.; McPherson, C. A. *J. Appl. Polym. Sci.* **1981**, *26*, 907.
7. Williams, R. J. J.; Borrajo, J.; Adabbo, H. E.; Rojas, A. J. In *Rubber-Modified Thermoset Resins;* Riew, C. K.; Gillham, J. K., Eds.; Advances in Chemistry 208; American Chemical Society: Washington, DC, 1984; p 195.
8. Williams, R. J. J.; Adabbo, H. E.; Rojas, A. J.; Borrajo, J. In *New Polymeric Materials;* Martuscelli, E.; Marchetta, C., Eds.; VNU Science: Utrecht, 1987.
9. Vazquez, A.; Rojas, A. J.; Adabbo, H. E.; Borrajo, J.; Williams, R. J. J. *Polymer* **1987**, *28*, 1156.
10. Matonis, V. A.; Small, N. C. *Polym. Eng. Sci.* **1969**, *9*, 90.
11. Broutman, L. J.; Agarwal, B. D. *Polym. Eng. Sci.* **1974**, *14*, 591.
12. Ricco, T.; Pavan, A.; Danusso, F. *Polym. Eng. Sci.* **1978**, *18*, 774.
13. Dekkers, M. E. T.; Dortmans, J. P. M.; Heikens, D. *Polym. Commun.* **1985**, *26*, 145.
14. Schlund, B.; Lambla, M. *Polym. Compos.* **1985**, *6*, 272.
15. Crammer, J. H.; Tesoro, G. C.; Uhlman, D. R. *Ind. Eng. Chem. Prod. Res. Dev.* **1982**, *21*, 185.
16. Sabra, A.; Pascault, J. P.; Seytre, G. *J. Appl. Polym. Sci.* **1986**, *32*, 5147–5160.

17. Montarnal, S.; Galy, J.; Sautereau, H.; Pascault, J. P.; Michel-Dansac, F. In *5eme Journees Nationales Composites JNC5;* Bathias, C.; Menkes, D., Eds.; Pluralis: Paris, 1986; pp 443–458.
18. Bartlet, P.; Pascault, J. P.; Sautereau, H. *J. Appl. Polym. Sci.* **1985,** *30,* 2955.
19. Merle, G.; O, Y. S.; Pillot, C.; Sautereau, H. *Polym. Test.* **1985,** *5,* 37–43.
20. Chan, L. C.; Gillham, J. K.; Kinloch, A. J.; Shaw, S. J. In *Rubber-Modified Thermoset Resins;* Riew, C. K.; Gillham, J. K., Eds.; Advances in Chemistry 208; American Chemical Society: Washington, DC, 1984; pp 235–260.
21. Wang, T. T.; Zupko, H. M. *J. Appl. Polym. Sci.* **1981,** *26,* 2391.
22. Vercheres, D.; Pascault, J. P.; Sautereau, H.; Moschiar, S.; Riccardi, C.; Williams, R. J. J. *Polymer* **1989,** *30,* 107–115.
23. Fox, T. G. *Bull. Am. Phys. Soc.* **1956,** *1,* 123.
24. Butta, E.; Levita, G.; Marchetti, A.; Lazzeni, A. *Polym. Eng. Sci.* **1986,** *26, 1,* 63–73.
25. Wu, S. *Polymer* **1985,** *26,* 1855–1863.
26. Kinloch, A. J.; Young, R. J. *Fracture Behavior of Polymers;* Applied Science: London, 1983.
27. Kinloch, A. J.; Shaw, S. J.; Hunston, D. L. *Polymer* **1983,** *24,* 1341–1354.
28. Yee, A. F.; Pearson, R. A. *J. Mater. Sci.* **1986,** *21,* 2462–2474, 2475–2488.
29. Adams, G. C.; Wu, T. K. *SPE (ANTEC)* **1981,** 185.
30. Siebert, A. R. In *Rubber-Modified Thermoset Resins;* Riew, C. K.; Gillham, J. K., Eds.; Advances in Chemistry 208; American Chemical Society: Washington, DC, 1984; p 179.
31. Kunz-Douglass, S.; Beaumont, P. W. R.; Ashby, M. F. *J. Mater. Sci.* **1980,** *15,* 1109–1129.
32. Bascom, W. D. Cottington, R. L.; Jones, R. L.; Peyser, P. *J. Appl. Polym. Sci.* **1975,** *19,* 2545–2562.
33. Ishai, O.; Cohen, L. J. *J. Compos. Mater.* **1968,** *2,* 302.
34. Bucknall, C. B.; Cote, F. F. P.; Partridge, I. K. *J. Mater. Sci.* **1986,** *21,* 301–313.
35. Broutman, L. H.; Panizza, G. *Int. J. Polym. Mater.* **1971,** *1,* 95.
36. Hong, S.; Chung, S. Y.; Neilson, G.; Fedors, R. F. In *Characterization of Highly Cross-linked Polymers;* Labana, S. S.; Dickie, R. A., Eds.; ACS Symposium Series 243; American Chemical Society, Washington, DC, 1984; pp 91–108.
37. DiBenedetto, A. T.; Nicolais, L. In *Interfaces in Composites;* Piatti, G., Ed.; Applied Science: London, 1978; Chapter 8, p 153.
38. Bell, J. P.; Chang, J.; Rhee, H. W.; Joseph, R. *Polym. Compos.* **1987,** *8,* 46–52.
39. Lin, Y. G.; Pascault, J. P.; Sautereau, H. In *Annales des Composites;* Lamicq, P., Ed.; Association des Matériaux Composites: Paris, 1985; Vol. 4, pp 323–338.
40. Lin, Y. G.; Gérard, J. F.; Cavaillé, J. Y.; Sautereau, H.; Pascault, J. P. *Polym. Bull.* **1987,** *17,* 97–102.
41. Gérard, J. F. *Polym. Eng. Sci.* **1988,** *28,* 568.
42. Pogany, G. A. *Br. Polym. J.* **1969,** *1,* 177.
43. Arridge, R. G. C.; Speak, J. H. *Polymer* **1972,** *13,* 450.

RECEIVED for review February 11, 1988. ACCEPTED revised manuscript September 12, 1988.

Morphology of Rubber-Toughened Polycarbonate

C. K. Riew and R. W. Smith

BF Goodrich Company, Research and Development Center,
Brecksville, OH 44141

Polycarbonate was toughened with preformed rubbery particles. A study was made of Izod impact strength at different toughener levels and at different test temperatures. Fractographs from the Izod impact test specimens by transmission and scanning electron microscopy were analyzed to elucidate rubber-toughening mechanisms. Polycarbonates deform through shear yielding with or without the presence of a discrete rubbery second phase. With rubbery domains, however, toughness is enhanced by simultaneous cavity formation (a dilatational process) and shear deformation (a deviatoric process). Cavities are formed in the rubber domains and/or in the matrix surrounding the rubber domains within the plastic zone ahead of the crack tip.

A DISCRETE RUBBERY PHASE IN A CONTINUOUS PLASTIC MATRIX provides significant improvements in toughness (i.e., crack and/or impact resistance) often without deterioration of the desirable inherent load-bearing-strength properties (*1*).

Because of this enhancement of performance properties, an ever-increasing number of articles in the literature has emphasized the importance of new commercial impact modifiers or rubber-toughened plastics. Merz et al. (*2*) proposed the first rubber-toughening mechanism for the high-impact polystyrene (HIPS) system in 1956. Since then many other researchers have presented valuable rubber-toughening mechanisms (*3–6*).

In recent years the automobile, aviation, and construction industries, which seek to develop tougher materials, have shown considerable interest in toughened plastics. Consequently, scientists are focusing on toughening mechanisms to develop new polymers or tailor existing polymers to meet new requirements. The work described in this chapter is concerned with morphological analyses that compare the crack-tip deformation and fracture behavior of a model rubber-toughened polycarbonate with that of unmodified polycarbonate.

Experimental Details

Materials. *Preformed Rubber Particles as Toughener.* The preformed rubber particles were experimental core-shell-type emulsion polymers. The emulsions were coagulated to produce free-flowing (dry) particles. The powders were further dried under vacuum at 60 °C for a minimum of 16 h to reduce or eliminate moisture prior to blending and injection molding. Number average diameters of two preformed rubber particles in latex form are as follows: latex A, 150.5 nm (core) and 161.5 nm (final); latex B, 30.7 nm (core) and 56.3 nm (final).

Polycarbonate. A low-molecular-weight polycarbonate resin (Lexan 141 from General Electric Co. or Calibre 300–10 from Dow Chemical Co.) was used. The granules of polycarbonate were dried at 120 °C for a minimum of 4 h to eliminate moisture prior to blending and injection molding.

Blending. The tougheners were added to the polycarbonate at the 2.5, 5.0, 7.5, and 10.0% levels. A mixer (Gelimat, Gl-s type, Werner & Pfleiderer Corp., Ramsey, NJ) was used to blend the mixture.

The mixer conditions were dumping temperature, 200 °C; screw speed, 6000 rpm; residence time, 10–12 s; batch size, ~250 g. The blended semimolten dough was further blended with a two-roller (25.4-cm diameter) mill maintained at 200 °C for 1.0–1.5 min.

Injection Molding. Test specimens of 6.35 × 12.7 × 127.0 mm or 3.175 × 12.7 × 12.7 mm were injection molded. An injection-molding machine (Arburg, 220E, Hydronica Allrounder, 40 ton, 2.2-oz. shot size, Polymer Machinery, Berlin, CT) was used. A hopper dryer (Novatec, model MD–25A) was maintained at 100 °C and was attached to the injection-molding machine to prevent moisture pickup during operation.

The nozzle temperature was set at 265–275 °C and barrel temperatures at 275–285, 265–275, and 255–265 °C for the first, second, and third zones, respectively. Mold temperature was maintained at 60–80 °C, depending on the thickness of the specimens.

Izod Test Specimen Preparation. Izod impact test specimens with dimensions of 3.175 or 6.35 × 12.7 × 63.5 mm were notched according to ASTM procedure D 256.

Testing. Six specimens for each sample were tested at 25, 10, 0, –20, and –40 °C. After they were notched, all specimens were placed in a desiccator for a minimum of 1 week at room temperature before they were tested to release thermal stresses possibly introduced during the injection-molding and notching operations. Izod impact testing was carried out according to Method A of ASTM D 256.

Scanning and Transmission Electron Microscopy. Scanning electron microscopy (SEM) was conducted on fractured specimen surfaces that had been sputter-coated with gold.

Transmission electron microscopy (TEM) was conducted on replicas of fractured surfaces. The replication procedure used was the gelatin–carbon system described by Andrews (7). Briefly, a 10% aqueous gelatin solution is applied to the fracture surface on the desired spot and allowed to dry to brittleness. This first-stage replica is peeled from the fracture surface and backed by evaporated carbon. The gelatin is then dissolved in a 0.1 N saline solution containing 50 mg of the enzyme trypsin and held at 80 °C. After the few minutes required to dissolve the gelatin, the floating carbon second-stage replica is picked up on a TEM grid, washed in distilled water, and shadowed with germanium at a 45° shadow angle. The completed replica is a positive replica of the fracture surface.

Particle-Size Analysis of Latexes. A latex was diluted to a total solid of about 0.01% and irradiated with ultraviolet light (rich in the 250–300-nm range) for 24 h to harden the particles. A carbon-coated 200-mesh copper grid was then dipped into the serum. After drying, the preparation was viewed in the transmission electron microscope and photographed at the appropriate magnification. Electron micrographs were used to count and measure particles with a particle analyzer (Zeiss TGZ–3) (Figure 1). Particle-size data were processed with a computer program that produces averages and distributions.

SEM Examination of Shear Banding. A fractured Izod bar was placed in a microtome in such a manner that the glass knife was able to shave material away from the fracture edge at a 45° angle. The shaved edge was then sputter-coated with gold and examined in the scanning electron microscope so that the shaved edge and fracture face could be viewed simultaneously.

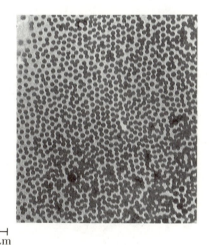

0.5 μm

Figure 1. Transmission electron micrographs of latexes; rubber phases of latexes A and B.

Plasma Etching.   The specimen was polished by using graded abrasive papers with 180–600 grits and completed with 50-nm alumina. Then the specimen was cleaned with distilled water in an ultrasonic bath. After thorough drying, the specimen was placed in a plasma reactor (Branson/IPC, Series 2000). The reactor has a barrel configuration with a 30.5-cm diameter and 50.8-cm length. The surface was etched by introducing pure oxygen to the reactor at flow rate of 100 sccm (standard cubic cm) per min, 300 m·Torr, and 100 W of applied power for 30 and 15 min (Figure 2). Both specimens were overetched. An etching time of 10 min or less would have been better.

Results and Discussion

Figure 1 shows transmission electron micrographs of the two latexes examined. The particle sizes are listed in the "Materials" section. After the latexes are dried, they are free-flowing particles.

During injection molding of the samples, uniaxial melt-drawing causes distortion of rubber particles and possibly nonuniform orientation. In addition, the degree of shell–polycarbonate compatibility may cause rheological changes. These factors contribute most to the so-called pearlescence or nacreous luster (i.e., silvery appearance). Therefore, consistent processing procedures and environment were carefully maintained. One major concern

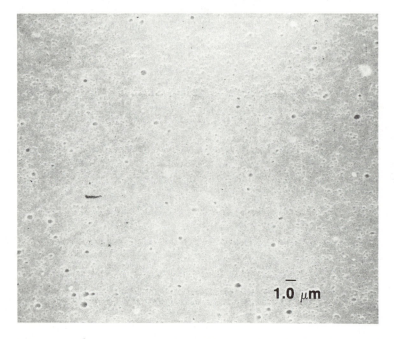

Figure 2. Scanning electron micrograph of rubber-toughened polycarbonate; plasma-etched surface etched for 15 min.

was whether the toughener particles retain their integrity after the high-shear blending processes.

Figure 2 shows a surface that was plasma-etched for 15 min after injection molding, as explained in the experimental section. It was slightly overetched and shows some large holelike black circles. However, the overall particle shapes and sizes are in the range of the original latex particle sizes shown in Figure 1. This result may indicate that the original size and shape of the preformed rubber particles are maintained (i.e., they are not agglomerated or disintegrated during the relatively high-shear melt-blending and injection-molding processes).

The impact strength of polycarbonate depends on its specimen thickness. Like many other ductile plastics, the polycarbonates are, in general, very notch-sensitive. Because of this notch sensitivity, Izod impact tests do not correlate very well with field performance. The industrial practice of using a specimen thickness of 3.175 mm (1/8 in.), often gives high Izod impact values (800–950 J/m tested at room temperature) for both modified and unmodified polycarbonate resins. With a specimen thickness of 6.35 mm, however, the Izod impact strength of Lexan 141 or Calibre 300–10 resin decreases to 100–250 J/m. The critical thickness for the brittle–ductile transition lies between 3.175 and 6.35 mm, depending on the type and molecular weight of the polycarbonate used.

In the present study, therefore, we used 6.35-mm-thick Izod specimens to evaluate the true toughening efficiency of the preformed tougheners.

Table I and Figures 3 and 4 show the effect of toughener level and test temperature on Izod impact strength. Impact strength is highest at 5.0 or 7.5% toughener. Izod impact strength begins to drop at 10% toughener, except at low temperatures.

Table I. Effect of Test Temperature and Toughener Level on Izod Impact Strength

Toughener Level, %	RT	10 °C	0 °C	−20 °C	−40 °C
Latex A: Toughener with Large Particle Sizes					
0.0	166	101	117	104	75
2.5	812	141	160	127	80
5.0	791	674	269	177	109
7.5	779	667	628	215	142
10.0	721	619	518	203	163
Latex B: Toughener with Small Particle Sizes					
0.0	166	101	117	104	75
2.5	785	533	251	133	105
5.0	776	253	211	141	115
7.5	684	606	350	173	137
10.0	678	544	534	187	117

NOTE: All results are in Joules per meter of notch. RT is room temperature.

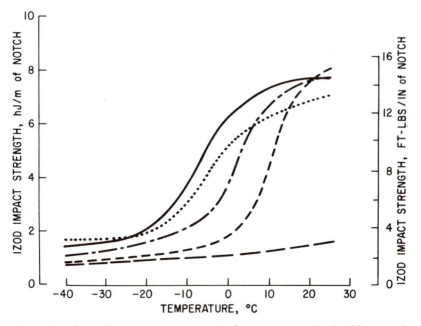

*Figure 3. Effect of test temperature on Izod impact strength of rubber-tough-
ened polycarbonate. Key: — —, 0%; - - -, 2.5%; — · —, 5.0%; ——, 7.5%;
and · · · ·, 10%; large particle toughener, latex A.*

The fractographs in Figure 5 show the effects of toughener level and
test temperature on fracture surfaces. In general, as the toughener level
increases, more stress whitening occurs. However, at 10% toughener, the
injection-molded part begins to show pearlescence from skin–core separation
in the mold, caused perhaps by incompatibility or molding characteristics
of the rubber-modified polycarbonates (8).

Table I and Figure 6 show the effects of toughener level and test temperature on fracture behavior and fracture surfaces. The transition from
plane-strain to plane-stress failure occurs at about 7.5% toughener with latex
A and B, at about 0 °C. A large amount of stress whitening occurs at these
conditions. These changes in toughener levels or temperatures bring about
changes in the fracture behavior from plane-strain to plane-stress conditions.
However, careful examination of the fractured Izod specimens showed two
regions in most of the rubber-toughened polycarbonate. One is the plane-
stress fracture region with shear lips and the other is the smooth plane-strain
fracture region, as depicted in Figure 7. The feature is quite different from
our study on compact tension specimen tests for critical stress factor, K_{Ic}
(ASTM E 399). The region with shear lips in the fractured compact specimens
under mixed plane-stress to plane-strain conditions is normally flat (9).

In principle, the state of triaxial stress at a crack tip near the surface
region of the specimen is low and approaching plane stress (in fact, ap-

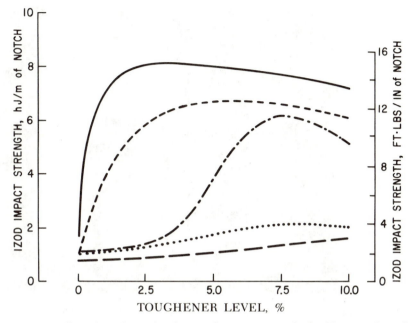

Figure 4. *Effect of toughener level on Izod impact strength of rubber-toughened polycarbonate. Key: ——, 25 °C; - - -, 10 °C; — · —, 0 °C; · · · ·, –20 °C; and — —, –40 °C.*

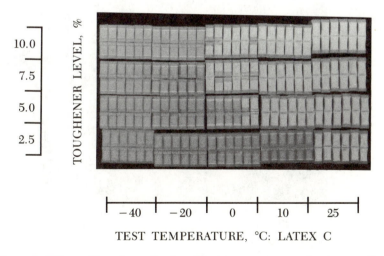

Figure 5. *Effects of toughener level and test temperature on fracture surfaces of Izod impact specimens.*

LATEX A

LATEX B

TOUGHENER
LEVEL

2.5   5.0   7.5   10.0%

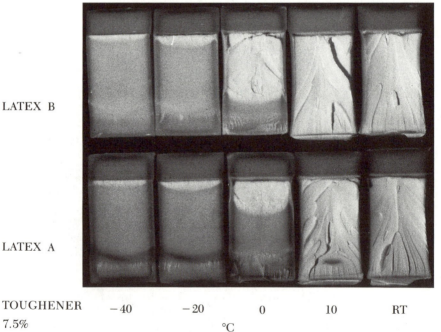

LATEX B

LATEX A

TOUGHENER   −40   −20   0   10   RT
7.5%
                              °C

Figure 6. Effects of toughener level and test temperature on failure mode transition, showing change from plane-strain to plane-stress fracture.

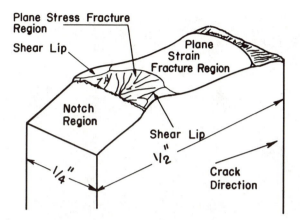

Plane Stress Fracture Region

Shear Lip

Plane Strain Fracture Region

Notch Region

Shear Lip

" \12

1/4 "

Crack Direction

Figure 7. Schematic crack plane of Izod impact specimen, showing plane-stress fracture region.

proaching biaxiality). At a central thick specimen region, the triaxiality is high and plane-strain. Because of this difference in stress state in neighboring regions, there must be differences in the kinetics of crack growth. The difference in crack-growth resistance gives various features to the fracture surface.

Figure 8 shows the fracture sites under plane-stress conditions. The parabolic multiple-fracture surfaces with flow of ridges always seemed to be associated with rubber particles at focal points. Formation of multiple-fracture surfaces with furrows seen under scanning electron microscopy is considered an additional energy-absorption mechanism because it can be seen often in tough plastics (*1*).

The unmodified polycarbonate in this study failed by shear yielding. No stress whitening was observed. At room-temperature testing, it failed under plane-strain conditions when the specimen thickness was 6.35 mm, whereas it failed under plane-stress conditions with a 3.175-mm-thick specimen and formation of shear lips or a plastic zone was observed. At −20 °C all failed under plane-strain conditions. Figure 9 shows the onset of shear lips and shear bands.

When a rubber-toughened polycarbonate fractures under plane-stress conditions, it always shows both shear yielding and cavitation with stress whitening, as shown in Figure 10. In the microtomed region (bottom half of the fractographs) there are no cavities, perhaps because compressive stress must have been applied with development of a plastic zone during fracture. In addition, shear bands may be started from or to the rubber particles. At the interfacial region between the fracture region (top half) and microtomed region, the cavities initiated by the rubber particles are aligned along the shear bands. The shear-yielding mechanism was reported to be largely

Figure 8. Parabolic fracture surfaces in plastic zone. Latex B, 10%, modified polycarbonate, tested at room temperature.

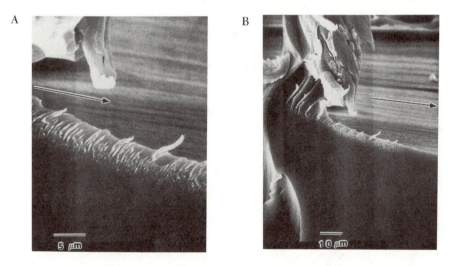

Figure 9. Fractured Izod impact specimen of unmodified polycarbonate that shows development of shear lips and shear bands. Magnification (A) 1000×, (B) 3000× (SEM). Arrow indicates fracture direction.

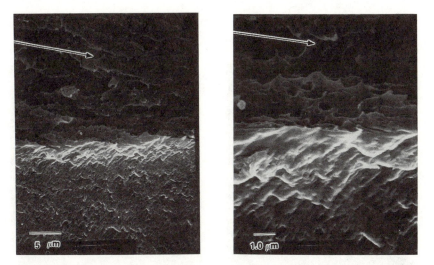

Figure 10. Fractured Izod impact specimen (at room temperature) of latex A (10%) modified polycarbonate showing fracture and plastic region (microtomed). Shear bands interact with cavitation (SEM). Arrow indicates fracture direction.

responsible for the enhanced toughness of polyethylene or methacrylate–butadiene–styrene (MBS) copolymer-modified polycarbonates (*10, 11*).

In Figure 11, the effect of the level of toughener B on fracture surface and cavity formation is shown. Here, as the toughener level increased, the sizes and population of cavities (i.e., volume fraction of cavities) decreased. The decrease in cavities in turn directly affected Izod impact strengths, which were 785, 776, 684, and 678 J/m (room-temperature testing) at 2.5, 5, 7.5 and 10% toughener levels, respectively. Growth of cavities is very much constrained at 7.5 and 10.0% toughener content. The same effect was observed for the latex-A-modified polycarbonates, as shown in Figure 12. The larger rubber particles of the latex A generated larger cavities with Izod impact strengths of 812, 791, 779, and 721 J/m, at 2.5, 5, 7.5, and 10% toughener levels, respectively. In other words, the cavity sizes and numbers (cavity volume fraction) correlate well with toughness properties.

In general, increasing the toughener level will raise the level of stress concentration around the equator of each rubber particle (*12*). However, as the particle population increases (i.e., as the volume fraction of rubber increases) the stress fields begin to interfere with each other. Because the rubber particles cavitate and enhance plastic flow of the continuous matrix resin phase, the larger number of rubber particles means increased cavitation sites. A cavity may form and grow, but the growth will soon be interrupted by adjacent growing cavities formed by neighboring particles. The limit that allows the volume fraction (whether cavity or rubber-phase population) and

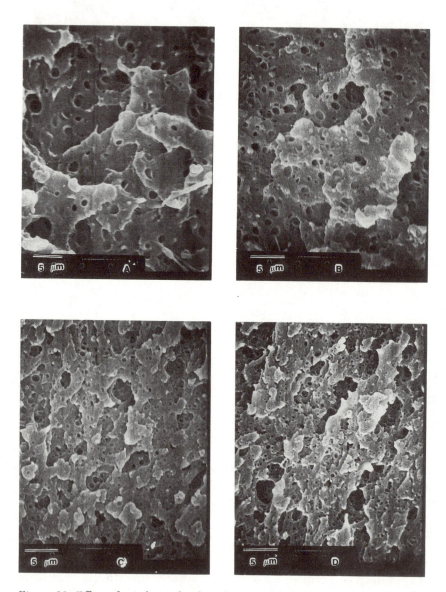

Figure 11. Effect of toughener level on fracture surface showing furrows and cavitations (SEM). Latex B-modified polycarbonate; toughener level: A, 2.5; B, 5.0; C, 7.5; and D, 10%.

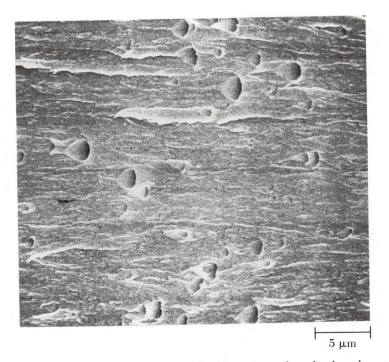

Figure 12. Scanning electron micrograph of latex-A-toughened polycarbonate showing extensive cavitation.

crack-growth resistance to be maximized is the most favorable kinetics of cavity growth.

Figure 13 shows that the multiple-fracture surface or furrows are all recoverable plastic deformation. The fractograph is two halves of the fractured Izod impact test specimen; the specimen on the right side shows the surface after the impact test. The left side shows a recovered (heat-healed or heat-relaxed) surface after the specimen was heated in an oven from room temperature to 180 °C at a heating rate of about 5 °C/min. About 2 min after the oven temperature reached 180 °C, the furrows and multiple-fracture sites returned to their original dimensions and relaxed to form a smooth surface.

Kinloch and co-workers (13) proposed a meniscus instability mechanism to describe ductile crack growth in rubber-modified epoxy resins. The fingerlike furrows are formed ahead of growing cracks when the smooth crack fronts break up as the cracks grow through the plastic zone, which is constrained by the adjacent elastically deformed material.

In Figure 14 a cryofractograph (made by fracturing the specimen at liquid nitrogen temperature) shows additional microstructure of latex-A-modified polycarbonates. The polycarbonate showed extensive plastic flow,

Figure 13. Effect of thermal treatment on fracture surface, showing recoverable plastic deformation.

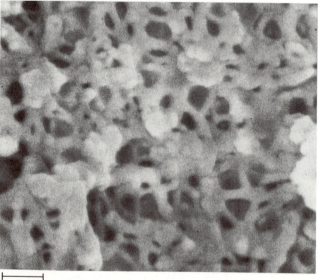

1.0 μm

Figure 14. Scanning electron micrograph of cryofracture surface. Latex A, 10% modified polycarbonate showing loose web structure.

even under cryogenic conditions in the presence of a rubber toughener. Cavity formation is rather substantial and extensive. The micrograph shows many nipples, which must have formed by polycarbonate strands that stretched, fractured, and then snapped back. The cavities are no longer spherical but irregularly shaped, perhaps because they are constrained by growing neighboring cavities. The latex-A-modified polycarbonate loose web-strand structure indicates maximized cavity formation and interaction among the cavities.

The replica micrographs of transmission electron microscopy in Figure 15 show the effect of preformed toughener particles on fracture surfaces. Latexes A and B again show the larger cavities. Figure 15A shows cavitated toughener particles within shear-yielded matrix shells (craters). The cavitated toughener particles range from 80 to 300 nm and the matrix-shell size ranges from 500 to 1500 nm. If the replica is accurate, as in the case of Figure 2, the micrographs indicate little or no distortion in toughener shapes or sizes during blending or injection molding. It also shows that much of the cavitation process resulted from the plastic flow at the toughener matrix interface rather than cavitation only within the toughener.

Conclusions

Morphology.   The original size and shape of the preformed particles are maintained during high-shear melt-blending and injection-molding processes. Two factors maximize toughness and enhance shear yielding and cavity formation ahead of crack tip. First, a critical toughener level exists at which the volume fraction of load-bearing matrix is reduced excessively by the volume fraction of cavities caused by toughener particles. Second, an excessive number of toughener particles (i.e., at a high toughener level such as 10%) constrains the shear-yielding and cavitation processes. In addition, the preferred particle size may be larger than 40 nm, but smaller than 200 nm.

Fracture Behavior.   Plane-strain–plane-stress transition occurs at a critical thickness, toughener level, and temperature. Thin samples, high toughener levels, and high temperature with the same notch dimensions favor plane-stress fracture.

Toughening Mechanism.   Enhanced toughness, crack resistance, and/or impact resistance result from greater energy-absorbing deformation processes in the unmodified or rubber-toughened polycarbonate. In unmodified polycarbonate, shear yielding is followed by with or without formation of small shear lips. A plastic zone develops as the shear yielding propagates ahead of the crack tip.

Rubber-toughened polycarbonate is enhanced through simultaneous deformation by shear yielding and cavitation of the matrix at the matrix

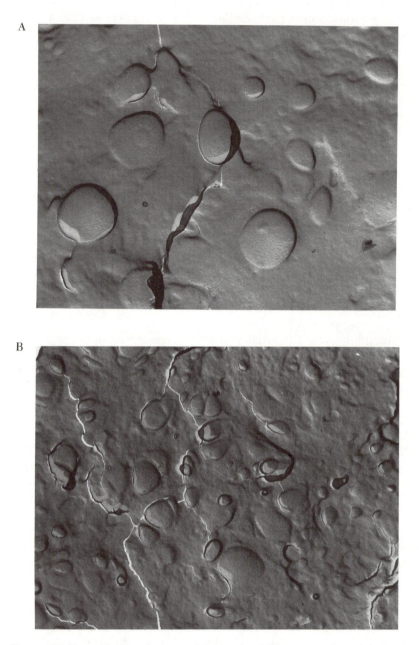

Figure 15. Transmission electron micrograph of fracture surface. (A) Large particle size; (B) small particle size.

resin–rubber-particle interface or within the rubber particles. The inter-action of shear yielding and cavities enhances formation of parabolic multiple-fracture sites and furrows that may be an additional energy-absorption mech-anism. A larger plastic zone than that found in unmodified polycarbonate develops ahead of the crack tip.

Acknowledgments

We acknowledge the contributions of the following persons: Roger E. Morris for his synthesis of tougheners, Frank H. Howard for his assistance in the development of products and processes and in test specimen preparation, Stan Prybyla for plasma etching, Ken Kight for his various assistance to Morris and Howard, and finally Frederick J. McGarry (MIT, Cambridge, MA) and Anthony J. Kinloch (Imperial College of Science and Engineering, London, England) for their useful discussion.

References

1. Riew, C. K.; Rowe, E. H.; Siebert, A. R. In *Toughness and Brittleness of Plastics*; Deanin, R. D.; Crugnola, A. M., Eds.: Advances in Chemistry 154; American Chemical Society: Washington, DC, 1976; p 326.
2. Merz, E. H.; Claver, G. C.; Baer, M. *J. Polym. Sci.* **1956**, *22*, 325.
3. *Toughness and Brittleness of Plastics*; Deanin, R. D.; Crugnola, A. M., Eds.; Advances in Chemistry 154; American Chemical Society: Washington, DC, 1976.
4. Bucknall, C. B. *Toughened Plastics*; Applied Science: London, 1977.
5. Kinloch, A. J.; Young, R. J. *Fracture Behavior of Polymers*; Elsevier-Applied Science: London, 1983.
6. *Rubber-Modified Thermoset Resins*; Riew, C. K.; Gillham, J. K., Eds.; Advances in Chemistry 208; American Chemical Society: Washington, DC, 1984.
7. Andrews, E. H. *J. Polym. Sci.* **1958**, *33*, 39.
8. Karger-Kocsis, J.; Csikai, I. *Polym. Eng. Sci.* **1987**, *27*, 241.
9. Riew, C. K., unpublished data.
10. Yee, A. F. *J. Mater. Sci.* **1977**, *12*, 757.
11. Yee, A. F.; Olszewski, W. V.; Miller, S. In *Toughness and Brittleness of Plastics*; Deanin, R. D.; Crugnola, A. M., Eds.; Advances in Chemistry 154; American Chemical Society: Washington, DC, 1976; p 97.
12. Goodier, J. N. *Trans. Am. Soc. Mech. Eng.* **1933**, *55*, 39.
13. Kinloch, A. J.; Gilbert, D.; Shaw, S. *J. Polym. Commun.* **1985**, *26*, 291.

RECEIVED for review March 25, 1988. ACCEPTED revised manuscript October 7, 1988.

Morphology and Mechanical Properties of Rubber-Modified Epoxy Systems

Amino-Terminated Polysiloxane

T. Takahashi[1], N. Nakajima[2], and N. Saito[2]

[1]Tsukuba Research Laboratory, Sumitomo Chemical Co., Ltd., Tsukuba, Ibaragi 300–32, Japan
[2]Ehime Research Laboratory, Sumitomo Chemical Co., Ltd., Niihama, Ehime 792, Japan

The morphology and mechanical properties of cured epoxy systems modified with amino-terminated polysiloxanes were investigated in relation to their compositions and the reaction conditions of the modification. The amino-terminated polydimethylsiloxane and its statistical copolymers with methylphenylsiloxane were used to chemically modify glycidyl ethers of o-cresol novolac. The compatibility between the epoxy resin and the silicone modifiers could be enhanced by increasing the percentage of methylphenylsiloxane units relative to dimethylsiloxane. This enhancement increased the rate of the modification reaction between terminal amino groups of the polysiloxanes and epoxy groups of the epoxy resin. Increasing the percentage of methylphenylsiloxane units raised the α-glass-transition temperatures of the polysiloxane from −128 to −17 °C. Morphology of cured states was subject to the reaction conditions. The low elastic modulus was achieved at the small spherical domains.

IMPROVEMENT OF FRACTURE PROPERTIES with minimal decrease in elastic modulus and mechanical strength in applications requiring high impact and fracture strengths is the primary objective of modification of thermoset resins by the addition of rubber. Such applications include reinforced plastics, matrix resins for composites, and coatings. On the other hand, the low modulus is preferable in applications requiring low internal stresses.

Such low-stress applications include packaging materials and epoxy molding compounds for encapsulation of semiconductor integrated circuit (IC) devices (1, 2). The thermal expansion coefficients of cured epoxy resins are almost 20 times higher than those of silicon chips, from which IC devices are made. The IC devices are stressed because of this thermal mismatch (3). This internal stress causes package cracks, deformation of aluminum patterns on IC devices, and passivation film cracks (4–6). It is, therefore, necessary to reduce the thermal expansion coefficient and the elastic modulus of epoxy molding compounds to reduce internal stresses.

The conventional epoxy molding compounds consist of phenol novolac (PN, ~10 wt %), glycidyl ethers of o-cresol novolac (ESCN, ~20 wt %), and inorganic fillers (~70 wt %). The inorganic fillers are used mainly to reduce the thermal expansion coefficient. The PN-cured ESCN system was used mainly because the system is characterized by low water absorption, high heat resistance, and high electrical resistance.

Addition of reactive liquid rubber to epoxy resins can reduce the elastic modulus of the cured states. Although liquid acrylonitrile–butadiene copolymers with carboxy or amine end groups have been widely used as epoxy modifiers (7–9), the relatively high glass-transition temperature (T_g) of acrylonitrile–butadiene copolymer limits their low-temperature applications. Their addition to epoxy molding compounds raises the thermal expansion coefficient. Polysiloxanes exhibit lower T_g values than conventional elastomers, along with very good thermal stability. Addition of amino-terminated polysiloxane to ESCN can reduce both the elastic modulus and the thermal expansion coefficient of the cured epoxy molding compound.

The solubility parameter of polydimethylsiloxane, 7.4–7.8, is much lower than that of epoxy resins, which is ~10.9 (10, 11). Polydimethylsiloxane is, therefore, not compatible with ESCN unless it has functional groups that react with ESCN epoxy groups. The solubility parameter of the polysiloxane, along with its reactive functional groups, is considered to be a key factor governing the morphology of cured states. Copolymerizing dimethylsiloxane with methylfluoropropylsiloxane or diphenylsiloxane enhances compatibility with glycidyl ethers of bisphenol A (10, 12–14).

The overall objective of this study was to investigate the morphology and mechanical properties of cured epoxy systems in relation to the compositions of amino-terminated polysiloxanes and reaction conditions of the modification.

Experimental Details

Materials.   Glycidyl ethers of o-cresol novolac (Sumi-epoxy ESCN–195X, Sumitomo Chemical Co., Ltd.) were used as epoxy resins. A phenol novolac (PN) was synthesized according to the conventional method. Analytical values of ESCN and PN are listed in Table I; their molecular structures are shown in Chart 1. Formulations were stoichiometric, with one epoxy group per phenolic hydroxy group. 1,8-Diazabicyclo[5.4.0]undec-7-ene (DBU) was used as an accelerator of the curing reaction.

Amino-terminated polysiloxanes were supplied by Wacker Chemie GmbH. Their average molecular structures are indicated in Chart 2. Epoxy-terminated polydimethylsiloxane, SF–8411, was supplied by Toray Silicone Co. Analytical values of both amino- and epoxy-terminated polysiloxanes are listed in Table II. It is difficult to synthesize block copolymers of dimethylsiloxane and methylphenylsiloxane by the conventional method. Graft copolymers were prepared by the addition of SF–8411 and excess amounts of the amino-terminated polymethylphenylsiloxane (V). The graft copolymers were then used as substitutes for the block copolymers.

Modification of ESCN with Amino-Terminated Polysiloxanes.   Two kinds of modified ESCN were prepared. One was formed by reacting ESCN with amino-terminated polysiloxanes in the melting state at 140 °C for 80 min. The other was formed in toluene solution at 105 °C for 5 h.

Specimen Preparation.   Two kinds of formulations were adopted for the measurement. The epoxy molding compound consisted of amino-terminated polysiloxane-modified ESCN, PN in stoichiometric amounts, DBU (1 wt % of total resins), and fused silica (70 wt % of the total molding compound). The other formulation was the neat resin system, which consisted of the modified ESCN, PN, and DBU. A precursor was made by melt-blending all of the ingredients. The precursor was placed between

Table I. Analytical Values of ESCN and PN

Characteristics	ESCN	PN
Softening point, °C	69.6	97.0
ICI viscosity, poise	5.1	6.0
Epoxy equivalent weight, g/eq	196.3	–
Phenolic hydroxy equivalent weight, g/eq	–	110
Number average molecular weight	1200	600

ESCN

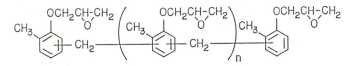

PN

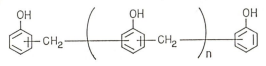

Chart 1. Molecular Structures of ESCN and PN.

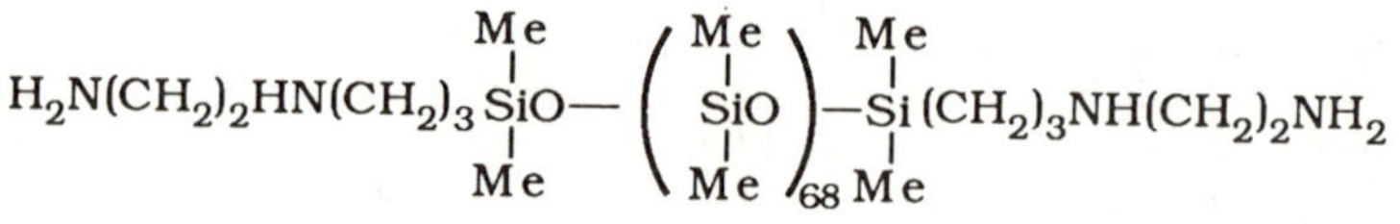

$$H_2N(CH_2)_2HN(CH_2)_3 \underset{\underset{Me}{|}}{\overset{\overset{Me}{|}}{Si}}O - \left(\underset{\underset{Me}{|}}{\overset{\overset{Me}{|}}{Si}}O \right)_{68} Si(CH_2)_3NH(CH_2)_2NH_2$$

I

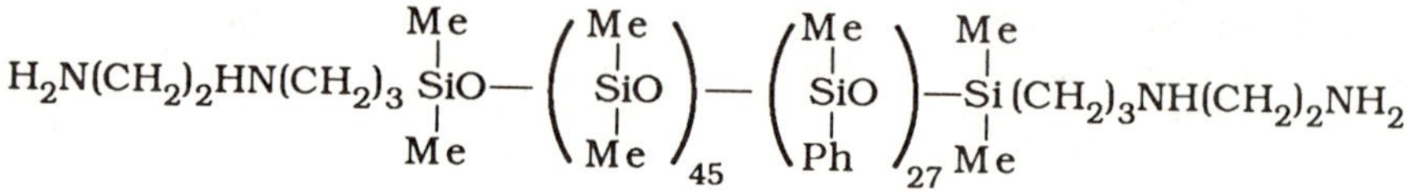

$$H_2N(CH_2)_2HN(CH_2)_3 \underset{\underset{Me}{|}}{\overset{\overset{Me}{|}}{Si}}O - \left(\underset{\underset{Me}{|}}{\overset{\overset{Me}{|}}{Si}}O \right)_{45} - \left(\underset{\underset{Ph}{|}}{\overset{\overset{Me}{|}}{Si}}O \right)_{27} Si(CH_2)_3NH(CH_2)_2NH_2$$

II

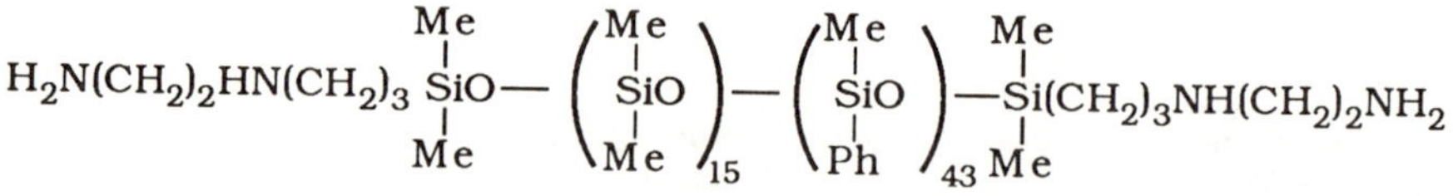

$$H_2N(CH_2)_2HN(CH_2)_3 \underset{\underset{Me}{|}}{\overset{\overset{Me}{|}}{Si}}O - \left(\underset{\underset{Me}{|}}{\overset{\overset{Me}{|}}{Si}}O \right)_{15} - \left(\underset{\underset{Ph}{|}}{\overset{\overset{Me}{|}}{Si}}O \right)_{43} Si(CH_2)_3NH(CH_2)_2NH_2$$

III

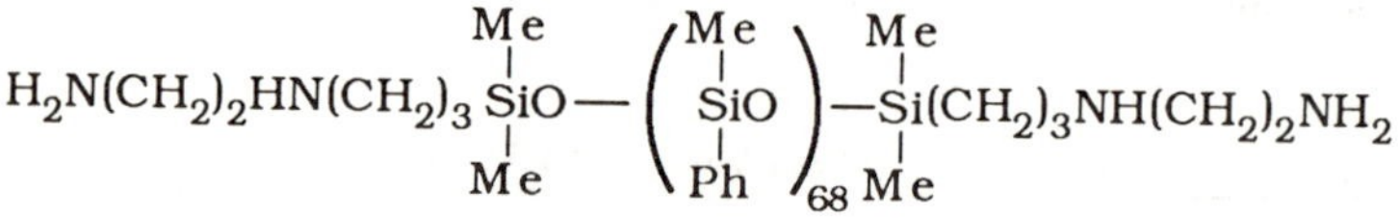

$$H_2N(CH_2)_2HN(CH_2)_3 \underset{\underset{Me}{|}}{\overset{\overset{Me}{|}}{Si}}O - \left(\underset{\underset{Ph}{|}}{\overset{\overset{Me}{|}}{Si}}O \right)_{68} Si(CH_2)_3NH(CH_2)_2NH_2$$

IV

$$H_2N(CH_2)_2HN(CH_2)_3 \underset{\underset{Me}{|}}{\overset{\overset{Me}{|}}{Si}}O - \left(\underset{\underset{Ph}{|}}{\overset{\overset{Me}{|}}{Si}}O \right)_{23} Si(CH_2)_3NH(CH_2)_2NH_2$$

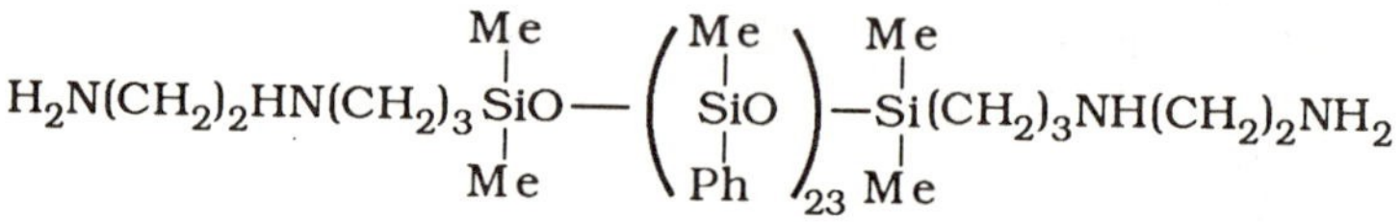

V

Chart 2. Molecular Structures of Amino-Terminated Polysiloxanes.

Table II. Analytical Values of Amino- and Epoxy-Terminated Polysiloxanes

Sample	Functional Group	Equivalent Weight, g/eq	Viscosity, 25 °C, cst
I	amine	1330	1100
II	amine	1780	560
III	amine	1660	1400
IV	amine	2200	7600
V	amine	820	2380
SF–8411	epoxy	3000	8000

steel plates, cured by a hot press at 160 °C and 50 atm for 10 min, and then postcured at 180 °C for 5 h.

Dynamic Mechanical Properties. Dynamic mechanical properties as a function of temperature were measured for cured sheets of the neat resin systems in a fully automated dynamic mechanical analyzer (DMA, Rheolograph Solid, Toyoseiki Co.) over the range from −150 to 250 °C at a heating rate of 2 °C/min in nitrogen atmosphere.

Transmission Electron Microscopy. The morphology of cured specimens of the neat resin systems was examined by using transmission electron microscopy (TEM). Specimens were stained with osmium tetroxide and microtomed by using the Kato method (*15, 16*).

Thermal Properties. The T_g values of cured sheets of epoxy molding compounds were determined by using thermomechanical analysis (TMA–100, Seiko I&E) at 5 °C/min. Thermal expansion coefficients were measured at 70 °C below T_g for cured sheets of the epoxy molding compounds and the neat resin systems.

Mechanical Properties. Flexural properties were measured from rectangular strips (2.0 × 10 × 80 mm) by using a three-point-bend assembly attached to a tensile tester (model 1122, Instron) according to ASTM D 790 (JIS K–6911).

Results and Discussion

Morphology: Effect of Reaction Conditions. ESCN was modified by reacting ESCN with the amino-terminated polysiloxanes. The end point of the modification reaction was determined by titrating tertiary amine, which was the final form of amino groups reacted with epoxy groups of ESCN, in the reaction mixture. The modification was performed by the two kinds of the reaction conditions, as described in the "Experimental Details" section. The phase of the reaction mixture was strongly dependent on the compatibility of amino-terminated polysiloxane with ESCN in the melting-state reaction. On the other hand, a homogeneous phase was obtained throughout the modification reaction by the addition of toluene as a solvent.

The modification reaction in the melting state was observed as follows. The low-molecular-weight amino-terminated polymethylphenylsiloxane (V in Chart 2) was completely soluble in ESCN, and the modified ESCN was a transparent pale yellow solid. The cured neat resin system of the modified ESCN, PN, and DBU was also a transparent pale yellow solid, and no rubber domains were observed. The high-molecular-weight amino-terminated polymethylphenylsiloxane (IV in Chart 2) was partially compatible with ESCN. The amino-terminated statistical copolymers (III and II in Chart 2) were less compatible with ESCN than IV. The amino-terminated polydimethylsiloxane (I in Chart 2) was not compatible with ESCN. The cloudiness of the reaction mixtures increased as the contents of methylphenylsiloxane in the statistical copolymers decreased.

The morphology of the cured neat resin system is shown in Figure 1. The content of the polysiloxanes in the neat resin systems was 16.7 wt %. White spots in the TEM micrographs were open holes in the specimens made during the microtoming process. The shapes and sizes of the domains differed from each other. There were no consistent trends in the shape and size of the domains, although the compatibility of the polysiloxanes with ESCN increased as the content of methylphenylsiloxane in the statistical copolymers increased.

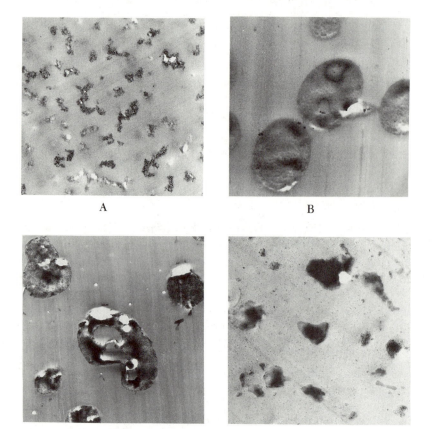

Figure 1. TEMs of cured neat resin systems for modified ESCNs. The modification reaction was conducted in the melting state of ESCN. Key: A, reacting with the amino-terminated high-molecular-weight polymethylphenylsiloxane (IV in Chart 2); B, the statistical copolymer of dimethylsiloxane (D) and methylphenylsiloxane (P) (III) (D:P = 26:74); C, the statistical copolymer (II) (D:P = 62:38); D, polydimethylsiloxane (I). There was no domain structure for ESCN modified with the low-molecular-weight amino-terminated polymethylphenylsiloxane (V).

   Contrary to our observation, it was reported in the literature that the enhancement of compatibility produced smaller rubber particle size (*10, 17*). The shape and size of the rubbery domains were dependent on the modification reaction conditions and the curing conditions. The reaction of the amino-terminated polysiloxanes with ESCN occurred in the heterogeneous phase for the polysiloxanes of I, II, III, and IV, but not for the low-molecular-weight amino-terminated polydimethylsiloxane (V). ESCN is a polyfunctional epoxy that averages about 6.1 epoxy groups per molecule, in contrast ˙ɔ diglycidyl ethers of bisphenol A, which are cited in the literature (*10, 17*) as having about 2 epoxy groups per molecule. These differences would explain the observed variations in morphology.

   The modification reaction was conducted in the homogeneous phase with the addition of toluene as a solvent. The modified ESCN did not include any gels, even after the completion of the modification reaction. The cured neat resin system of the modified ESCN had a rubbery domain of about 0.01 μm (Figure 2). There was no difference in morphology between ESCN modified with the polydimethylsiloxane (I) and the statistical copolymer (II). The shape and size of the rubbery domains were considerably dependent on the phase of the reaction mixture, in which amino groups of the polysiloxane reacted with epoxy groups of excess molar amounts of ESCN. The compatibility of polysiloxanes with ESCN did not affect the morphology of the cured state of ESCN modified with polysiloxanes that had reactive functional groups such as an amino group, as long as the reaction was conducted in the homogeneous phase.

   The compatibility of polysiloxanes with ESCN enhanced the rate of modification reactions in the homogeneous phase (Figure 3). The secondary and tertiary amines were serially produced in the reaction of amino groups of polysiloxanes with epoxy groups of ESCN. The concentration of primary amine in the reaction mixture decreased rapidly for both the polydimethylsiloxane (I) and the statistical copolymer (II). The final product of tertiary amine increased gradually. The end point of the modification was determined as the stage when the molar ratio of tertiary:total amine in the reaction mixture became >80 mol %. The modification reaction was completed within 5 h for the statistical copolymer, which was more compatible with ESCN than the polydimethylsiloxane.

   Dynamic Mechanical Properties.   The dynamic mechanical properties of the cured neat resin systems are summarized in Table III. The dynamic mechanical spectra are shown in Figures 4, 5, and 6. The sample designations correspond to those in Chart 2. The designation in Table III of V/SF–8411–A, –B, and –C represents the graft copolymer synthesized by additional reaction of the amino-terminated polymethylphenylsiloxane (V) with the epoxy-terminated polydimethylsiloxane (SF–8411) at the V/SF–8411 composition ratios of 15:85, 30:70, and 50:50, respectively.

0.1 µm

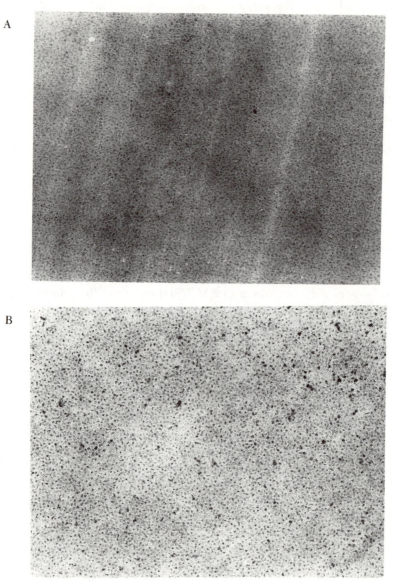

Figure 2. TEMs of cured neat resin systems for modified ESCNs. The modification reaction was conducted in toluene solution of ESCN. Key: A, with the amino-terminated polydimethyl-siloxane (I in Chart 2); B, the statistical co-polymer (II).

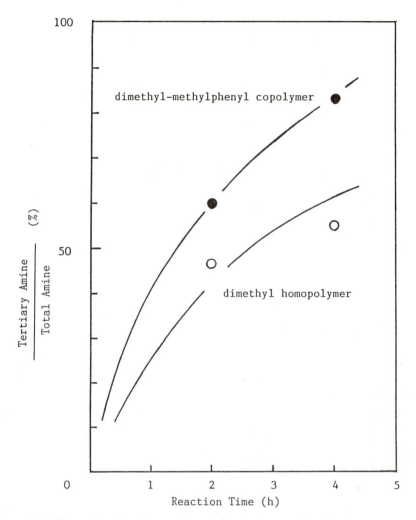

Figure 3. Molar ratios of tertiary amine–total amine in reaction mixture versus reaction time. Key: ●, *molar ratios for the amino-terminated statistical co-polymer (II in Chart 2) reacting with ESCN in toluene solution;* ○, *molar ratios for the polydimethylsiloxane (I) reacting with ESCN in toluene solution.*

The T_g due to the polysiloxane rose as the percentage of methylphen-ylsiloxane relative to dimethylsiloxane increased in the statistical copolymer, because the steric hindrance of the bulky phenyl group restrained molecular motion (Figures 5 and 7). The small peak showing the α-transition of polymethylphenylsiloxane was observed at –18 °C for the cured neat resin system of ESCN modified with low-molecular-weight polymethylphenylsi-loxane (V), in spite of the normal peak height at –24 °C for the α-transition of high-molecular-weight polymethylphenylsiloxane (IV) (Figure 4). This

Table III. Dynamic Mechanical Properties and Thermal Expansion Coefficients of Cured Neat Resin Systems

Polysiloxanes		Dynamic Mechanical Properties			Thermal Expansion Coefficient, $\times 10^{-5}/°C$
Sample Designation	Content, vol %	T_g, °C	Elastic Modulus, 20 °C, $\times 10^9$ N/m^2	T_g, Si linkage, °C	
I	19.0	178	1.53	−128	6.2
II	18.3	176	1.57	−81	6.9
III	17.7	172	1.96	−42	6.9
IV	17.2	168	1.89	−24	7.8
V	17.2	166	1.88	−18	7.5
V/SF−8411−A	18.7	178	1.68	−128	7.1
V/SF−8411−B	18.5	176	1.81	−128, −24	7.0
V/SF−8411−C	18.1	174	1.81	−126, −24	6.4
Unmodified	0	176	2.39	none	6.5

temperature difference is caused by the difference in morphology between the cured resin systems of V and IV. The former did not have any rubbery domains, but the latter had the rubbery domains shown in Figure 1.

The peak at about −60 °C was β-transition because of the hydroxy ether linkage produced by the cross-linking reaction of ESCN with PN (Figure 4) (9). The graft copolymers of the V/SF−8411 series exhibited two T_gs that indicated the polysiloxane units at −126 and −24 °C (Figure 6). The former T_g was assigned to polydimethylsiloxane and the latter to polymethylphenylsiloxane. This result implies that the graft copolymer of V/SF−8411 type could be substituted for amino-terminated block copolymers of dimethylsiloxane and methylphenylsiloxane, as far as dynamic mechanical properties are concerned.

The Gordon–Taylor copolymer equation was adopted to determine the fraction of dissolved polysiloxane (rubber) in the cured neat resin systems (17).

$$\frac{1}{T_g} = \frac{1}{V_1 + K \cdot V_2}\left(\frac{V_1}{T_{g1}} + \frac{K \cdot V_2}{T_{g2}}\right)$$

where V_1 and V_2 are volume fractions of epoxy and rubber, respectively; T_{g1} and T_{g2} are the T_gs of unplasticized epoxy and pure rubber, respectively; T_g is the glass-transition temperature of the homogeneous epoxy-rich matrix phase; and K is a normalization constant. K could be calculated from the low-molecular-weight polymethylphenylsiloxane (V). The cured neat resin system of ESCN modified with V did not contain any phase-separated polysiloxane domains, hence $V_1 = 0.172$ and $V_2 = 0.828$. K was determined to be 0.15. The volume fraction of dissolved polysiloxane could, then, be calculated from the plasticized epoxy T_g, as shown in Table IV. The volume fractions of dissolved polysiloxane increased as the percentage of methyl-

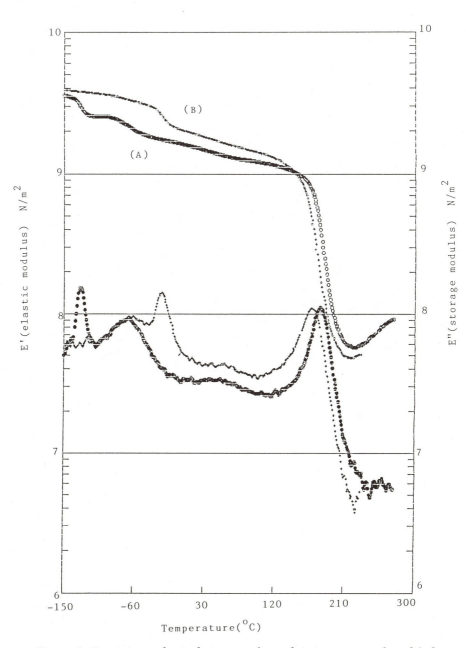

Figure 4. Dynamic mechanical spectra of cured resin systems of modified ESCN. Key: A, with the amino-terminated polydimethylsiloxane (I) in melting state; B, high-molecular-weight polymethylphenylsiloxane (IV).

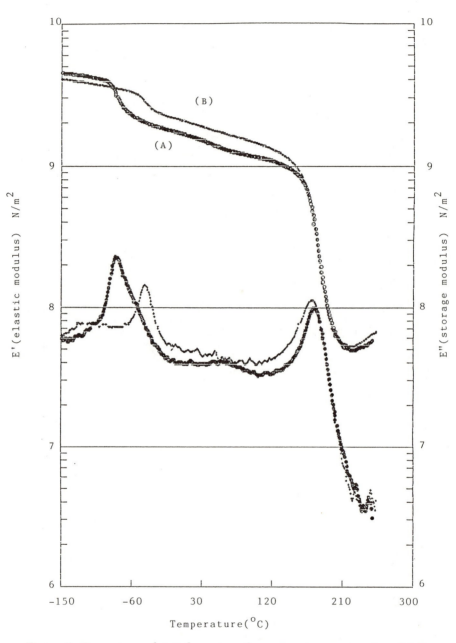

Figure 5. Dynamic mechanical spectra of cured resin systems of modified ESCN. Key: A, with the low-methylphenyl (P) statistical copolymer content (II) in the melting state; B, with the high-P statistical copolymer (III).

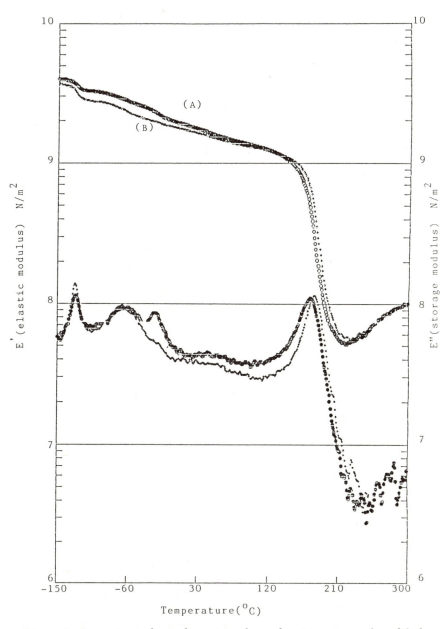

Figure 6. Dynamic mechanical spectra of cured resin systems of modified ESCN. Key: A, with the graft copolymer of 1 and V/SF–8411–A (V:SF–8411 = 15:85) in the melting state; B, that of V/SF–8411–C (V:SF–8411 = 50:50).

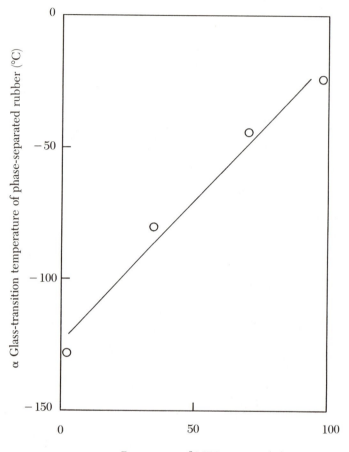

Figure 7. α-Glass-transition temperatures of rubbery domain versus methylphenylsiloxane contents relevant to dimethylsiloxane in the statistical copolymers.

phenylsiloxane increased relative to dimethylsiloxane in the statistical co-polymer, as predicted from the experimental observations.

Incorporation of methylphenylsiloxane in polydimethylsiloxane en-hanced compatibility with ESCN. This modification had, however, several disadvantages when amino-terminated polysiloxane-modified ESCN was ap-plied to the epoxy molding compounds for encapsulation of IC devices. It increased the volume fraction of polysiloxane dissolved into the cured neat resin system and consequently decreased both the T_g of the cured system and the volume fraction of phase-separated rubbery domains so as to limit the extent of the modulus reduction. It raised the T_g of the polysiloxane to limit low-temperature flexibility. Its advantage, from an industrial point of

Table IV. Volume Percent of Dissolved Polysiloxane Calculated from Gordon–Taylor Equation

Variable	II	III	IV
Volume fraction of polysiloxane (VI)	18.3	17.7	17.2
Volume fraction of dissolved polysiloxane	0	6.0	13.3

view, would be the reduced time for modification of the amino-terminated polysiloxane with ESCN in toluene solution. The mechanical properties of the cured epoxy molding compounds that were related to the industrial application for encapsulation of IC devices were measured for ESCN modified with the amino-terminated polydimethylsiloxane (I) and its statistical copolymer with methylphenylsiloxane (II).

Mechanical Properties.   The internal stress on IC devices encapsulated by epoxy molding compounds is closely related to the mechanical properties and thermal expansion coefficient of the cured epoxy molding compounds (*4, 18, 19*). Such stress tends to decrease when the elastic modulus and the thermal expansion coefficient are reduced.

The mechanical properties, T_gs, and thermal expansion coefficients of the cured epoxy molding compounds are indicated in Table V. The sample designations correspond to those in Chart 2 and Table III. The mechanical properties of the cured compound of ESCN modified with the amino-terminated polydimethylsiloxane (I) in toluene solution were equivalent to those modified with its statistical copolymer. The elastic modulus for I and II was reduced to about 1050 kg/mm^2 from 1450 kg/mm^2 for unmodified ESCN, at the polysiloxane content of 4.8 wt % in the epoxy molding compounds and 16 wt % in the neat resin systems. The thermal expansion coefficients of I and II were reduced to about 1.8×10^{-5} °C from 2.4×10^{-5} °C for unmodified ESCN, although it remained constant or slightly increased for the cured neat resin systems, as shown in Table III. This aspect, one of the most advantageous properties of the amino-terminated polysiloxane modification over the other technologies, remains an issue to be investigated through further studies.

The elastic modulus decreased linearly as the polysiloxane content in-

Table V. Mechanical Properties of Cured Epoxy Molding Compounds

Property	Unmodified	I, in Solution	II in Solution	in Melt
Polysiloxane content, wt %	0	16.0	16.0	16.0
Flexural modulus, kg/mm^2	1450	1050	1070	1330
Flexural strength, kg/mm^2	15.0	15.0	15.7	17.2
TMA T_g, °C	146	147	146	143
Thermal expansion coefficient, $\times 10^{-5}$/°C	2.4	1.8	1.9	2.2

creased (Figure 8). The reduction of the modulus was much larger in the cured compounds of ESCN modified in toluene solution than in those in the melting state. There was a considerable difference in morphology between the modification reaction in toluene solution and in melting state (Figures 1 and 2). The ESCN modified in the melting state contained gels that were not dissolved in tetrahydrofuran, so that the rubbery domains

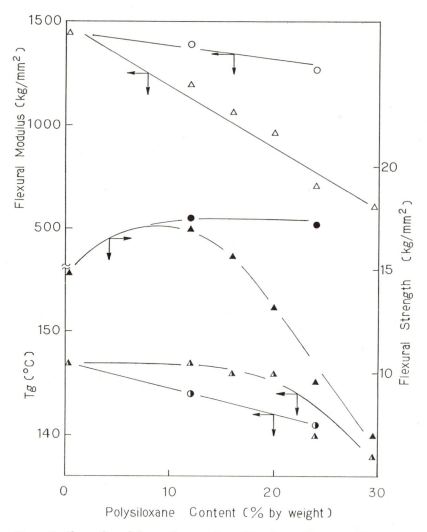

Figure 8. *Flexural modulus and strength and* T$_g$ *of cured epoxy molding compounds versus polysiloxane content indicated in percent by weight of neat resin system. Circles indicate flexural properties for modification in the molten state and triangles indicate those in toluene solution.*

were possibly estimated to exhibit higher molecules than those of the ESCN modified in toluene solution. This difference in the stiffness of the rubbery domains between two modification reaction conditions would affect the extent of the modulus reduction.

Hedrick et al. (*20, 21*) recently investigated mechanical properties in relation to the alternating sequence distributions of hard and soft segments (i.e., diglycidyl ethers of bisphenol A and amino-terminated polysiloxanes). They indicated that the elastic modulus of the cured materials increased with the content of the hard segments made from diglycidyl ethers of bisphenol A. Their result implies that the rubbery domains in ESCN modified in the melting state exhibit a higher elastic modulus than those in toluene solution because it is likely that the ESCN content in the rubbery domain is larger for melting-state modification than in solution.

The flexural strength increased slightly in the region of low polysiloxane content as the content increased, probably because the polysiloxane would have acted as a coupling agent for fused silica, so that the interfacial bond strength was increased by its addition. The flexural strength, then, decreased linearly with the polysiloxane content from about 12 wt % of the neat resin system. This decrease corresponded to the transformation in morphology, where the rubbery domains of about 0.01 μm began to aggregate and construct bridgelike structures, as shown in Figure 2. On the other hand, the flexural strength remained up to 24 wt % polysiloxane for modification in the melting state. The domains for this case stayed dispersed in the matrix, as shown in Figure 1. The minimum allowable flexural strength in encapsulant materials for IC devices is 10 kg/mm^2. Therefore, the epoxy molding compound of ESCN modified with either polydimethylsiloxane (I) or its statistical copolymer (II) in toluene solution must reduce the flexural modulus to half of that of unmodified ESCN.

The decrease in T_g of the cured compound corresponded to that of the elastic modulus, as expected from the transformation in morphology (Table V). There was, however, a noteworthy difference in the case of modification in toluene solution. The threshold of T_g decrease was at the polysiloxane content of 24 wt % in the neat resin system, in spite of 12% for the flexural strength. The extent of aggregation of submicrometer domains was considered the dominant influence on flexural strength. T_g was less affected by the morphological transformation.

The modification of ESCN with amino-terminated polysiloxane is the most effective technology to reduce internal stress. It can reduce not only the modulus, but also the thermal expansion coefficient. It decreased gradually as the polysiloxane content increased (Figure 9). Modification in toluene solution could reduce the thermal expansion coefficient more effectively than that in the melting state. Reduction of the thermal expansion coefficient would be related to particular distribution of polysiloxane-rich domains. Further study will provide an analysis of this phenomenon.

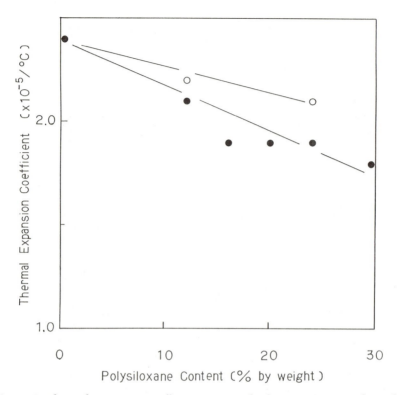

Figure 9. Thermal expansion coefficient versus polysiloxane content, indicated in wt % of the neat resin system, in relation to the two modification conditions of ESCN. Key: ○, molten state; ●, in toluene solution.

Conclusions

The morphology and mechanical properties of the cured neat resin system and the cured molding compound have been investigated in relation to the molecular structures of amino-terminated polysiloxane and the reaction conditions of modification. The modification reaction in the melting state occurred in the heterogeneous phase except for the low-molecular-weight polymethylphenylsiloxane. In toluene solution the modification occurred in the homogeneous phase.

Increasing the percentage of methylphenylsiloxane units relative to dimethylsiloxane enhanced compatibility of polysiloxanes with the epoxy resin. This enhancement increased the volume fraction of polysiloxanes dissolved in the epoxy matrix. It did not affect the morphology of the cured neat resin system so much as the modification reaction. However, it increased the rate of the modification reaction between terminal amino groups of the polysiloxanes and epoxy groups of the epoxy resin. T_g values of the statistical copolymers increased with methylphenylsiloxane content.

The modification in solution exhibited the minimum size of 0.01 μm for rubbery domains. This morphology resulted in effective reduction of the elastic modulus and the thermal expansion coefficient of the cured molding compound.

Acknowledgments

We are indebted to C. K. Riew for offering us the opportunity to participate. Financial support by Sumitomo Chemical Co., Ltd., is also appreciated.

References

1. Wyble, C. W.; Parr, F. T. *SPE J.* **1970**, *28*, 281.
2. DiGiacomo, G. M. *SPE J.* **1970**, *28*, 273.
3. Natarajan, B.; Bhattacharyya, B. *IEEE Electron. Compon. Conf.* **1986**, *36*, 544.
4. Nakamura, Y. et al. *IEEE Trans. Compon. Hybrids Manuf. Technol.* **1987**, *CHMT-12*, 502.
5. Segawa, M. et al. Jpn. Patent, Tokkyo Kokai 1982-180626, 1982.
6. Crabtree, D. J.; Park, E. Y. U.S. Patent 4 476 285, 1984.
7. Riew, C. K.; Rowe, E. H.; Siebert, A. R. In *Toughness and Brittleness of Plastics*; Deanin, R. D.; Crugnola, A. M., Eds.; Advances in Chemistry 154; American Chemical Society: Washington, DC, 1976; p 326.
8. Chan, L. C. et al. In *Rubber-Modified Thermoset Resins*; Riew, C. K.; Gillham, J. K., Eds.; Advances in Chemistry 208; American Chemical Society: Washington, DC, 1984; p 117.
9. Saito, N. et al. *Polym. Mater. Sci. Eng.* **1986**, *55*, 54.
10. Yorkgitis, E. M. et al. In *Rubber-Modified Thermoset Resins*; Riew, C. K.; Gillham, J. K., Eds.; Advances in Chemistry 208; American Chemical Society: Washington, DC, 1984; p 137.
11. Burrell, H. *Encycl. Polym. Sci. Technol.* **1970**, *12*, 618.
12. Yorkgitis, E. M. et al. *Adv. Polym. Sci.* **1985**, *72*, 79.
13. Riffle, J. S.; Yilgor, I.; Tran, C.; Wilkes, G. L.; McGrath, J. E.; Banthia, A. K. In *Epoxy Resin Chemistry II*; Bauer, R. S., Ed.; ACS Symposium Series 221; American Chemical Society: Washington, DC, 1983; p 21.
14. Riffle, J. S. et al. *Org. Coat. Appl. Polym. Sci. Proc.* **1982**, *46*, 397.
15. Kato, K. *J. Electron Microsc.* **1965**, *14*, 219.
16. Kato, K. *Polym. Eng. Sci.* **1967**, *7*, 38.
17. Manzione, L. T.; Gillham, J. K. *J. Appl. Polym. Sci.* **1981**, *26*, 907.
18. Oizumi, S.; Imamura, N.; Tabata, H.; Suzuki, H. In *Polymers for High Technology*; Bowden, M. J.; Turner, S. R., Eds.; ACS Symposium Series 346; American Chemical Society: Washington, DC, 1987; p 537.
19. Thomas, R. E. *IEEE Trans. Compon. Hybrids Manuf. Technol.* **1985**, *CHMT-8*, 427.
20. Hedrick, J. L. et al. *Macromolecules* **1988**, *21*, 67.
21. Hedrick, J. L. et al. *Polym. Bull.* **1988**, *14*, 573.

RECEIVED for review February 11, 1988. ACCEPTED revised manuscript January 25, 1989.

Phase-Separation and Transition Phenomena in Toughened Epoxies

Wai H. Lee[1], Kenneth A. Hodd[2], and William W. Wright[3]

[1]Ciba-Geigy Plastics, Bonded Structures, Duxford CB2 4QD, England
[2]Department of Materials Technology, Brunel University, Uxbridge UB8 3PH, England
[3]Department of Materials and Structures, The Royal Aircraft Establishment, Farnborough GU14 6TD, England

Three epoxy resins with different cross-link densities were modi-fied with low-molecular-weight carboxyl-terminated butadiene–acrylonitrile (CTBN) rubbers. The phase-separation process was investigated by using light microscopy and transmission electron mi-croscopy. It was found that the phase-separation process is favored by increasing the cross-link density of the epoxy matrix, but that the rubber particles in the more highly cross-linked epoxies undergo cavitation upon cooling from their final curing temperatures. Possible explanations for these observations were investigated. The measured volume fractions of the phase-separated rubber particles were com-pared with the theoretical volume fractions. Dynamic mechanical thermal analysis was used to characterize the transition phenomena of the cured toughened epoxies. The heights, positions, and areas of their tan δ peaks were related to the microstructures of the rubber-toughened resins.

IMPACT RESISTANCE AND FRACTURE TOUGHNESS of thermoset epoxy resins were improved by the incorporation of liquid rubber, as demonstrated in the late 1960s by Sultan and McGarry (1). Since then, these modified resins have been subjected to intensive investigations (2–7). Because of their en-hanced fracture properties, they have found wide application as adhesives and matrix resins in the aerospace and other industries.

The properties of rubber-modified epoxy are strongly dependent on the

0065-2393/89/0222-0263$07.25/0

morphology generated during its cure (5–7). The morphology in turn is determined by factors such as the before-cure compatibility of the rubber–resin system, the choice of curing agent, and curing time and temperature (5, 7).

The effect of acrylonitrile content on the phase-separated particles was demonstrated by Rowe et al. (7). Generally, the higher the acrylonitrile content of the carboxyl-terminated butadiene–acrylonitrile (CTBN) rubber, the smaller the particle size. Siebert and Riew (8) also described the chemistry of the particle formation in an epoxy system involving a CTBN rubber, a diglycidyl ester of bisphenol A epoxy resin, and piperidine catalyst. According to them, three reactions can occur between CTBN rubber and epoxy resins. The three reactions are (1) epoxy–acid (rubber) reaction (esterification); (2) acid–aliphatic hydroxyl (from reaction 1) reaction (esterification); and (3) epoxy–aliphatic hydroxyl reaction (etherification). Reactions 1 and 2 are the important reactions, and reaction 1 is termed the "chain extension reaction" (i.e., building up of molecular weights by reaction of carboxyl and epoxy groups).

Riew et al. (9) later showed the superiority of carboxyl terminal groups over others such as phenol, epoxy, hydroxyl, and mercaptan. We now report an investigation of the influence of both the compatibility of the resins with the rubber and the cross-link density of the cured resin on the morphology of the transitions of rubber–epoxy system networks.

Experimental Details

Materials. Diglycidyl ether of bisphenol A (DGEBA) resin (Ciba-Geigy MY750) and tetraglycidyldiaminodiphenylmethane (TGDDM) resin (Ciba-Geigy MY720) were modified with CTBN rubbers from BF Goodrich. The types and properties of the CTBN rubbers are given in Table I, and the structures of the resins and hardeners are shown in Chart 1.

The epoxy and epoxy–rubber systems were cured with either piperidine or diaminodiphenylsulfone (DDS) and boron trifluoromonoethylamine (BF₃MEA).

For DDS-cured systems, a rubber–epoxy prereaction was prepared at 80 °C for MY750 resin and at 100 °C for MY720 resin. Triphenylphosphine catalyst was added in the case of rubber–MY750 resin. The reaction was carried out under nitrogen atmosphere and followed by determination of the carboxyl content at regular intervals (10).

Table I. Properties of Hycar CTBN Elastomers

Property	CTBN 1300×13	CTBN 1300×8
Molecular weight	3500	3500
Functionality	1.85	1.85
Acrylonitrile content, %	27	17
Solubility parameter, Cal cm	9.14	8.77
Specific gravity at 25 °C/25 °C	0.96	0.948

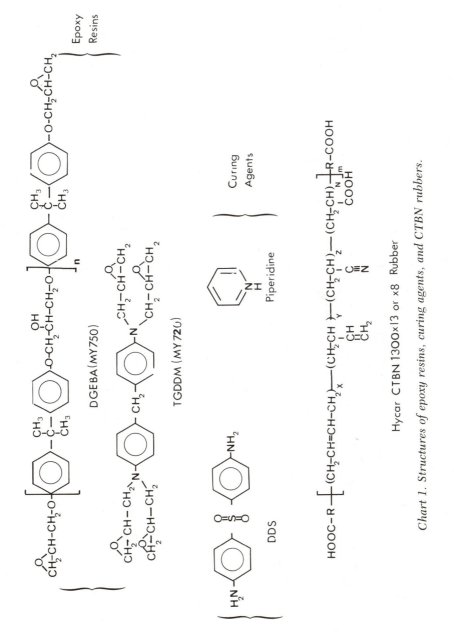

Chart 1. Structures of epoxy resins, curing agents, and CTBN rubbers.

The reaction was stopped when the acid value had been reduced to 0.05 mmol/g or below (usually 3–4 h). The prereaction was cooled and stored at 5 °C.

The prereaction was used in combination with additional resin to obtain the desired rubber content. The mixing was carried out by heating the mixture (prereaction, resin, and DDS as curing agent) to 100 °C and stirring for 15 min. The temperature was then raised to 135–140 °C, and the mixture was degassed until all the DDS had dissolved. The mixture was cooled to 110 °C, and BF$_3$MEA catalyst was added. After 3 min more of stirring, the mixture was poured into preheated glass molds and cured in a heated oven for the required time.

For the rubber-modified piperidine cure system, the desired rubber content was added to 100 parts of resin. Then the mixture was heated and degassed to 80 °C for 30 min before being cooled to 30–40 °C. Piperidine was added, and the mixture was stirred for 5 min before it was poured into a preheated glass mold and cured.

For cast resin without rubber, the premixing stage with the rubber was omitted. All the formulations and curing schedules used are given in Table II. Also included are the notations used to describe the different formulations.

Characterization Techniques. Optical microscopes (OM, Zeiss or Microstar) equipped with polarizers were used to examine phase-separated particles. Specimens were prepared by placing a speck of uncured sample on precleaned glass slides or cover slips and curing in situ.

Submicrometer specimens (<0.5 μm) were stained with osmium tetroxide (OsO$_4$) (11) before being analyzed with 100–C transmission electron microscope (TEM), with a spectrometer (Kevax 5100) for energy-dispersion X-ray analysis (EDAX). The analysis was carried out across particles by using the osmium peak

Table II. Formulations of the Unmodified and Rubber-Modified Epoxies

Type of Rubber	Rubber Content, %	Notation
MY750 and Piperidine[a]		
–	0	P750
1300 × 13	10	P750–13 (10)
1300 × 13	20	P750–13 (20)
1300 × 13	30	P750–13 (30)
1300 × 8	10	P750–8 (10)
1300 × 8	20	P750–8 (20)
1300 × 8	30	P750–8 (30)
MY750 and DDS[b]		
–	0	D750
1300 × 13	10	D750–13 (10)
1300 × 13	20	D750–13 (20)
MY720 and DDS[c]		
–	0	D720
1300 × 13	10	D720–13 (10)
1300 × 13	20	D720–13 (20)
1300 × 13	30	D720–13 (30)

[a]Curing schedule: 120 °C, 16 h.
[b]Curing schedule: 130 °C, 1 h + 200 °C, 2 h.
[c]Curing schedule: 135 °C, 1.5 h; 177 °C, 2 h + 205 °C, 2.5 h.

height (osmium concentration) as a measure of the rubber concentration (12). A similar technique was used to determine the presence of sulfur on the rubber particles in systems cured with DDS.

Distributions of the rubber-particle sizes were determined from TEM micrographs using a Quantimat 520. The volume fractions of the phase-separated rubber particles measured (V_p) were compared with the volume fractions calculated (V_R) from density and weight fraction considerations (13).

Transition phenomena were investigated with a dynamic mechanical thermal analyzer (DMTA, Polymer Laboratory), temperatures from −100 to 200–300 °C, and a heating rate of 5 °C/min. The samples (2 mm thick, 4 mm wide, and 25 mm long) were clamped in dual cantilever mode and scanned at a frequency of 3 Hz, which gave the best resolution of both low- and high-temperature transitions and a strain corresponding to a maximum displacement of 62.6 μm. The temperatures at which damping (tan δ) peaks were observed are quoted as the transition temperatures.

Differential scanning calorimetry (DSC, Perkin-Elmer) was also used to identify transitions. Samples were scanned at a rate of 20 °C/min for the higher temperature scanning thermograms (above 20 °C) and at 40 °C/min for lower temperature ones (below 20 °C).

Results and Discussion

The results of rubber-particle fractions and transitions are summarized in Tables III and IV.

Relative Cross-Link Density. The molecular weight between cross-links, $\overline{M}_c$, can be estimated by using the empirical equation based on the shift of T_g that was proposed by Nielsen (14).

$$\overline{M}_c = \frac{3.9 \times 10^4}{T_g - T_{g_0}} \tag{1}$$

where T_g is the glass-transition temperature of the cured resin and T_{g_0} is the glass-transition temperature of the uncured resin. The estimated values for the respective unmodified epoxy resins are given in Table IV. These values have been found to be comparable to experimental $\overline{M}_c$ values (15),

Table III. Estimated Volume Fraction (V_p) of the Rubber Particles for Rubber-Modified Epoxies

Material	Rubber Content	W_R	V_R	V_P
P750–8	10	8.7	11.2	12
	20	16	20.8	20
D750–13	10	7.1	11.5	11
	20	13.2	20.0	17
D720–13	10	7.1	11.9	18
	20	13.3	21.3	25
	30	18.7	28.95	30

NOTE: All values are given in percents.

Table IV. Transition Temperatures of Different Resin Systems

	DSC		DMTA			$\overline{M}_c$	
Composition	T_g Rubber	T_g Epoxy	T_β Epoxy	T_{gR} Rubber	T_{gE} Epoxy	Expt.	Estimated
P750	–	–	−58	–	100	2416	1696
P750–8 (10)	−68	84	−58	−48	100		
P750–8 (20)	−66	83	−58	−43	98		
P750–8 (30)	−63	84	−58	−36	98		
P750–13 (10)	–	–	−58	−12	93		
P750–13 (20)	–	–	−58	−11	93		
P750–13 (30)	−35	–	−58	−9	−90		
D750	–	190	−33	–	201	463	317
D750–13 (10)	−59	186	−33	−25	192		
D750–13 (20)	−44	179	−33	−23	179		
D750–13 (30)	–	180	−33	−15	190		
D720	–	–	−40	–	253	–	246
D720–13 (10)	−66	–	−40	−29	253		
D720–13 (20)	−63	–	−40	−23	253		
D720–13 (30)	−50	–	−40	−20	253		

NOTE: All T_g values are in degrees Celsius. T_g of CTBN 1300×8 determined from DSC is −60 °C. T_g of CTBN 1300×13 determined from DSC is −40 °C. $\overline{M}_c$ is molecular weight between two cross-link points.

determined by thermomechanical techniques similar to that used by Gillan (16) and Lau et al. (17), as shown in Table IV.

Aspects of Morphology. For the P750–8 system (with 10 and 20% rubber content) an initially clear solution became cloudy after 60 min at 120 °C, at which point rubber particles formed, as shown in Figure 1. The average diameter of these particles, determined from OsO_4-stained samples (Figure 2), ranged from 0.6 to 4 μm, with the mean diameter increasing with increasing rubber content (Figure 3). This phase separation of the particles occurred via the building up of molecular weight, as described previously.

At the highest rubber content, 30%, the rubber was the continuous phase. Figure 4 shows a TEM micrograph of this system. The phase-inversion process was also observed under optical microscopy, as may be seen in Figure 5.

In contrast, no rubber particles were observed under OM or TEM for the P750–13 systems. The higher acrylonitrile content of the CTBN 1300×13 rubber (see Table I) increases the compatibility of the rubber with the MY750 epoxy, and consequently the phase separation of the rubber particles is suppressed (18, 19).

When the curing agent was changed from piperidine to DDS–BF₃MEA (D750–13 systems), rubber particles again formed. However, the phase separations in these cases were slightly different from those of P750–8 systems.

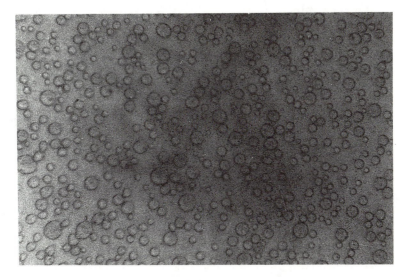

Figure 1. Optical micrograph of P750–8 rubber-modified epoxy resin (×400).

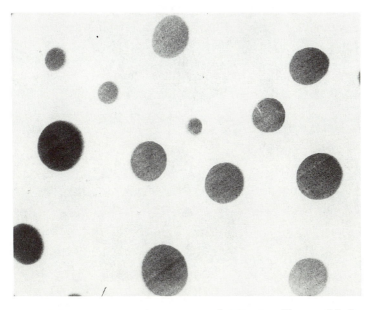

Figure 2. TEM micrograph of OsO_4-stained P750–8 rubber-modified epoxy resin (×4000).

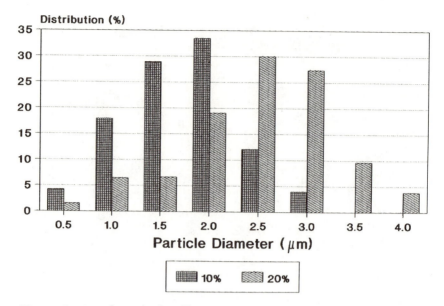

Figure 3. Distribution of rubber-particle diameters for rubber-modified P750–8 epoxies with 10 and 20% rubber content.

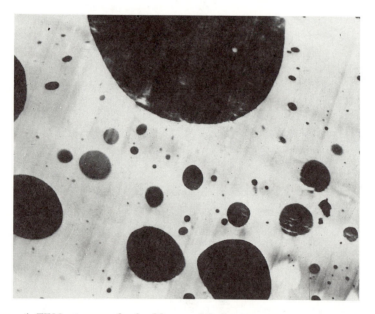

Figure 4. TEM micrograph of rubber-modified D750–13 epoxy resin with 30% rubber content, revealing bimodal distribution of rubber-particle sizes (×3600).

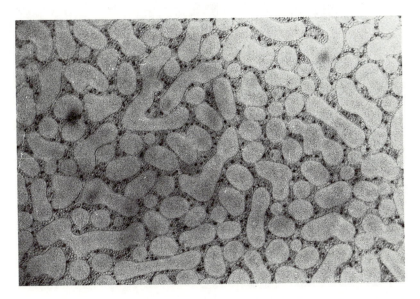

Figure 5. Phase inversion of 30% rubber-modified P750–8 observed under optical microscope (×400).

For the DDS-cured systems, both the epoxy–rubber prereaction and the epoxy–DDS mixture by themselves gave clear solutions at 130–140 °C. However, their combinations always produced cloudy solutions, even at high temperatures. Under OM, the systems showed a turbid or "snowy" appearance, unlike the clear solutions seen with the P750–8 systems. An example of the turbidity seen the in DDS systems is shown in Figure 6.

Spherical particles formed (Figure 7) as the cure of the D750 or D720 systems proceeded at 130 °C. Upon completion of the cure (Table II gives the schedules), these particles were observed to contain dark central spots (Figure 8). These spots appeared in samples at temperatures below their cure temperature, but disappeared if the cured samples were held at temperatures above their final cure temperature. Also, when viewed under polarized light, birefringence was observed around the rubber particles (light and dark areas around the particles in Figure 8).

The dark spots have previously been identified as cavities (20). These cavities and the birefringence are both results of stress concentration in and around the rubber particles. The explanation for both phenomena derives from the stresses developed within a cured resin as it cools from the curing temperature. The rubber particles are well bonded to the matrix. Therefore, when the rubber-modified resin is cooled to room temperature after completion of cure, the difference between the coefficients of thermal expansion of the rubber particles and the epoxy matrix imposes significant triaxial

Figure 6. Optical micrograph of rubber-modified D750–13 epoxy at zero curing time (×400).

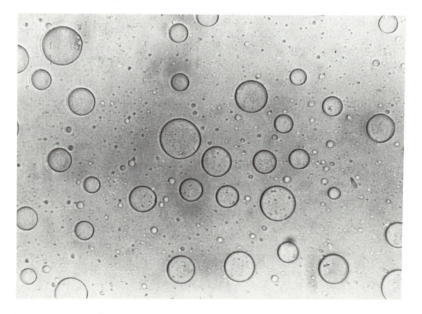

Figure 7. Optical micrograph of rubber-modified D750–13 epoxy at intermediate stage (130 °C/h) of cure (×400).

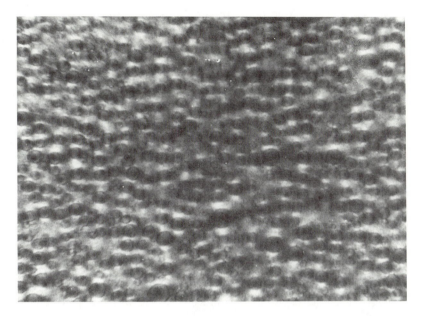

Figure 8. Fully cured rubber-modified D750–13 epoxy resin observed under polarized light microscopy (×200).

tensile stresses upon the rubber particles. The magnitude of these stresses can be estimated from Goodier's equation (*21*):

$$\sigma_{rr} = \sigma_{\sigma\sigma} = \sigma_{\psi\psi} = \frac{4 \, (\beta_p - \beta_m)(1 + \nu_p) \, G_m G_p \Delta T}{6 \, (1 - 2\nu_p) \, G_m + 3 \, (1 + \nu_p) \, G_p} \tag{2}$$

where σ_{rr}, $\sigma_{\sigma\sigma}$, and $\sigma_{\psi\psi}$ are stress tensors; β_p and β_m are the coefficients of thermal expansion of the rubber and the matrix, respectively; G_p and G_m are the shear moduli of rubber and matrix, respectively; ν_p is the Poisson ratio of the rubber; and ΔT is the temperature drop causing the thermal stresses.

The triaxial stresses within the rubber particles were estimated (*15*) from equation 2 and are as follows: for P750–8, 0.18 GPa; for D750–13, 0.26 GPa; and for D720–13, 0.35 GPa. The stresses increase with increasing cross-link density (smaller $\overline{M}_c$ value) of the matrix. Consequently, the rubber particles cavitated (*20, 22*), a result that explains the appearance of the dark spots. Similarly, the larger differences in the shear modulus between the rubber and the higher cross-link epoxy matrices resulted in a stress concentration induced by the rubber particles on the matrix; hence the observation of the birefringence stress pattern around the particles.

The change of curing agent from piperidine (P750–13) to DDS (D750–13) caused a clearly discernible phase separation in the latter system. Although no phase separation was observed for the P750–13 systems, both rubber and resin glass transitions were detectable (see DMTA results)

Two factors that can cause phase separation are precipitation induced by the curing agent and buildup of the molecular weight of the rubber–resin molecules. Phase separation induced by the curing agent has been reported by Sayre et al. (23) for a DGEBA epoxy modified with CTBN 1300 × 8 rubber cured with diethanolamine. In the D750–13 systems such a precipitation of rubber would explain the immediate turbidity previously referred to, for at this early stage of curing, little or no reaction has occurred between the epoxy end-capped rubber and the bulk of the resin. As the cure proceeds, the rubber precipitates and aggregates to form larger spherical particles. At the same time more particles of different sizes may be formed through either precipitation or buildup of molecular weight.

The size range of the particles formed may be generated by the balance of two complementary separation processes, aggregation of particles and the chain extension processes described by Siebert and Riew (8).

For the 10 and 20% rubber-modified D750–13 systems, the size of the rubber particles ranged from less than 0.6 μm to about 5 μm; the mean diameter increased with the increased rubber content (Figure 9). For the 30% rubber system, however, a bimodal distribution was observed with particle sizes ranging from less than 1 μm in diameter to 10 μm and larger. Figure 10 shows this distribution.

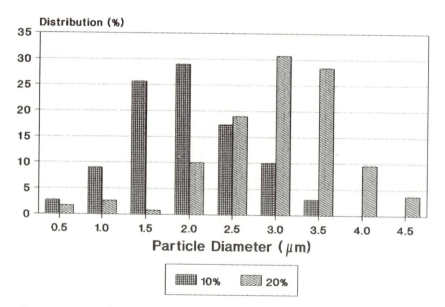

Figure 9. Distribution of rubber-particle diameters for rubber-modified D750–13 epoxies with 10 and 20% rubber content.

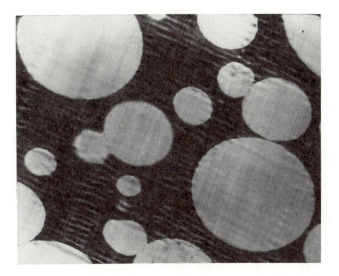

Figure 10. TEM micrograph of OsO_4-stained P750–8 rubber-modified epoxy resin with 30% rubber content showing phase inversion (×2600).

The curing schedules of D720–13 systems were slightly different from those used for D750–13. The D720–13 systems phase-separated in a similar manner, but showed a much greater particle-size distribution, ranging (for the systems with 10 and 20% rubber content) from particles of close to 0.5 µm to some above 20 µm in diameter (Figure 11). Particles above 30 µm were seen in the system of 30% rubber content, and many of the larger rubber particles contained epoxy occlusions (Figure 12).

The smaller rubber particles of D750–13 and D720–13 systems, which had been stirred with OsO_4, were analyzed for their sulfur content by using EDAX. Sulfur was expected to originate from the curing agent, DDS. The absence of a sulfur peak in these analyses indicated that the epoxy-end-capped rubber did not react with the curing agent.

The smaller rubber particles of all the systems were also microanalyzed by X-ray across their diameter for their OsO_4 concentration (Figure 13). An even distribution of OsO_4 was obtained from these measurements, which suggested that there was no significant variation in the concentration of rubber across the diameters of such particles.

For the 30% rubber content D720–13 system, the occlusion of epoxy resin, which was mentioned previously, increased the size of the rubber particles. Consequently, the apparent volume fraction (V_p) of the rubber phase was also increased.

The amount of epoxy incorporated during the formation of the rubber particles (according to the mechanism suggested by Siebert and Riew (8))

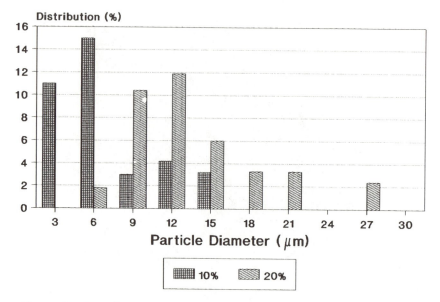

Figure 11. *Distribution of rubber-particle diameters for rubber-modified D720–13 epoxies with 10 and 20% rubber content.*

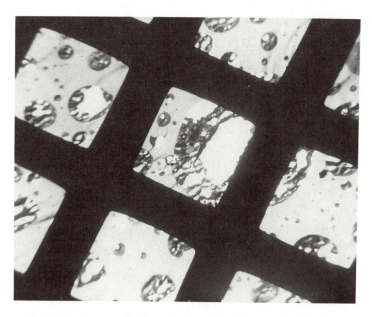

Figure 12. *TEM micrograph of OsO₄-stained rubber-modified D720–13 epoxy resin showing epoxy occlusions (×300).*

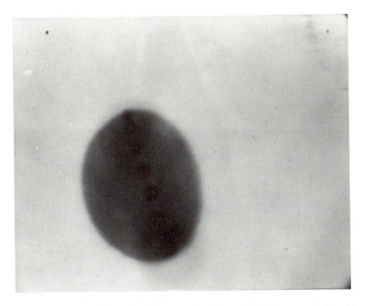

Figure 13. TEM micrograph of an OsO$_4$-stained rubber particle, showing sampling location for energy-dispersive X-ray analysis (×18,000).

will, therefore, have a significant influence on the V_p value of a system. This influence is shown in Figure 14, which is a plot of V_p versus V_R for the three systems. The line drawn at 45° is for V_p values equal to V_R. The V_p values of the D720–13 systems are greater than the corresponding values of V_R because of the large amounts of occluded resin (*see* Figure 12). The slight leveling off of the V_p value at higher rubber content reflects the need to take micrographs at a lower magnification to accommodate the larger rubber particles. This setting rendered the small particles undetectable. This problem has been referred to by Sayre et al. (*23*).

The V_p and V_R values of P750–8 systems followed the 45° line closely. Lower V_p values were observed for D750–13 systems, a finding that suggests that some rubber had dissolved in the epoxy matrix in this case.

Transitions in the Rubber–Resin Systems.   Despite the absence of rubber particles detectable by OM, SEM, or TEM from P750–13 systems, a tan δ peak was observed at around –10 °C (Figure 15) for each rubber content. These peaks are attributable to the glass transition of the rubber, T_{g_R}, in the P750–13 system. Also at lower temperature, the β peak decreased as the tan δ peak increased with increasing rubber content. The glass-transition temperature of the epoxy, T_{g_E}, decreased in like manner (Table IV and Figure 16).

The high T_{g_R} and the reduction of the T_{g_E} are indications of high com-

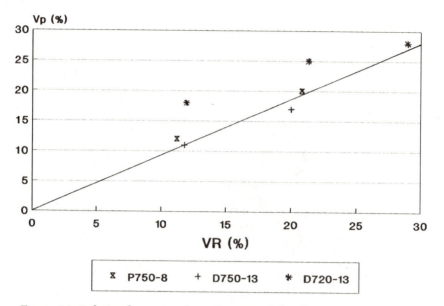

Figure 14. Relation between volume fraction of the phase-separated rubber particles (V_p) *and the calculated volume fraction* (V_R) *for the rubber-modified P750–8, D750–13, and D720–13 epoxies.*

patibility between MY750 and the 1300 × 13 rubber, as noted by other workers (*18, 19*). This compatibility caused the rubber to remain in the epoxy matrix and plasticized the system; hence the reduction of T_{g_E}.

The T_{g_R} and T_{g_E} values for the P750–8, D750–13, and D720–13 systems are collected in Table IV. Figures 16, 17, and 18 show typical thermograms in which transition peaks are clearly delineated. The T_{g_E} values of the P750–8 and D720–13 systems changed very little with the incorporation of rubbers, but a reduction in T_{g_E} was observed for the D750–13 system. This reduction increased from 10 to 20% with the addition of rubber (192 and 179 °C), but was less marked at 30% rubber (190 °C).

The reduction in T_{g_E} for the D750–13 system may be attributed to the good compatibility of the 1300 × 13 rubber with the MY750 epoxy resin as found for the P750–13 system. On completion of the cure, some rubber remained dissolved in the matrix and hence plasticized the cured resin. The reduction in the T_{g_E} for the 10 and 20% rubber contents correlated well with the smaller V_p/V_R values observed for these systems, as was discussed earlier.

The increase in T_{g_E} at 30% rubber content may be explained by phase separation of the dissolved rubber. This suggestion is confirmed by the appearance of smaller particles (<1 μm) in the dispersed phase, as shown in Figure 10.

Interestingly, the T_{g_E} tan δ peak heights (Figures 16 and 17) of the

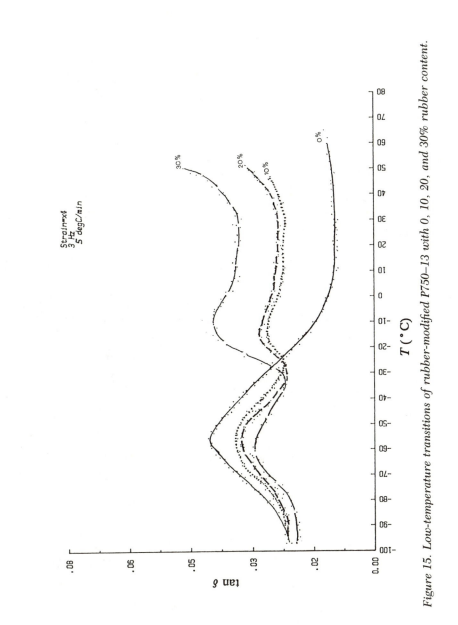

Figure 15. Low-temperature transitions of rubber-modified P750–13 with 0, 10, 20, and 30% rubber content.

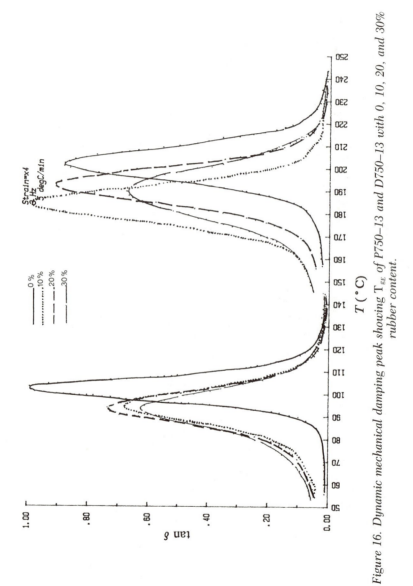

Figure 16. Dynamic mechanical damping peak showing T_{gE} *of P750–13 and D750–13 with 0, 10, 20, and 30% rubber content.*

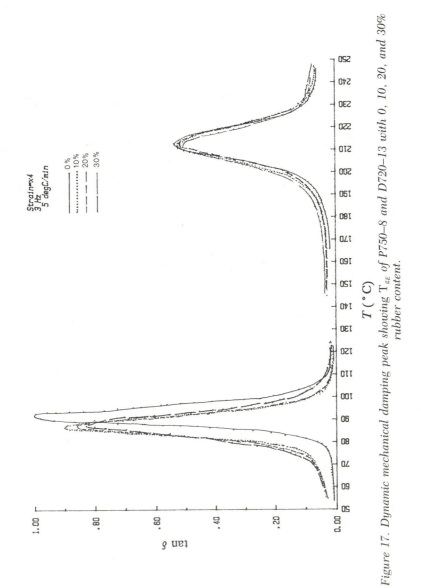

Figure 17. Dynamic mechanical damping peak showing T_{gE} of P750–8 and D720–13 with 0, 10, 20, and 30% rubber content.

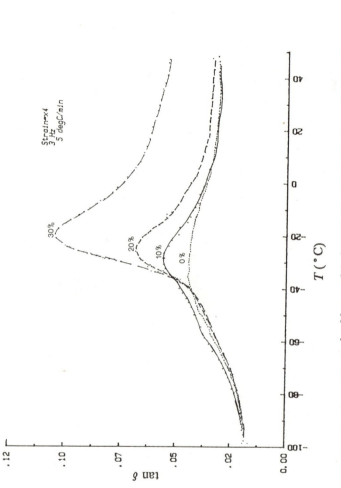

Figure 18. Low-temperature transitions of rubber-modified D720–13 with 0, 10, 20, and 30% rubber content.

P750–8, P750–13, and D720–13 systems decreased with increasing rubber modification, although an increase was observed for the D750–13 system (except at 30% rubber content). However, these reductions in tan δ peak heights were usually accompanied by broadening of the bases of the peaks. The T_{g_E} tan δ peak area reflects the volume fraction of the epoxy phase. Rubber incorporated in the epoxy phase will contribute to the peak areas observed by expanding this phase and so increasing the peak area. The extent of broadening may also indicate a slight extension in distribution of cross-link density because of incorporation of the rubbers.

Figure 19 gives the relative tan δ peak area estimated for the various systems. Greater increases in T_{g_E} peak area were observed for the P750–13 and D750–13 systems, compared with those for P750–8 and D720–13. This result is a strong indication of lower solubility of rubber in the resin matrix in the latter two systems, especially in the D720.

This reasoning is consistent with the observation that for the D750–13 system at 30% rubber content, a bimodal phase separation occurred. This separation resulted in small peak areas compared to 10 or 20% rubber content of the same system.

The position of the T_{g_R} is known to be associated with compatibility between the epoxy and the rubber. Figure 20 is a plot of T_{g_R} versus rubber content. The significantly higher T_{g_R} of 1300×13 rubber-modified system, compared with 1300×8 system, has been attributed by Romanchick et al. (*24*) to the good compatibility of the resin and the rubber. However, a

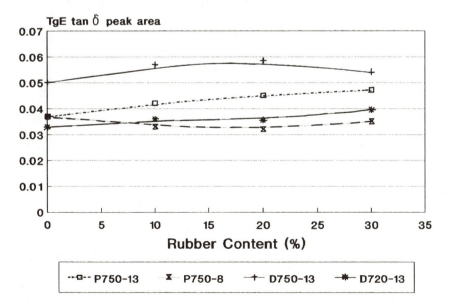

Figure 19. Relation between T_{g_E} *tan δ peak area of rubber-modified epoxies and rubber content for P750–8, P750–13, D750–13, and D720–13 systems.*

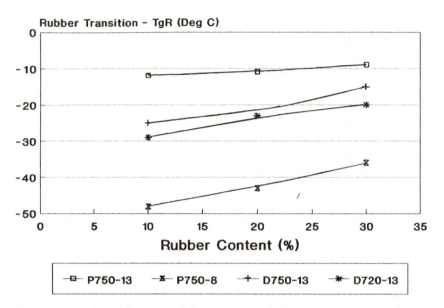

Figure 20. Relation between rubber content and the tan δ peak of T_{g_R} *for the rubber-modified P750–8, P750–13, D750–13, and D720–13 epoxies.*

comparison of the collected T_{g_R} values for the various systems as determined from DSC (Table IV) with the respective pure T_{g_R}s shows that the T_{g_R}s of the rubber in the epoxies are higher in all cases.

Because rubber particles are cross-linked by the epoxy (8, 25), one would expect the T_{g_R} to increase to values above that observed for the respective T_{g_R} of pure rubber, yet this increase was not observed. This lowering of T_{g_R} was explained by Manzionne and Gillham (26) in terms of stresses induced by thermal shrinkage. This triaxial thermal stress, as described earlier, will increase the free volume of the rubber particles and thus depress the T_{g_R} below that of the neat rubber. If this hypothesis is correct, it would explain the greater depression of T_{g_R} of the D720–13 or D750–13 rubber-modified epoxies, compared to the P750–8 system. In addition, it would suggest that the cavitation of the rubber particles in the former two systems did not fully relieve the triaxiality of the rubber particles.

The T_{g_R} tan δ peak height has been related to the V_p values of the rubber particles (27, 28). For example, Bucknall and Yoshii (28) observed a linear relationship between V_p and the T_{g_R} tan δ peak height for various rubber-modified epoxy systems after subtracting the contribution by the β transition peak of the epoxy. No simple linear relationship was obtained from the results of the present work, as may be seen in Figure 21.

The β relaxation of epoxies occurs at temperatures at which the molecular segments in the polymer network are frozen. It is attributed to vibration of segment chains by the so-called crank shaft mechanism (29). For amine-

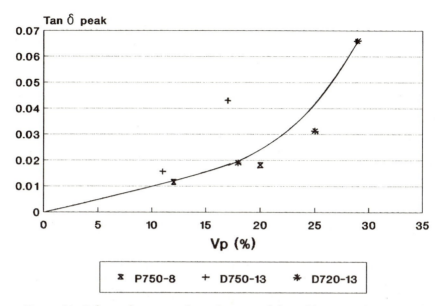

Figure 21. Relation between volume fraction of the rubber particles (V_p) and the tan δ peak of T_{gR} for the rubber-modified 750–8, D750–13, and D720–13 epoxies.

cured epoxies, β relaxation is usually associated with the molecular motion of segments resulting from the amine–epoxy and hydroxy–epoxy reactions. This transition has been reported to be influenced by the cross-link density of the epoxy network (30–35). No definite relationship of the β transition to the cross-link density of the epoxy could be derived from the present work. Probably, differences in either the curing agent or the epoxy used produce different network structures. In turn, the structures will have a large influence on the degree of mobility of the hydroxy ether group.

Influence of Cross-Link Density on Phase Separation of Rubber.   As noted in Table IV, the cross-link density of the piperidine-cured MY750 system was much lower than that in DDS-cured systems. The highest cross-link density was measured in the MY720–DDS system. The incorporation of various levels of rubber was not expected to influence the cross-link density significantly. In general terms, it was observed that increasing the cross-link density (decreasing $\overline{M}_c$ values) favored the phase separation of the rubber. No rubber particles were observed for the P750–13 system, whereas large and small particles were observed for the D750–13 system. Furthermore, increasing the cross-link density also affected the particle-size distribution, a result that is an indication of phase-separation process.

Finally, the T_{g_E} of the highest cross-linked epoxy (D720–13 system) remains constant regardless of the rubber content. However, as each system involved a different resin or curing agent, a simple interpretation in terms of $\overline{M}_c$ alone, excluding consideration of the changing solubility parameters of the systems, must be viewed with caution.

Summary and Conclusions

For this study the phase-separation processes were followed by OM; the phase-separated systems were examined by OM, SEM, TEM, DSC, and DMTA. This combination of techniques gave useful insights into the nature of the phase separation.

The microscopic methods allowed an identification of the continuous and dispersed phases, and the particle size and size distribution of the dispersed phase. Such examinations provided values for V_p; for the four rubber-modified systems studied, it was generally noted that $V_p < V_R$. This difference was attributed to rubber dissolved in the epoxy matrix.

The thermoanalytical techniques, especially DMTA, provided information that confirmed this attribution. As the dissolved rubber content of the matrix increased, it was generally observed that its T_{g_E} peak (DMTA tan δ thermogram) broadened and its area increased.

The complete absence of dissolved rubber from the TGDDM systems (MY720) was demonstrated by the consistency of the T_{g_E} peak, regardless of rubber content.

The sensitivity of thermomechanical means of detecting phase separation was further demonstrated by the finding that glass transitions for both rubber and resin were observable in the P750–13 formulations for which no separation was detectable microscopically.

Phase separation in the rubber-toughened epoxy systems studied was influenced by the nitrile content of the CTBN, the amount of rubber added to the resin, the hardener, and the cross-link density of the cured resin. For the systems examined, there seems to be no simple relationship between any one of these factors and phase separation. Rather, a subtle combination of these factors is operating.

Acknowledgments

The material is based upon work supported by the Ministry of Defense. We are grateful to the BF Goodrich Chemical Company for generously donating the rubber and also to Ciba-Geigy Plastics, Duxford, for the epoxy resins.

References

1. Sultan, J. N.; McGarry, F. J. MIT Research Report No. R69–59, 1969.
2. Pocius, A. U. *Rubber Chem. Technol.* **1985**, *58*, 622.
3. Siebert, A. R. *Org. Coat. Plast. Chem.* **1983**, *49*, 427.

4. Ting, R. Y. In *The Role of Polymeric Processing and Structural Property of Composite Material;* Seferis, J. C.; Nicolais, L., Eds.; Plenum: New York, 1983; p 171.
5. Sultan, J. N.; McGarry, F. J. *Polym. Eng. Sci.* **1973**, *13*, 29.
6. Rowe, E. H.; Riew, C. K. *Plast. Eng.* **1975**, **45**.
7. Rowe, E. A.; Siebert, A. R.; Drake, R. S. *Mod. Plast.* **1970**, *47*, 110.
8. Siebert, A. R.; Riew, C. K. *Org. Coat. Plast.* **1971**, *31*, 552.
9. Riew, C. K.; Rowe, E. H.; Siebert, A. R. In *Toughness and Brittleness of Plastics;* Deanin, R. D.; Crugnola, A. M., Eds.; Advances in Chemistry 154; American Chemical Society: Washington, DC, 1976; p 326.
10. B. F. Goodrich Co. *Toughen Epoxy Resin with Hycar RLP;* Product Data RLP–2.
11. Riew, C. K.; Smith, K. W. *J. Polym. Sci. Part A–1* **1971**, *9*, 2737.
12. Kato, K. *Polym. Eng. Sci.* **1967**, *7*, 35.
13. Kunz-Douglas, S.; Beaumont, P. W. R.; Ashby, M. F. *J. Mater. Sci.* **1980**, *15*, 1109.
14. Nielsen, L. E. *J. Macromol. Sci. Rev. Macromol. Chem.* **1969**, *C3*, 77.
15. Lee, W. H. Ph.D. Thesis, Brunel University, 1987.
16. Gillan, K. T. *J. Appl. Polym. Sci.* **1978**, *22*, 1291.
17. Lau, C. H.; Hodd, K. A.; Wright, W. W. *Br. Polym. J.* **1986**, *18*, 316.
18. Sohn, J. E. *Org. Coat. Plast. Chem.* **1982**, *46*, 977.
19. Sohn, J. E.; Emerson, J. A.; Chen, J. K.; Siegel, A. F.; Koberstein, J. T. *Org. Coat. Plast. Chem.* **1983**, *48*, 447.
20. Lee, W. H.; Hodd, K. A.; Wright, W. W. *ASE 85 Conference Proceedings;* London, 1985, Day 1, p 147.
21. Goodier, J. N. *J Appl. Mech.* **1933**, *55*, 39.
22. Gent, A. N.; Lindley, P. B. *Proc. R. Soc. London A* **1958**, *249*, 195.
23. Sayre, J. A.; Assink, R. A.; Lagasse, R. R. *Polymer* **1981**, *22*, 87.
24. Romanchick, W. A.; Sohn, J. E.; Geibel, J. F. In *Epoxy Resin Chemistry II;* Bauer, R. S., Ed.; ACS Symposium Series 221; American Chemical Society: Washington, DC, 1985; p 85.
25. Brown, H. P. *Rubber Chem. Technol.* **1963**, *36*, 931.
26. Manzionne, L. T.; Gillham, J. K. *J. Appl. Polym. Sci.* **1981**, *26*, 889.
27. Keskkula, H.; Turley, S. G.; Boyer, R. F. *J. Appl. Polym. Sci.* **1971**, *15*, 351.
28. Bucknall, C. B.; Yoshii, T. *Br. Polym. J.* **1978**, *10*, 53.
29. Shatzki, T. F. *J. Polym. Sci.* **1962**, *57*, 496.
30. Kline, D. E. *J. Polym. Sci.* **1960**, *47*, 237.
31. Pogany, P. A. *Polymer* **1970**, *11*, 66.
32. Kreahling, R. P.; Kline D. E. *J. Appl. Polym. Sci.* **1969**, *13*, 2411.
33. Hata, N.; Kumanotoui, J. *J. Appl. Polym. Sci.* **1971**, *15*, 2371.
34. Takahama, T.; Geil, P. H. *J. Polym. Sci., Polym. Lett. Ed.* **1982**, *20*, 453.
35. Yee, A. F.; Pearson, R. A. NASA Contract Report No. 3852, 1984.

RECEIVED for review February 11, 1988. ACCEPTED revised manuscript August 31, 1988.

Optimum Rubber Particle Size in High-Impact Polystyrene

Further Considerations

Henno Keskkula

Department of Chemical Engineering and Center for Polymer Research, University of Texas, Austin, TX 78712

The role of submicrometer rubber particles in polystyrene toughening is reviewed, and it is shown that cross-linked latex rubber particles may be quite effective in toughening polystyrene. Their effectiveness may be further enhanced by the presence of some particles in excess of 1 μm in diameter. The role of submicrometer rubber particles in toughening polystyrene and styrene–acrylonitrile copolymers is compared. Their reduced effectiveness in toughening polystyrene is related to a high propensity for molecular orientation in polystyrene and a concomitant weakness of moldings perpendicular to the flow orientation. Rubber-modified styrene–acrylonitrile copolymers show a lesser tendency to anisotropy and a higher strength perpendicular to the flow direction. This strength leads to a high practical toughness of fabricated parts. Discussion of possible mechanisms of toughening of brittle styrene polymers with a large concentration of small particles leads to a suggestion that they are crucial in the control of growth and termination of crazes in rubber-modified polystyrene.

T OUGHENING OF POLYSTYRENE REQUIRES RUBBER PARTICLES ≥ 1–2 μm in diameter, distributed throughout the polystyrene (PS) matrix (1–7). Although most authors suggest a 1–6-μm particle size range as the most desirable, some early reports (5) indicate that even larger particle sizes up to 10 μm are required. Moore (5) reported that rubber particles in high-impact polystyrene (HIPS) should be larger than 2 μm for effective toughening. On

0065–2393/89/0222–0289$06.00/0

the basis of microscopic examination of stressed thin films of HIPS, Donald and Kramer (7) concluded that particles <1 μm were ineffective in toughening PS. Bucknall (4) reported that 0.8 μm was the minimum critical particle size for toughening PS. Most commercial HIPS, indeed, contain complex rubber particles in the range of one to several micrometers in diameter. Uzelmeier (8) reports that commercial HIPS are characterized by number-average particle sizes of 1–2 μm and weight-average particle sizes of 3–6 μm.

More recently, however, a number of investigators have shown that toughening of PS is effective with a dual rubber-particle-size population if most particles are below 1 μm (9, 10). Such HIPS are particularly desirable when improved gloss and reduced opaqueness are required. Also, they are desirable in blends with polyphenylene oxide. As rubber particles below 1 μm have been reported to be ineffective in craze initiation or termination, the mechanism by which such polymers are toughened is of great interest.

The role of very small rubber particles in toughening styrene–acrylonitrile (SAN) copolymers and other more ductile polymers is associated with their ability to initiate shear bands. Shear band formation has not been reported in HIPS, however. Hobbs (10) has suggested that in HIPS, with a dual particle-size population, all particles initiate crazes but only large particles are effective in terminating crazes. Although the initiation of the majority of crazes by large particles is expected (11), the role of small particles remains to be elucidated in HIPS and rubber-modified SANs. It has been reported (12, 13) that styrene–butadiene–styrene (SBS) block copolymer particles and small (0.1–0.2 μm) grafted rubber latex particles in combination are effective in toughening SAN and that HIPS can be further toughened by an introduction of 0.1–0.2-μm methyl methacrylate grafted rubber particles (13).

However, direct comparisons of HIPS from a mass process and a mechanical blend of cross-linked latex rubber and PS should not be considered. The size of the complex rubber particles in HIPS cannot be independently controlled without affecting other variables such as rubber-phase volume fraction, internal particle morphology, molecular weight, and adhesion to the matrix. The desired particle size in commercial HIPS is achieved by controlling polymerization recipe variables such as initiator, chain-transfer agent, and solvent concentrations, as well as agitation rate. Size and other characteristics of the solid latex rubber particles are determined during the polymerization of rubber. Accordingly, the characteristic features of the rubber and the matrix phases are controlled separately for a mechanically blended product.

It is important not to compare the two products directly, as the rubber phases have significantly different mechanical properties. For instance, shear modulus of the rubber phase from a mass product is about 40 times as high as that of a mildly cross-linked polybutadiene (14). In addition to crazing, cavitation may occur within the solid latex particles, as is often observed in acrylonitrile–butadiene–styrene (ABS) polymers. Future studies on the mechanisms of toughening of blends that contain large populations of small particles must answer whether they initiate shear yielding or contribute to the control of craze growth and termination by some interparticle association mechanism.

This chapter covers some of the early background of the commercial development of HIPS and reports on the reasons for the success of the mass polymerization approach in comparison to mechanically blended rubber–PS products that were under development in the early 1950s, but failed to become commercial products. Furthermore, this chapter speculates on the mechanisms of toughening blends that consist primarily of submicrometer rubber particles.

History

Since the early 1950s, HIPS has been produced by a mass or solution process (*1, 15*). In this process, 3–10% of a butadiene-based rubber is dissolved in styrene, and the solution is polymerized. A critical feature of this process requires shearing agitation during the phase inversion, in which the morphology of the dispersed rubber phase is established (*1, 16*). If the shearing agitation and a number of other recipe variables are controlled, a variety of HIPS products is possible. Indeed, over the years, the basic mass processes have been fine-tuned to give a large variety of commercially attractive products that are used for injection molding, vacuum formed and stamped sheet, blow molding, and similar applications. During the early period of HIPS development, the variables that controlled surface texture and its relationship to toughness and other mechanical properties were not fully understood. Accordingly, problems existed in producing extruded sheets of HIPS for vacuum forming of refrigerator door inner liners that would be impact-resistant and have a smooth glossy surface. The goal of a smooth glossy surface was particularly elusive. The presence of undesirable macrogels or "fish eyes" was often conspicuous.

The period of refrigerator market development for HIPS generated a need for a product with an improved surface on the extruded sheet. Mechanical blends of latex rubber with PS were extensively studied at the Dow Chemical Co. for the replacement of mass-made HIPS in this application.

Discussion

Mechanical Blends of Rubber and Polystyrene. In an attempt to obtain tough blends of rubber and PS, it was discovered early that the uncross-linked styrene–butadiene rubber (government rubber–styrene,

GRS) was not satisfactory in producing toughened polystyrene. Later publications (17–20) report that up to 20% gum rubber mixed with PS produced blends with an insignificant increase in impact strength. Emulsion-polymerized styrene–butadiene copolymers, however, produce tough blends with many desirable features.

Styron 480, an early Dow product, was based on a spray-dried butadiene–styrene (45/55) copolymer latex. About 35% of the rubber was blended with a styrene–α-methylstyrene (75/25) copolymer (21). Injection moldings of the resulting product had excellent impact strength at room temperature, satisfactory tensile strength (3000–4000 psi), and high elongation at failure (>20%). However, because of the high glass-transition temperature (T_g) of the rubber, the blend became brittle at low temperatures. The rubber used in this product was developed as a latex paint, and it was substantially cross-linked during polymerization. Because of the state of manufacturing technology, an increase in the butadiene content was not readily possible then, and the product was short-lived.

A GRS–2003 rubber latex with a high content (70%) of butadiene was available from Firestone. It was a high-solids (~60%) latex made by a hot emulsion polymerization recipe, and it had a broad particle-size distribution with cross-linked rubber particles. A blend of PS with about 8–12% GRS–2003 rubber, along with appropriate antitoxidants and butyl stearate or mineral oil as lubricants, gave a rubber-toughened product with an excellent notched Izod impact strength and high surface gloss (15, 22). No morphological characterization of these products was possible, because only phase-contrast light microscopy was available at the time. Rubber latex particles, even agglomerated ones, were below the resolving capacity of light microscopy. The characteristics of the product were cross-linked rubber particles, maximum particle size well below 1 μm, and excellent surface characteristics (high gloss).

Notched Izod impact strength of blends based on the Styron 480 development, as well as those derived from the use of GRS–2003, are shown in Figure 1. The impact values, obtained from injection-molded specimens, show clearly the effectiveness of these rubbers in toughening PS. Blends of PS with uncross-linked butadiene–styrene copolymers, however, are not effective. Impact values of less than 0.5 ft-lbs/in. are obtained with blends containing up to 20% rubber (17–20). Although it appeared that any desired impact strength can be achieved by the use of an appropriate level of rubber and modifiers in the blend, it became clear that other mechanical properties were of crucial importance in the performance of the vacuum-formed (from extruded sheet) refrigerator door liners.

In this development, the critical property that proved fatal to the product was determined by a door-slam test. Door liners fabricated from the mechanically blended product failed in several orders of magnitude fewer cycles than the ones made from a mass-made product (22). After extensive testing, it was concluded that elongation at failure of compression-molded specimens

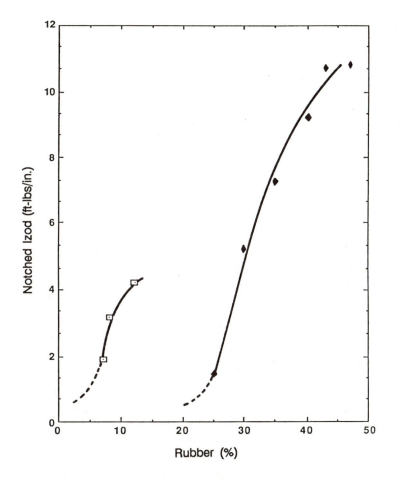

Figure 1. Notched Izod impact strength of mechanically blended cross-linked styrene–butadiene (SB) copolymer rubbers with styrene polymers. Key: ⊡, *70/30 SB blend with PS;* ◆, *45/55 SB blend with styrene–α-methylstyrene (75/25) copolymer (15, 21, 22).*

correlated with the door-slam test. The mechanically blended product had comparable impact strengths with those of the mass product and a high elongation (>20%) when tested on injection-molded specimens, but with compression-molded specimens it failed shortly after yielding, without significant stress-whitening. Because of this experience, compression-molding of test specimens became widely used in research and in quality control testing of HIPS.

HIPS from a Mass Process with a Bimodal Rubber Particle-Size Distribution.   An early indication of the potential advantages of a probable bimodal rubber particle-size distribution was reported in a 1965 Monsanto patent (23). It described a mass process where a high rate of agitation was used for only a short period of time during the critical period of phase inversion. A product with good impact strength and high gloss resulted. However, if this high rate of agitation was maintained during the remainder of the polymerization, impact strength was not satisfactory. In the absence of appropriate morphological data, it may be speculated that during the short period of intensive mixing only a small portion of the rubber particles remained large, with the majority of particles reduced to submicrometer levels. Intensive mixing during the remainder of the polymerization produced only small particles, which were ineffective for toughening PS.

This speculation may be reasonable in light of a subsequent Monsanto patent (9). Lavengood (9) showed that blends of two HIPS with significantly different particle sizes permit optimization of impact strength and gloss. Some of his data are summarized in Table I. It shows that the introduction of only 5% HIPS containing large particles (3 μm) into a HIPS with an average particle size of 0.6 μm produces a blend with excellent properties. The addition of large particles was sufficient to increase the impact strength to its full potential and to produce a major improvement in gloss. The estimated volume fraction of the rubber phase increases from 20 to only 21.3%.

More recently, Hobbs (10) also showed the desirability of blending HIPS resins with two different particle sizes of 0.92 and 3.9 μm. In his work, the blends containing 10–20% HIPS with large particles showed about a threefold increase in impact strength. In fact, the blends had a slightly higher impact strength than the HIPS with 3.9-μm particles. On the basis of these experimental observations, Hobbs proposed a mechanism of toughening of HIPS with a bimodal rubber particle-size distribution. The assumptions included the requirements that all particles act as craze initiators, that the probability of initiation is independent of particle size, and that only large particles are effective in terminating crazes. Possible toughening mechanisms in various rubber-modified PSs will be discussed in the next section.

Speculation on Toughening Mechanisms.   High-impact polystyrene stressed in tension deforms only by a crazing mechanism first proposed

Table I. Impact Strength and Gloss of HIPS with Dual-Size Rubber Particles

Rubber Particle Size, μm[a]	Large Particles in Blend, %[b]	Rubber-Phase Volume Fraction[c]	Notched Izod, ft-lbs/in.	Gloss
0.6	0	0.20	0.9	100
3.0	100	0.45	1.85	50
0.6 and 3.0	5	0.213	1.85	96.3
0.6 and 3.0	30	0.275	1.90	81.3

NOTE: Partially reprinted from ref. 9.
[a]Weight-average diameter.
[b]Weight percent of HIPS containing large particles in the blend.
[c]Estimate based on data in ref. 29.

by Bucknall and Smith (*24*). Numerous studies of the volume change during deformation have confirmed the exclusive nature of crazing in HIPS (*1, 25*). Catastrophic cracks do not develop because rubber particles, as stress concentrators, initiate crazes as well as participate in their termination. Apparently, the large concentrations of crazes that develop on impact or straining, which are associated primarily with large particles, are responsible for the ductility of commercial HIPS (*1, 26*).

Kramer (*7*) has studied the micromechanical deformation of ultrathin films of HIPS and concluded that only particles with diameters >1 μm are capable of initiating crazes. Because it is commonly believed that smaller particles do not participate in the termination process, their presence seems unnecessary. The conclusion that small particles do not participate in the termination process is apparently derived from the fact that there is no indication of shear yielding in commercial HIPS, and no other mechanisms have been proposed.

In light of the preceding discussion, it is surprising indeed that polystyrenes with large populations of submicrometer particles are tough. It seems most probable that the larger particles in HIPS with a bimodal particle-size distribution or in the mechanically blended latex rubber and PS are responsible for the initiation of crazes, as there are only limited data indicating the possibility of the initiation of crazes at submicrometer particles (*11, 26*). As the rubber particle size is reduced, the craze initiation stress is increased (*27*). This correlation suggests that it is unlikely that HIPS with a dual particle-size distribution will have a significant number of crazes initiating at other than large particles. Therefore, it is most important to consider the role of small particles in the toughening mechanism of polystyrene.

Any mechanism for the participation of small particles in the HIPS deformation process should consider their ability to control craze growth as well as termination. Mechanical dilatometry experiments should reveal whether or not small particles are capable of initiating shear bands in the polystyrene matrix, clarify the requirements of particle size and concentration, and verify if the change of crazing to shear yielding does take place. If shear banding does not occur, however, the role of small particles may be even more elusive. Interactive stress fields may blunt or divert craze growth (*12, 13*), but ideas for conclusive experiments have not been proposed.

Additional questions arise from an attempt to examine the ductility of toughened PS with submicrometer particles. Although HIPS with a bimodal rubber particle population has good adhesion between the grafted rubber particles and the PS matrix, the level of adhesion between a random styrene–butadiene (SB) copolymer is likely to be lower. Good adhesion between the matrix and the rubber is necessary for toughening. Good adhesion is considered particularly important in HIPS with a bimodal particle distribution (*28*).

In addition to the consideration of the critical particle size for toughening of PS, a required level of the rubber-phase volume fraction for toughening has been considered (11, 29, 30). Although it is reasonable that there exists a critical rubber-phase volume fraction in HIPS with a narrow particle-size distribution or a uniform particle type, such a criterion for a HIPS with a bimodal particle-size distribution requires a more careful analysis. The data in Table I show that an increase of rubber-phase volume from 20 to 21.3% may double the impact strength.

As more questions are raised in connection with the attempts to understand the reasons for the ductility of the cross-linked rubber–PS blends, the importance of interparticle distances in the control of craze growth and termination should be considered. Michler (11) suggested that both the interparticle distance and the mean free path between the particles are important in controlling fracture in HIPS. With a high concentration of small particles, the growth of cracks may be impeded by the small particles in the vicinity of the crack tip as they become effective for initiating crazes within the stress field of the crack tip (31). Such cooperative processes may lead to a formation of thin PS ligaments controlled by the small-particle population. Such a possibility could be investigated by the use of in situ scanning electron microscopy. The fracture of HIPS was recently demonstrated by Beaumont (32), who used such a technique. It showed a significant plastic flow near the crack tip and the ductile deformation of the PS ligaments.

Small Rubber Particles in HIPS vs. ABS. It has been demonstrated in this chapter that cross-linked latex rubber particles are effective in toughening PS. Excellent impact strength and high elongations at failure are achieved with injection-molded test specimens. Grafted rubber particles of similar size are not only effective in toughening SAN copolymer matrix, but they form a basis for a most successful family of commercial ABS polymers. Is the success of ABS based only on the good adhesion between grafted rubber particles and on the ability of the SAN matrix to shear yield, or are there other important differences between the similarly modified PS and ABS?

The differences in the response of HIPS and ABS injection moldings to a biaxial stress in dart-drop testing may offer some insight into the difference in ductility of the two polymer types. In 1959, Keskkula and Norton (33) reported on the anisotropy in HIPS moldings and on their impact strength. They concluded that the injection moldings are highly anisotropic when molded at low melt temperatures close to "short-shot" conditions. These moldings are brittle when tested by a dart-drop test. However, tough moldings are produced by molding at higher temperatures or result from relatively isotropic compression moldings or extruded sheet.

A similar study was carried out with a number of ABS polymers (34). Both emulsion and mass-made ABS were compared with HIPS. Some of these data are summarized in Table II. ABS moldings are less sensitive to

Table II. Tensile and Impact Properties of Injection-Molded Polymer Disks

Polymer	Molding Temp, °F	Tensile, lbs/in.²				Rupture Elong., %		Notched Izod impact, ft-lbs/in.		Dart Drop, ft-lbs
		Yield		Rupture						
		‖	⊥	‖	⊥	‖	⊥	‖	⊥	
HIPS	400	3600	3050	4250	2800	30.55	19.4	1.9	0.8	2.1
	450	3300	3050	3800	2900	33.9	27.3	1.6	1.0	4.2
	500	3150	3050	3500	2850	32.4	24.5	1.4	1.1	20.0
	550	3150	3200	3450	3050	27.4	24.4	1.2	1.0	15.0
Mass ABS	400	5450	4800	4650	4150	5.8	3.0	4.1	1.4	9.7
	450	4900	4550	4150	4100	5.3	2.0	4.3	2.8	16.0
	500	4800	4600	4050	4000	2.7	3.2	3.7	2.5	23.9
	550	4400	4350	3850	3650	3.3	2.9	2.5	2.0	25.2
Emulsion ABS	400	5600	5550	4750	4750	3.2	2.3	5.1	2.5	11.4
	450	5550	5400	4650	4750	3.6	2.1	4.5	3.0	26.9
	500	5500	5550	5000	5050	2.4	2.2	3.9	3.0	23.6
	550	5650	5600	4750	5000	2.8	2.1	3.7	2.9	23.4

NOTE: Disks were 6-in. diameter, 1/8 in. thick. Specimens were cut parallel (‖) and perpendicular (⊥) to the flow direction (adapted from ref. 34).

molding temperature variations when determined by a dart-drop test. The tensile strength ratio ($\parallel / \perp$) is closer to 1, as compared to similar data on HIPS. ABS appears to be less sensitive to molded-in orientation effects than HIPS, and ductile moldings result over a variety of molding conditions. Rupture elongation of HIPS remains high at low molding temperatures, even though the disk moldings have a low impact strength. On the basis of this observation, it may be speculated that the PS–cross-linked rubber blends (Figure 1) are also brittle when tested by dart drop. These blends showed an excellent elongation of tensile specimens and a low elongation when tested as compression-molded specimens. This result seems to indicate a relatively poor strength perpendicular to the flow-orientation direction, as well as in more isotropic compression moldings and extruded sheet.

Differences such as the flow-orientation effects on the practical toughness of moldings and the property variations obtained on compression- and injection-molded specimens may illustrate the fundamental differences responsible for the early success of the emulsion-produced ABS and the mass-made HIPS. However, small-particle cross-linked rubbers can be used for toughening PS, and there may be applications where these blends are useful. Combination of such toughened PS with other polymers may offer additional possibilities.

Summary

In HIPS from a mass process, rubber particles ≥ 1–2 μm are required for satisfactory toughening. Submicrometer cross-linked latex rubber particles are also effective in toughening polystyrene. High impact strength and high tensile rupture elongations are obtained on injection-molded samples.

Bimodal rubber particle distributions are useful in producing HIPS with an excellent impact strength. They have a low rubber-phase volume fraction, and the majority of the particles are below the minimum critical rubber particle-size diameter of 0.8 μm (4).

Limitations to the toughening of polystyrene, particularly with small rubber particles, seem to relate to its high propensity to molecular orientation and to the weakness of the moldings across the flow direction. Rubber-modified SANs, however, show a lesser tendency to anisotropy and a higher strength in the across-the-flow direction, which leads to a high practical toughness of fabricated parts.

It needs to be determined whether small rubber particles under some circumstances are capable of initiating shear yielding in stressing the PS matrix.

The role of the small particles in the control of craze growth rate and termination requires study. A tentative suggestion is that a high concentration of small particles (small interparticle distance) in HIPS controls the growth and termination of crazes. The crazes are initiated at large particles, and the small particles, by a cooperative mechanism, control the ductile PS ligament thickness in the fracture zone. An in situ video observation of scanning electron microscopy may prove useful in these studies.

Acknowledgments

I am indebted to D. R. Paul for many valuable discussions.

References

1. Bucknall, C. B. *Toughened Plastics;* Applied Science: London, 1977.
2. Donald, A. M.; Kramer, A. J. *J. Mater. Sci.* **1982**, *17*, 2351.
3. Boyer, R. F.; Keskkula, H. *Encycl. Polym. Sci. Technol.* **1970**, *13*, 392.
4. Bucknall, C. B. In *Polymer Blends;* Paul, D. R.; Newman, S., Eds.; Academic: New York, 1978; Vol. 2, p 99.
5. Moore, J. D. *Polymer* **1971**, *12*, 478.
6. Angier, D. J.; Fettes, E. M. *Rubber Chem. Technol.* **1965**, *38*, 1164.
7. Donald, A. M.; Kramer, E. J. *J. Appl. Polym. Sci.* **1982**, *27*, 3729.
8. Uzelmeier, C. W. *SPE Reg. Tech. Conf. (Mid-Ohio Valley)* **1976**, 49–65.
9. Lavengood, R. E. (Monsanto) U.S. Patent 4 214 056, 1980.
10. Hobbs, S. Y. *Polym. Eng. Sci.* **1986**, *26*, 74.
11. Michler, G. H. *Plaste Kautsch.* **1986**, *26*, 680.
12. Fowler, M. E.; Keskkula, H.; Paul, D. R. *Polymer* **1987**, *28*, 1703.
13. Keskkula, H.; Paul, D. R.; McCreedy, K. M.; Henton, D. E. *Polymer* **1987**, *28*, 2063.
14. Keskkula, H.; Turley, S. G. *Polym. Lett.* **1969**, *7*, 697.
15. Amos, J. L. *Polym. Eng. Sci.* **1974**, *14*, 1.
16. Amos, J. L.; McCurdy, J. L.; McIntire, O. R. (Dow) U.S. Patent 2 694 692, 1954.
17. Turley, S. G. *J. Polym. Sci. Part C* **1963**, *1*, 101.
18. Haward, R. N.; Mann, J. *Proc. R. Soc. London A* **1964**, *282*, 120.
19. Keskkula, H. *Appl. Polym. Symp.* **1970**, *15*, 51.
20. Keskkula, H.; Turley, S. G.; Boyer, R. F. *J. Appl. Polym. Sci.* **1971**, *15*, 351.
21. Keskkula, H.; Price, R. M.; Roche, A. F. (Dow) U.S. Patent 2 958 671, 1960.
22. Keskkula, H. (Dow), unpublished information, 1956.
23. Monsanto Company Br. Patent 1 005 681, 1965.
24. Bucknall, C. B.; Smith, R. R. *Polymer* **1965**, *6*, 437.
25. Coumans, W. J.; Heikens, D. *Polymer* **1980**, *21*, 957.
26. Keskkula, H.; Schwarz, M.; Paul, D. R. *Polymer* **1986**, *27*, 211.
27. Arends, C. B. (Dow), private communication, 1988.
28. Yee, A. F. *International Conference on the Toughening of Plastics, II;* Plastics and Rubber Institute: London, 1985, 19/1.
29. Turley, S. G.; Keskkula, H. *Polymer* **1980**, *21*, 466.
30. Bucknall, C. B.; Davies, P.; Partridge, I. K. *J. Mater. Sci.* **1987**, *22*, 1341.
31. Gilbert, D. G.; Donald, A. M. *J. Mater. Sci.* **1986**, *21*, 1819.
32. Beaumont, P. W. R. *Polym. Mater. Sci. Eng.* **1987**, *57*, 422.
33. Keskkula, H.; Norton, J. W., Jr. *J. Appl. Polym. Sci.* **1959**, *2*, 289.
34. Keskkula, H.; Simpson, G. M.; Dicken, F. L. *SPE Prepr. Antec* **1966**, *12*, XV-2.

RECEIVED for review February 11, 1988. ACCEPTED revised manuscript August 30, 1988.

CHEMISTRY AND PHYSICS

Formation of Epoxy Networks, Including Reactive Liquid Elastomers

Karel Dušek and Libor Matějka

Institute of Macromolecular Chemistry, Czechoslovak Academy of Sciences, 162 06 Prague 6, Czechoslovakia

Formation and structure of epoxy networks based on diglycidyl ether of bisphenol A and on N,N-diglycidylaniline and its derivatives are discussed. The effect of reaction mechanism on the structure of epoxy networks cured with amines, acids, anhydrides, and amine-terminated elastomers is shown. The second part of this chapter discusses the formation and properties of rubbery networks prepared from carboxyl-terminated elastomers and polyepoxides. Both homogeneous and two-phase networks can be obtained. Improvement of mechanical properties is achieved mainly by physical reinforcement (i.e., by chemically bound hard polyester segments).

T HE PROPERTIES AND APPLICATIONS OF CURED EPOXY RESINS are determined by the structure of the resins and curing agents, conditions of curing, and possible modifications. An understanding of the relationship among network formation, structure, and properties is necessary to properly select the components and the curing regime. Understanding of the molecular and supermolecular structure and its changes plays a key role in this selection.

Branching theories can be used to describe structural changes occurring in the course of curing for a number of epoxy resin–curing agent systems (1). The results obtained on systems like diglycidyl ether of bisphenol A (DGEBA) diamines or on curing with polyacids and cyclic anhydrides can be mentioned as examples. However, systems based on polyfunctional epoxy resins derived from N,N-diglycidylamines are rather complicated with re-

0065–2393/89/0222–0303$06.00/0
© 1989 American Chemical Society

spect to the buildup of the network structure (2). Much more experimental and theoretical work must be done before network structure can be quantitatively understood and predicted.

Special attention has been paid recently to epoxy systems modified with reactive liquid polymers (RLP) in general and telechelic polymers in particular (cf. ref. 3 and this volume). One of the main aims of such studies is to toughen materials while retaining their temperature resistance. This aim is achieved by separation of a rubbery or tough phase during curing. Morphology and toughness depend on the stage of curing at which the new phase separates. One of the main factors controlling the onset of phase separation is the network buildup in terms of such factors as molecular weight increase, gel point, and cross-linking density. Branching theory also applies here. RLPs are a minor component in these modifications.

In another field of modification, RLP, the major component, is crosslinked or end-linked with epoxy resins. If the RLPs are elastomeric, the products resemble vulcanized rubbers or semielastic materials. The fields of application (like sealants and technical rubber) are becoming wider. Applications such as producing cast low-pressure tires for agricultural use seem promising. Here the study of cross-linking and combination with the branching theories is even more important, because one of the main properties of the systems—rubber elasticity—can be well correlated with network structure.

In this chapter, based mainly on our results, we summarize the present state of knowledge about the mechanism of curing and network buildup in epoxy–amine, epoxy–polycarboxylic acid, and epoxy–cyclic anhydride systems. The second part is devoted to elastomeric systems like epoxy–polyether amines or epoxy–carboxyl-terminated elastomers. For many systems discussed here, the experimental data on formation and properties were compared with the predictions of the theory of branching processes. The basis of the theory and its application are explained in ref. 1.

Curing with Amines

Cross-linking with amines is the most common way of curing epoxy resins. The DGEBA–diamine reaction proceeds by an alternating polyaddition. Cyclization is negligible. Etherification takes place in the reaction of aliphatic amines with an excess of epoxide, but not in systems with aromatic amines (1). The reactivity of the epoxy groups in DGEBA is independent, but the reactivity of hydrogens in the primary and secondary amino groups may differ.

$$RCH-CH_2 \quad + \quad R^1NH_2 \quad \xrightarrow{\ k_1\ } \quad RCHCH_2NHR^1$$
$$\underset{O}{\diagdown\diagup} \qquad\qquad\qquad\qquad\qquad\qquad \underset{OH}{\big|}$$

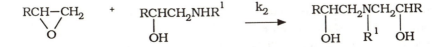

The ratio of rate constants for the reaction of primary (k_1) and secondary (k_2) amino groups $(\rho = k_2/k_1)$ is decisive for the kinetics and network structure. This ratio is equal to ½ for an ideal reaction (i.e., equal reactivity of hydrogens in the primary and secondary amino groups). Real systems have a negative substitution effect: $\rho = 0.1$–0.3 for aromatic and 0.3–0.5 for aliphatic amines (*4*).

The influence of the substitution effect on network structure can be seen in Figure 1. The figure illustrates the dependence of the sol fraction w_s on the molar ratio of functional groups $(r = 2[NH_2]/[epoxy])$ of nonstoichiometric networks. Theoretical curves for different values of ρ were calculated under the assumptions that epoxy groups in DGEBA are independent and that cyclization and etherification do not occur (*5*). The sol fraction is zero in stoichiometric networks $(r = 1)$ and increases with increasing excess of

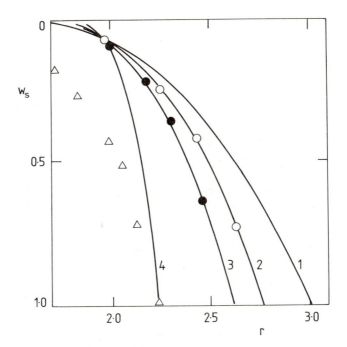

Figure 1. Dependence of the sol fraction w_s *of the nonstoichiometric networks on the ratio of functionalities* $r = 2[NH_2]/[epoxy]$. *Key:* ○, *DGEBA–OMDA;* ●, *DGEBA–DDM;* △, *DGA–DDM; 1–4, theoretical curves for* $\rho = 0.5, 0.32, 0.22,$ *and* 0.05, *respectively.*

the amine. At the critical molar ratio, r_c, the system does not gel any more and $w_s = 1$. In an ideal case, $r_c = 3$. The negative substitution effect results in a decrease in r_c and an increase in the sol fraction at the given r. The substitution effects for the systems under study—DGEBA–4,4'-diaminodiphenylmethane (DDM) and DGEBA–octamethylenediamine (OMDA)— were calculated from the experimental r_c values (Figure 2) (6): ρ (DGEBA–DDM) = 0.22; ρ (DGEBA–OMDA) = 0.32.

A very good agreement between theory and experiments was found (Figure 1). The critical ratio r_c for these systems is independent of dilution (Figure 2), a result that indicates the absence of cyclization. The ρ values determined directly from the reaction kinetics of model compounds are as follows: ρ (phenylglycidyl ether (PGE)–aniline) (7) = 0.24; ρ (PGE–dodecylamine) (8) = 0.41.

Polyepoxides based on N,N'-diglycidylamine derivatives represent an important class of epoxy resins used in high-performance composites. N,N,N',N'-tetraglycidyl-4,4'-diaminodiphenylmethane (TGDDM) is commonly used. Curing of these epoxides with amines has some special features attributable to the presence of a tertiary nitrogen atom in the molecules and sterical vicinity of the two epoxy groups. The formation of networks was

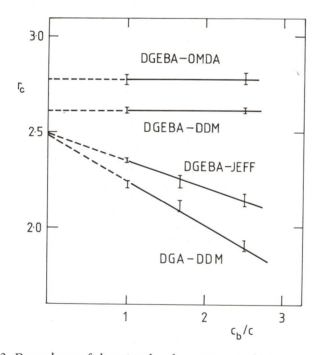

Figure 2. Dependence of the critical molar ratio r_c *on dilution.* c *and* c_b *are concentration of functional groups in the diluted sample and in the bulk, respectively (adapted from ref. 6).*

studied by using the bifunctional model N,N'-diglycidylaniline (DGA). (Figure 1 shows that the DGA–DDM networks have a much higher content of the sol at a given r than the DGEBA–DDM system. The simple theory developed for the cross-linking of DGEBA cannot account for such a deviation from the ideal case, even if the very negative substitution effect is taken into account (Figure 1). Consequently, studies of the more complicated DGA–DDM reaction mechanism and a modification of the theory were necessary.

Unlike DGEBA, the reactivities of epoxy groups in DGA are strongly dependent and show a positive substitution effect (9). An analysis of the kinetics and gelation experiments indicates that the reactivities of all functional groups in the system are interdependent (10). This interdependence means that the substitution effect in amine depends on the epoxide structure, and vice versa. Consequently, the reaction can be described by not two but four rate constants (6) in the absence of etherification.

Unlike the DGEBA–DDM network, the critical ratio r_c in the DGA–DDM system decreases with dilution (Figure 2). This decrease is a result of the reduced effective functionality of the reagents because of cyclization. The formation of small cycles has been demonstrated in the model systems DGA–aniline and DGA–N-methylaniline. Seven- and eight-membered cycles were isolated from the reaction mixture and identified (11).

Etherification takes place during the reaction of DGA with aniline, even in a stoichiometric mixture (9). The etherification was taken into account in the theory as a reaction initiated by the OH group arising during the addition step (12).

Knowledge of the rheology of the reacting system is important for processing. The increase in viscosity is determined by the increase in size of the molecules and by the increase in T_g. In addition, a temperature change can be programmed or determined by reaction heat and heat transfer (13). In the case of an isothermal reaction, the overall change of viscosity η with time t can be expressed as changes of η with conversion α and of α with time t:

$$\frac{d\eta}{dt} = \left(\frac{d\eta}{d\alpha}\right)\left(\frac{d\alpha}{dt}\right)$$

From the dependence of η on α, the dependences of η on molecular weight and on $(T - T_g)$ can be factored out: $\eta(\alpha) = \eta(M) \cdot \eta(T - T_g)$ (14, 15). The $\eta(M)$ dependence is still to be found; for the $\eta(T - T_g)$ dependence, the free-volume or alternative theories of segmental mobility can be used.

Control of the curing rate by both reaction mechanism and diffusion is very important when the reacting system enters the glass-transition region (16, 17). The increase of T_g in conventional systems during cure is of the order of 10^2 K, and for a majority of epoxy–curing-agent systems the final cure phases are diffusion-controlled by vitrification. Theoretical treatments have been developed that describe relatively well the passage from control by chemical kinetics to segmental mobility (diffusion) control below T_g. Both theory and experiments show that the fall of the apparent rate constant is relatively steep and commences rather abruptly when T_g of the reacting systems approaches the reaction temperature (Figure 3) (17). However, the reaction proceeds slowly even several tens of Kelvins below T_g.

Cross-Linking of Amine-Terminated Elastomers

Bi- and polyfunctional amine-terminated liquid elastomers (polyethers, polybutadienes, butadiene–acrylonitrile copolymers) can be cross-linked by polyepoxides to yield semiflexible to flexible products. Network formation in-

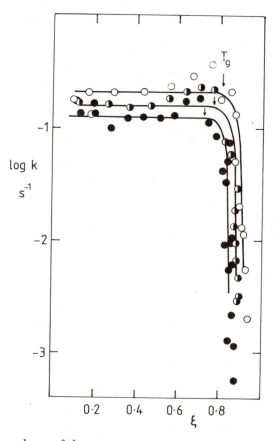

Figure 3. Dependence of the rate constant on conversion for the reaction of DGEBA with diaminopropane at t_{cure}. Key: ●, 50 °C; ◐, 60 °C; ○, 70 °C.

volving poly(oxypropylene) (POP) diamines and triamines (Jeffamine) have been studied in great detail (*18–20*).

Model reactions with phenylglycidyl ether have revealed that the reactivity of the terminal primary amino groups differs from that of NH_2 groups of aliphatic amines. The reaction rate of the former is generally lower because of competition between the epoxy groups and ether groups of the polymer backbone in an interaction with proton donors. Moreover, the substitution effect in the Jeffamine amino group is more negative than in aliphatic amines; $\rho = 0.2$ (*20*). Gelation of the DGEBA–Jeffamine systems is affected by this feature, as well as by the fact that cyclization is no longer negligible (Figure 2).

An analysis of equilibrium elasticity data reveals that these POP amine–epoxy networks are similar to POP triol 4,4'-diphenylmethane diisocyanate polyurethanes of comparable cross-linking density (cf. ref. 21).

Acid Curing

Curing with acids and cyclic anhydrides is widely used, especially in coatings. The carboxyl–epoxy reaction is also utilized in end-linking or cross-linking of carboxyl-terminated elastomers, mainly in modifying thermosets and thermoplastics.

The main esterification carboxyl–epoxy reaction is accompanied by a number of side reactions (polyetherification, hydroxyl–acid condensation), depending on whether carboxyl or epoxide is in excess and on a catalyst.

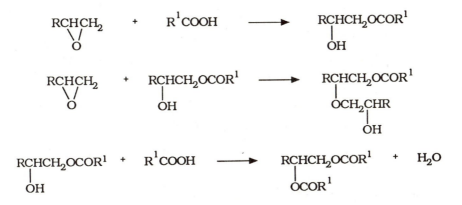

The transesterification reaction, in which an addition product—hydroxyester—is transformed into diester and diol (*22, 23*), is very important.

By this reaction, branch points and nonreactive ends are formed in the network. Theoretical analysis has shown that gelation caused by transesterification is possible even in a diacid–diepoxide system, which yields linear polyester in the first step. This prediction was confirmed by experiments (23). The tertiary amine works as a catalyst in this system (i.e., it does not become chemically bound). On the contrary, in an acid-free anhydride–epoxy system, tertiary amine becomes an initiator and is chemically bound to the structure (24).

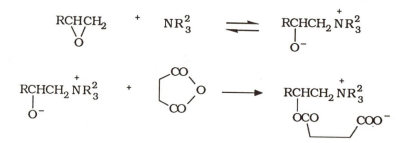

Differences between the stepwise noninitiated carboxyl–epoxide and initiated anhydride–epoxide reactions can be detected experimentally in several ways (25):

1. The molecular weight distribution of the polyesters formed from a monoepoxide and cyclic anhydride is much narrower than that of hydroxy polyesters formed from a diepoxide and dicarboxylic acid, even in the region where transesterification is negligible.

2. The number-average molecular weight M_n of cyclic anhydride–epoxide polyesters depends on the relative concentration of the tertiary amine, whereas that of diacid–diepoxide polyesters does not.

3. The gel-point conversion of epoxide–cyclic anhydride systems is dependent on the relative concentration of the tertiary amine (26).

Cross-Linking of Carboxytelechelic Polymers with Polyepoxides

This reaction has been used for a long time to increase the toughness of epoxy resins (27). In this case, the weight fraction of the telechelic polymer does not exceed 20–25%. The network has a two-phase structure, and one can talk about rubber-modified epoxide. Phase inversion occurs above the upper limit and yields rubber reinforced with a cross-linked epoxide. In the segregated epoxy domains, the epoxide units are cross-linked by etherification.

Phase separation and the supermolecular structure of the cross-linked rubbers depend very much on the nature of the telechelic polymers. For example, some of the butadiene–acrylonitrile copolymers exhibit a relatively good compatibility with epoxies; carboxyl-terminated polybutadienes are less compatible. Although carboxyl-terminated systems are cross-linked and reinforced by the polyepoxy phase in situ, their ultimate properties are not satisfactory.

For example, the carboxyl-terminated polybutadienes (CTB) cross-linked with a molar excess of DGEBA in the presence of a tertiary amine as a catalyst exhibit a two-phase structure with a rubbery matrix and glassy hard domains (28). The glass transition of the CTB phase occurs at around −80 °C, but the temperature of the hard-phase transition increases with increasing epoxy content to reach a value +90 °C for the polyDGEBA (Figure 4). The shift of this transition reflects a partial solubility of CTB in epoxy domains. The epoxy domains are highly cross-linked by etherification. The permanent cross-links reduce the mobility of domains and make the hard phase relatively brittle. This brittleness results in a reduction of elongation at break, and mechanical properties are poor (Table I) (28).

Several attempts have been undertaken to improve the ultimate properties of these systems:

1. Hard polyester blocks acting as a physical reinforcement are chemically attached to the chain ends of a carboxyl-terminated butadiene–acrylonitrile copolymer (CTBN) containing 30% acrylonitrile (28). The blocks of different lengths were prepared by reaction of CTBN with a monoepoxide (phenylglycidyl ether) and a cyclic anhydride (hexahydrophthalic anhydride, HHPA). Because these components are bifunctional, linear chains are formed. Some diepoxide (DGEBA) is added to make chemical cross-links within these blocks and prevent viscous flow at higher temperatures. The networks with longer blocks are heterogeneous (Figure 5). The hard phase from segregated blocks is softer and tougher than the hard phase composed of polyDGEBA. The ultimate properties are substantially improved, as can be seen from Table II. The disadvantage of these systems consists in very slow stress relaxation at elevated temperatures (above 60 °C) because of a low T_g of the polyester blocks (Figure 5).

2. The CTBN (30% acrylonitrile)–diepoxide system is modified by addition of the cyclic anhydride HHPA. Diepoxide is tetrafunctional and cyclic anhydride is difunctional, so that hard segments as well as cross-links are introduced and the

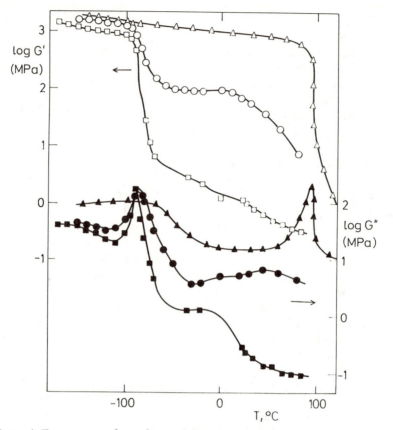

Figure 4. Temperature dependence of the storage, G', and loss, G'', dynamic shear moduli for CTB–DGEBA networks. Key: □ and ■, sample 1 (Table I); ○ and ●, sample 5 (Table I); △ and ▲, polyDGEBA.

Table I. Composition and the Ultimate Properties of CTB–DGEBA Networks

Sample	q^a	$V_{PB}{}^b$	$\sigma_b,{}^c$ MPa	$\epsilon_b{}^d$	$\sigma_b\lambda_b,{}^e$ MPa
1	0.7	0.89	1.49	8.44	14
2	0.47	0.84	3.04	3.96	15
3	0.35	0.80	3.16	1.91	9
4	0.18	0.67	5.76	0.36	8
5	0.12	0.57	9.62	0.25	12

[a] q is COOH–epoxy.
[b] V_{PB} is volume fraction of CTB ($M_n = 3350$).
[c] σ_b is stress at break.
[d] ϵ_b is elongation at break.
[e] $\sigma_b\lambda_b$ is true strength ($\lambda = 1 + \epsilon$).

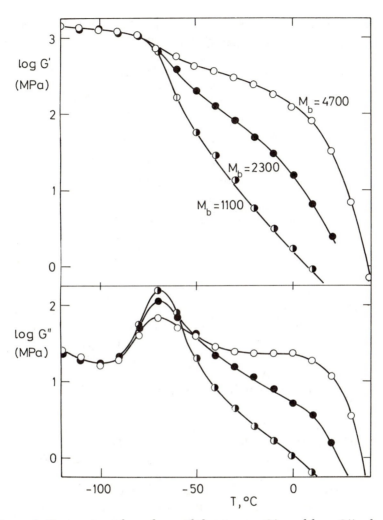

Figure 5. Temperature dependence of the storage, G', and loss, G'', shear moduli for CTBN–polyester block networks. M_b is molecular weight of the polyester block.

system is reinforced both physically and chemically. The networks were prepared by using a metal catalyst (Cr(III) chelate) that selectively catalyzes esterification carboxyl–epoxy groups. The networks are homogeneous, as evidenced by a single maximum on the temperature dependence of the loss modulus (Figure 6). The reason is a higher compatibility between CTBN with a relatively high content of acrylonitrile and polyester blocks. By varying the carboxyl:epoxide:anhydride ratio, the mechanical properties are varied in a wide range.

Table II. Composition and the Ultimate Properties of CTBN–Polyester and
CTBN–DGEBA–Hexahydrophthalic Anhydride (HHPA) Networks

Sample	$M_b{}^a$	$V_{PB}{}^b$	DGEBA, phr^c	HHPA, phr	$\sigma_b{}^d$ MPa	$\epsilon_b{}^e$	$\sigma_b\lambda_b{}^f$ MPa
6	4300	0.49	20.4	–	4.50	3.41	20
7	4900	0.48	20.4	–	7.77	2.25	25
8	5050	0.47	30.6	–	10.90	1.71	30
9	5450	0.45	40.8	–	16.33	0.78	29
10	–	0.86	15.3	1.27	1.95	8.6	19
11	–	0.79	20.4	5.06	2.08	4.0	10
12	–	0.62	40.4	20.0	4.74	1.4	11
13	–	0.83	15.7	5.06	2.16	6.2	15
14	–	0.84	13.6	5.06	2.30	9.4	24

$^a M_b$ is molecular weight of the polyester block.
$^b V_{PB}$ is volume fraction of CTBN (30% AN, M_n = 2930).
^{c}phr is parts per hundred parts of rubber.
$^d \sigma_b$ is stress at break.
$^e \epsilon_b$ is elongation at break.
$^f \sigma_b\lambda_b$ is true strength.

CTBN tensile strength values of 2 MPa at 400–900% elongation at break or 4.7 MPa at 140% elongation were reached (cf. Table II). With increasing DGEBA and HHPA content, T_g shifts to higher temperatures and the modulus in the rubbery region increases (Figure 6). Because the networks are homogeneous, their formation can be treated by the branching theory (29).

3. A polyfunctional epoxide is used as the cross-linking agent to increase the cross-linking density (30, 31). A detailed study has been made as to the cross-linking of CTB and CTBN with the tetrafunctional epoxide TGDDM. The networks were prepared by using Cr(III) octoate, which selectively catalyzes esterification. Figure 7 shows that all the networks are homogeneous except that with the highest epoxide content, $q = [\text{COOH}]/[\text{epoxy}] = 0.3$. Specifically, the effect of varying molar ratio q on the sol fraction and mechanical properties was investigated. The gel fraction, elastic modulus, and tensile strength all pass through a maximum; the elongation at break and the degree of swelling pass through a minimum at the ratio $q = 0.9$–1.0 (Figure 8). Because most of networks are homogeneous even with CTB, the experiment can be correlated with theory (31). An analysis of the sol fraction and equilibrium modulus dependences for the ratio $q < 1$ shows that some etherification takes place even with the selective metal catalysts. The critical molar ratio necessary for gelation is close to 3 at the carboxyl excess, but substantially less than

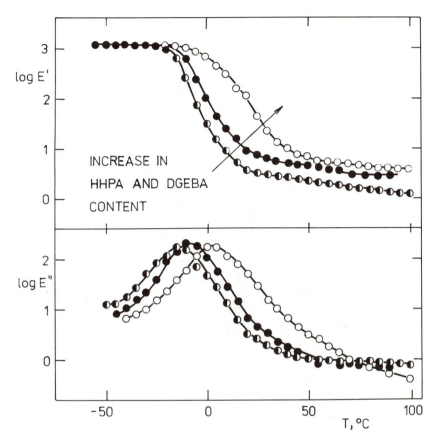

Figure 6. Temperature dependence of the storage, E', and loss, E'' (MPa). Young moduli for CTBN–HHPA–DGEBA networks. Key: ◑, *sample 10;* ●, *sample 11;* ○, *sample 12 (Table II).*

0.33 (0.1–0.15) for the epoxide excess, which indicates some additional cross-linking by etherification.

The experimental reduced equilibrium modulus is

$$G_r = \frac{G_e}{RTw_g} = A\nu_{eg}$$

where G_e is the equilibrium shear modulus, w_g is the gel fraction, ν_{eg} is the cross-linking density, and A is the front factor in the Flory rubber elasticity theory. This equilibrium modulus is higher than that calculated by using the branching theory under the assumption of $A = \frac{1}{2}$ for phantom networks or $A = 1$ for affine networks (Figure 9). This higher equilibrium modulus points to an important contribution by permanent topological constraints.

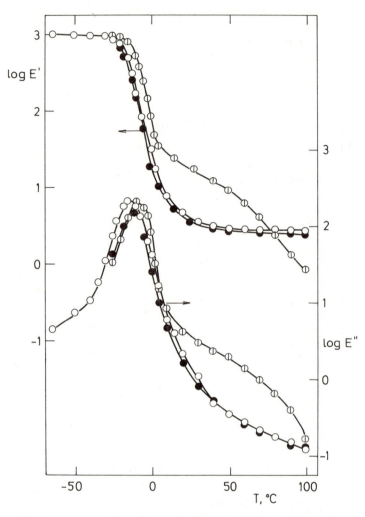

Figure 7. Temperature dependence of the storage, E′, and loss, E″. Young moduli for the CTBN–TGDDM network. q = [COOH]/[epoxy]. *Key:* ①, q = 0.3; ○, q = 0.8; ●, q = 1.2. *(Reproduced with permission from ref. 31. Copyright 1987 de Gruyter.)*

Conclusions

Curing of epoxy resins with amines, acids, or acid anhydrides may be adequately described by the branching theory under an assumption of a good knowledge of the reaction mechanism and characteristics of a system. Effects like unequal reactivity of functional groups, cyclization, or an initiation reaction mechanism can be taken into account.

Also, the cross-linking of amine or COOH-terminated liquid polymers by epoxides can be treated theoretically if the system is homogeneous. Poor

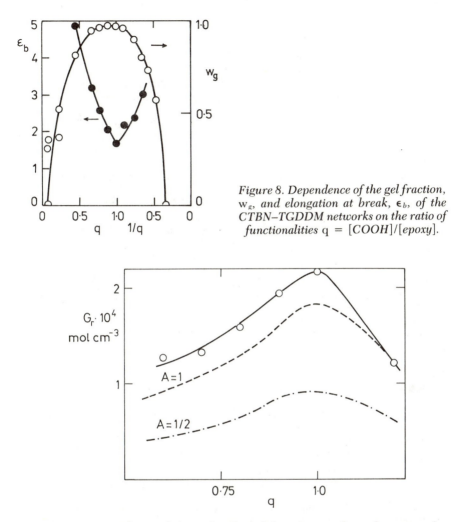

Figure 8. *Dependence of the gel fraction, w_g, and elongation at break, ϵ_b, of the CTBN–TGDDM networks on the ratio of functionalities* q = [COOH]/[epoxy].

Figure 9. *Dependence of the reduced modulus, G_r, on the molar ratio of reactive groups* q = [COOH]/[epoxy] *of CTBN–TGDDM networks. Key:* —— *and* -----, *theoretical curves for A = 1 and ½, respectively. (Reproduced with permission from ref. 31. Copyright 1987 de Gruyter.)*

mechanical properties of epoxy cross-linked carboxyl-terminated RLPs may be improved by polyester blocks attached to the chain ends of RLP, formation of a hard segment by the addition of a cyclic anhydride, and use of a poly-functional epoxide. Halatotelechelic polymers represent another route to improving the mechanical properties of these systems (32). Generally, the ultimate properties are better if the system is only physically reinforced and if the cross-linking in epoxy domains is suppressed by using the selective catalyst. However, some chemical cross-links are necessary to prevent vis-cous flow in the system at a higher temperature.

References

1. Dušek, K. *Adv. Polym. Sci.* **1986,** *78,* 1.
2. Dušek, K; Matějka, L. *Polym. Mater. Sci. Eng.* **1987,** *56,* 356.
3. *Rubber-Modified Thermoset Resins;* Riew, C. K.; Gillham, J. K., Eds; Advances in Chemistry 208; American Chemical Society: Washington, DC, 1984.
4. Rozenberg, B. A. *Adv. Polym. Sci.* **1985,** *75,* 113.
5. Dušek, K.; Ilavský, M.; Luňák, S. *J. Polym. Sci., Polym. Symp.* **1975,** *53,* 29.
6. Matějka, L.; Dušek, K.; Dobáš, I. *Polym. Bull.* **1985,** *14,* 309.
7. Johncock, P.; Porecha, L.; Tudgey, G. F. *J. Polym. Sci., Polym. Chem. Ed.* **1985,** *23,* 291.
8. Dušek, K.; Bleha, M.; Luňák, S. *J. Polym. Sci., Polym. Chem. Ed.* **1977,** *15,* 2393.
9. Matějka, L.; Pokorný, S.; Dušek, K. In *Crosslinked Epoxies;* Sedláček, B.; Kahovec, J., Eds.; de Gruyter: West Berlin, 1987; p 241.
10. Matějka, L.; Dušek, K. In *Crosslinked Epoxies;* Sedláček, B.; Kahovec, J., Eds.; de Gruyter: West Berlin, 1987; p 231.
11. Matějka, L.; Pokorný, S.; Tkaczyk, M.; Dušek, K. *Polym. Bull.* **1986,** *15,* 389.
12. Dušek, K. *Polym. Bull.* **1985,** *13,* 321.
13. Nicolais, L. *Polym. Mater. Sci. Eng.* **1987,** *56,* 434.
14. Apicella, A.; D. Amore, A.; Kenny, J.; Nicolais, L. *SPE Annu. Tech. Conf. (ANTEC 86)* **1986, 557.**
15. Macosko, C. W. *Br. Polym. J.* **1985,** *17,* 239.
16. Gillham, J. K. *Makromol. Chem., Makromol. Symp.* **1987,** *7,* 67.
17. Havlíček, I.; Dušek, K.; Štokrová, S.; Biroš, J. *Eur. Symp. Polym. Mater. Lyon* **1987,** preprint CA 03.
18. Morgan, J. R.; Kong, F. M.; Walkup, C. M. *Polymer* **1984,** *24,* 375.
19. Dušek, K.; Ilavský, M.; Luňák, S., Jr. In *Crosslinked Epoxies;* Sedláček, B.; Kahovec, J., Eds.; de Gruyter: West Berlin, 1987; p 269.
20. Dušek, K. et al. In *Crosslinked Epoxies;* Sedláček, B.; Kahovec, J., Eds.; de Gruyter: West Berlin, 1987; p 279.
21. Ilavský, M.; Dušek, K. *Polymer* **1983,** *24,* 981.
22. Matějka, L.; Pokorný, S.; Dušek, K. *Polym. Bull.* **1982,** *7,* 123.
23. Dušek, K.; Matějka, L. In *Rubber-Modified Thermoset Resins;* Riew, C. K.; Gillham, J. K., Eds; Advances in Chemistry 208; American Chemical Society: Washington, DC, 1984; p 15.
24. Matějka, L. et al. *J. Polym. Sci., Polym. Chem. Ed.* **1983,** *21,* 2873.
25. Matějka, L.; Pokorný, S.; Dušek, K. *Makromol. Chem.* **1985,** *186,* 2025.
26. Dušek, K.; Luňák, S.; Matějka, L. *Polym. Bull.* **1982,** *7,* 145.
27. Siebert, A. R. *Makromol. Chem., Makromol. Symp.* **1987,** *7,* 115.
28. Ilavský, M. et al. *Int. Rubber Conf. Moscow* **1984, preprint A68.**
29. Ilavský, M.; Dušek, K.; Vaněk, P.; Svoboda, P. *Plasty Kauč.* **1985,** *22,* 270.
30. Heřman, J. Diploma Thesis, Institute of Chemical Technology, Prague, 1986.
31. Ilavský, M. et al. In *Crosslinked Epoxies;* Sedláček, B.; Kahovec, J., Eds.; de Gruyter: West Berlin, 1987; p 347.
32. Broze, G.; Jerome, R.; Theyssie, P.; Marco, C. *Macromolecules* **1986,** *18,* 1376.

Received for review March 25, 1988. Accepted revised manuscript September 23, 1988.

Effect of Polydispersity on the Miscibility of Epoxy Monomers with Rubbers

Julio Borrajo, Carmen C. Riccardi, Stella M. Moschiar, and Roberto J. J. Williams

Institute of Materials Science and Technology (INTEMA), University of Mar del Plata–National Research Council, J. B. Justo 4302, (7600) Mar del Plata, Argentina

The effect of polydispersity on the miscibility of epoxy monomers with carboxyl-terminated butadiene–acrylonitrile random copolymers (CTBN) is analyzed by using a Flory–Huggins lattice model. Calculations lead to cloud-point curves in χ (interaction parameter) versus composition coordinates, composition of the segregated phase at the cloud point ("shadow" curve), and distribution of molecular weights in the minority phase. The effect of polydispersity of CTBN is to produce a precipitation threshold at a low rubber volume fraction (0.07–0.08), in agreement with experimental observations. When the polydispersity of the epoxy monomer is also considered, a flattening of the cloud-point curve near the critical point is observed. Although the effect of polydispersity must be taken into account to analyze experimental cloud-point curves, the monodisperse assumption gives a reasonable approximation when macroscopic separation is considered. The CTBN segregated in the minority phase is rich in higher-molecular-weight fractions, an effect that is enhanced at low initial rubber concentrations.

C<small>ARBOXYL-TERMINATED</small> <small>BUTADIENE–ACRYLONITRILE</small> random copolymers (CTBN rubbers) are used for the toughening of epoxy networks (*1*). In these systems, the miscibility of both components is the key point in attaining the desired morphology. Initially the mixture must remain homogeneous. How-

0065–2393/89/0222–0319$06.00/0

ever, at a certain reaction extent well before gelation, generation of rubber-rich domains must begin. This change leads to the final structure when the matrix gels (2).

Experimental cloud-point determinations of mixtures of epoxy monomers with CTBNs show an upper critical solution temperature behavior (3, 4). This behavior is to be expected because of the low average molecular weight of both components. Moreover, the thermodynamic description of the system through a Flory–Huggins lattice model, taking both components as monodisperse, led to a reasonable simulation of the observed phase separation process (5, 6). However, polydispersity may have a significant effect on the location of the miscibility gap for low-molecular-weight polymer mixtures (7). This possible effect is indicated by the appearance of a precipitation threshold (maximum in the temperature–composition cloud-point curve) located at a CTBN volume fraction of about 0.07 (3) (i.e., far away from the predicted critical point).

We propose to ascertain the effect of the polydispersity of one or both components on the thermodynamic behavior of CTBN–epoxy mixtures. The analysis will focus on the location of the miscibility gap. Once the system reaches this region, the nature of the demixing process will depend on the competition between rates of polymerization and phase separation (5, 6).

Molecular Weight Distributions

According to data specified by the supplier (BF Goodrich), most of the CTBNs have number-average molecular weights between 3000 and 4000, with polydispersity indexes close to 2. These values arise from free-radical polymerization with chain transfer as the main termination mode. The use of a convenient amount of a transfer agent determines the required average molecular weight. For these polymerizations, the normalized molecular weight distribution (by weight) is given (8) by

$$w_1(x) = x\, q^{x-1}\, (1 - q)^2 \tag{1}$$

where $w_1(x)$ is the mass fraction of the x-mer of component 1 (CTBN), and q is the probability that a growing chain will add a monomer unit rather than transfer (i.e., it represents the overall probability of propagation). Different averages of this distribution may be calculated as

$$x_n = \frac{1}{1 - q} \tag{2}$$

$$x_w = \frac{1 + q}{1 - q} \tag{3}$$

$$x_z = \frac{1 + 4q + q^2}{(1 + q)(1 - q)} \tag{4}$$

where x_n, x_w, and x_z, respectively, are the number-, weight-, and z-average degrees of polymerization. For illustration purposes, we will take a particular carboxyl-terminated butadiene–acrylonitrile random copolymer (CTBN 1300×8, BF Goodrich), characterized by M_n = 3600 g/mol and ρ = 0.948 g/cm^3. By assigning the repetitive unit a molar mass equal to 54 g/mol (butadiene (BU): 54, acrylonitrile (AN): 53) and neglecting the effect of chain ends, x_n = 66.67 and q = 0.985. This result in turn leads to x_w = 132.33 and x_z = 198.5. The polydispersity index is x_w/x_n = 1.985; this value compares well with the experimental value of 1.94, which was determined by size exclusion chromatography (4). The molar volume of the repetitive unit of CTBN is V_1 = 56.96 cm^3/mol. Because this value is smaller than the molar volume of the epoxy monomer, it will be taken as the unitary volume of the lattice size in the Flory–Huggins model. No effect of polydispersity in copolymer composition will be considered, although it must have a bearing on the observed experimental behavior. Epoxy monomers based on bisphenol A diglycidyl ether are selected:

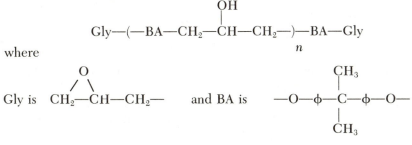

where

Gly is   and BA is

and ϕ is an aromatic ring.

Two different types of epoxy monomers will be considered. The first one, E(340), represents the first term of the series (n = 0, M = 340 g/mol) and has a molar volume V_2 = 283.3 cm^3/mol. The second one, E(382), represents a typical commercial product with $\bar{n}$ = 0.148, for which a molecular weight distribution is available (9). Table I shows the normalized molecular weight distribution (by weight); $w_2(y)$ represents the mass fraction of the y-mer of component 2 (epoxy). The molar volume of any y-mer is given by the product yV_2.

The following averages of the distribution shown in Table I can be calculated:

$$y_n = 1.123 \tag{5}$$
$$y_w = 1.334 \tag{6}$$
$$y_z = 1.950 \tag{7}$$

Table I. Molecular Weight Distribution of E(382)

n	M	y	w_2
0	340	1	0.807
1	624	1.835	0.113
2	908	2.671	0.029
3	1192	3.506	0.016
4	1476	4.341	0.015
5	1760	5.176	0.009
6	2044	6.012	0.005
7	2328	6.847	0.006

where y_n, y_w, and y_z, respectively, are the number-, weight-, and z-average degrees of polymerization.

Thermodynamics

The thermodynamic analysis of polymer mixtures, taking into account the effect of polydispersity, has been presented by Roe and Lu (7). The analysis starts by stating the Flory–Huggins free energy of mixing in the form

$$\frac{\Delta G_M}{RT} = \sum_x \left(\frac{\Phi_{1x}}{x}\right) \ln \Phi_{1x} + \sum_y \left(\frac{\Phi_{2y}}{Vy}\right) \ln \Phi_{2y} + \chi \, \Phi_1 \, \Phi_2 \qquad (8)$$

where ΔG_M is Gibbs free energy of mixing per unit cell, R is the universal gas constant, T is temperature, and $V = V_2/V_1 = 4.974$ represents the number of unit cells of volume V_1, occupied by the smallest mer $(y = 1)$ of component 2. Φ_{1x} and Φ_{2y} represent the volume fractions of the x-mer of polymer 1 and the y-mer of polymer 2. Then

$$\Phi_1 = \sum_x \Phi_{1x}, \quad \Phi_2 = \sum_y \Phi_{2y}, \quad \Phi_1 + \Phi_2 = 1 \qquad (9)$$

The interaction parameter χ is assumed to be a function of temperature, but independent of composition.

The determination of cloud-point curves, at which an infinitesimal volume of phase A has just separated out (the composition of phase B being essentially the same as the initial one), is detailed by Roe and Lu (7). For a selected value of Φ_1, calculations lead to the value of χ at which phase separation starts, the composition of the incipient phase A, Φ_1^A ("shadow" curve), and the molecular weight distributions in the new phase. In our case, consideration of 2000 x-mers of CTBN was satisfactory to achieve values that were not modified by adding extra terms (i.e., up to $x = 3000$).

On the other hand, spinodal curves and critical points can be calculated from expressions derived by Koningsveld and Kleintjens (*10*) by taking the volume ratio, V, into account. This leads to the spinodal curve

$$\frac{1}{x_w \Phi_1} + \frac{1}{V y_w \Phi_2} = 2\chi \tag{10}$$

and the critical point

$$\frac{1}{\Phi_{2c}} = 1 + \frac{V^{1/2} y_w x_z^{1/2}}{x_w y_z^{1/2}} \tag{11}$$

$$2\chi_c = \left(\frac{1}{\chi_z^{1/2}} + \frac{1}{V^{1/2} y_z^{1/2}}\right)\left(\frac{x_z^{1/2}}{x_w} + \frac{y_z^{1/2}}{V^{1/2} y_w}\right) \tag{12}$$

Location of the Critical Point

Two different mixtures will be considered: CTBN–E(340) and CTBN–E(382). For comparison purposes, both CTBN and E(382) will be taken as monodisperse, with all the averages equal to x_n and y_n, respectively. Then the six possibilities described in Table II arise.

The following trends are observed:

- Changing E(340) by E(382) decreases the miscibility of the system (i.e., decreases χ_c; compare 1 with 4 or 2 with 6). This decrease is caused by the increase in the molecular weight of epoxy monomer and the corresponding decrease of the entropic contribution to miscibility.

- The consideration of the actual polydispersity shifts the miscibility gap to lower χ values (i.e., the system becomes more incompatible than is expected by assuming monodisperse components). This shift results from the fact that the largest mers of both distributions are more readily separated.

Table II. Location of Critical Points for Two Mixtures and Different Approximations

System	ϕ_{1c}	χ_c
CTBN(poly)–E(340)	0.192	0.144
CTBN(mono)–E(340)	0.214	0.163
CTBN(poly)–E(382)(poly)	0.185	0.113
CTBN(poly)–E(382)(mono)	0.201	0.131
CTBN(mono)–E(382)(poly)	0.207	0.131
CTBN(mono)–E(382)(mono)	0.224	0.149

- Both the consideration of CTBN and epoxy polydispersities affect the results (compare 4 with 6 and 5 with 6). In spite of the low polydispersity index of E(382), $y_w/y_n = 1.19$, its significance is comparable to that of CTBN because of its lower molecular weight and the correspondingly higher entropic contribution. Both polydispersities shift Φ_{1c} to lower values.

Cloud-Point Curves

Figure 1 shows the cloud-point curve for the system CTBN (polydisperse)–E(340) and binodal for the system CTBN(monodisperse)– E(340). The latter was vertically shifted so that its critical point intercepts the cloud-point curve for the polydisperse system. This condition enables us to compare the effect of CTBN polydispersity on the shape of the cloud-point curve, apart from the expected decrease in miscibility (decrease in the χ values). The horizontal lines tie pairs of points, giving the composition of the continuous phase Φ_1 (in the cloud-point curve of the polydisperse system) and the segregated phase $\Phi_1{}^A$ (in the "shadow" curve) that coexist at the cloud point. The curves intercept each other at the critical point.

Branches of the cloud-point curve for the polydisperse CTBN and its "shadow" curve represent extreme compositions that may be obtained in both phases at equilibrium. Thus, when a differential amount of phase A is

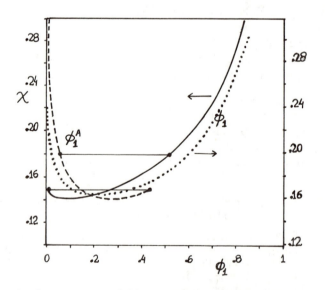

Figure 1. Cloud-point curve (solid line, labeled Φ_1) and "shadow" curve (dashed line, labeled $\Phi_1{}^A$) for the mixture CTBN(polydisperse)–E(340) compared with the binodal for the system CTBN(monodisperse)–E(340) (dotted line, labeled Φ_1). Values of χ are different for mono- and polydisperse systems.

segregated, it has a composition lying in the "shadow" curve; the principal phase composition lies in the cloud-point curve. Therefore, when there is a macroscopic phase separation, points representing the composition of both phases at equilibrium are located anywhere between both branches, at each side of the critical point, and joined by a horizontal line. If the system is assumed to be monodisperse and values of χ are adjusted to fit the experimental cloud-point curves at the critical point (4), the resulting binodal is located in a region where equilibrium compositions are expected for macrophase separation. This result may explain the reasonable predictive capability of a model that includes a simple thermodynamic calculation neglecting the effect of polydispersity on the particle-size distribution of dispersed domains produced in a CTBN–epoxy system (6).

Figure 2 shows an amplification of the region near the critical point for a polydisperse system. The composition at which a minimum solubility is predicted (precipitation threshold) is $\Phi_{1th} = 0.07$–0.08, close to experimental values reported in the literature (3).

The composition of the minority phase predicted by assuming that CTBN is monodisperse is significantly different from the one that results from taking the effect of polydispersity into account (Figure 3). For compositions in a range of commercial interest ($\Phi_1 = 0.1$–0.2), polydispersity decreases the rubber concentration in the segregated domains at the beginning of phase separation.

Figure 4 shows normalized weight distributions (on a weight basis) for

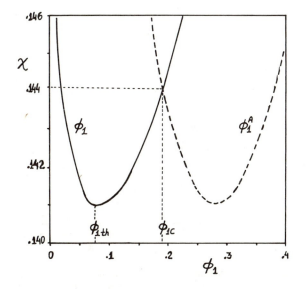

Figure 2. Amplification of Figure 1 in a region near the critical point (only for the polydisperse system).

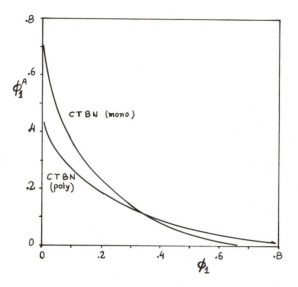

Figure 3. Predicted composition of the minority phase, $\Phi_1{}^A$, as a function of the composition in the principal phase, Φ_1, by assuming a monodisperse CTBN and by taking the effect of polydispersity into account.

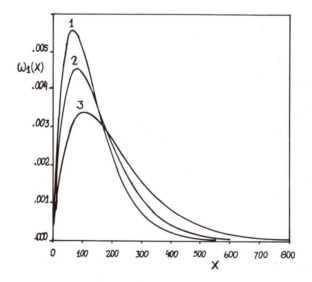

Figure 4. Normalized molecular weight distributions (on a weight basis) for CTBN. Curve 1, distribution in the principal phase; Curve 2, distribution in the segregated phase at the cloud point for $\Phi_1 = 0.15$; Curve 3, distribution in the segregated phase at the cloud point for $\Phi_1 = 0.10$.

CTBN. Curve 1 represents the initial distribution in the homogeneous mixture. Curves 2 and 3 indicate the corresponding distributions in the segregated phase for two different values of Φ_1. CTBN in the minority phase consists predominantly of higher-molecular-weight fractions, an effect that is enhanced for low initial rubber concentrations. This fractionation, which favors the higher-molecular-weight end of the distribution, is responsible for the decrease in the χ values at which phase separation takes place (compared to the assumption of a monodisperse polymer). For practical purposes, the polydispersity index of the segregated polymer does not change with respect to the initial index. Roe and Lu (7) showed the same behavior for polymers with Schulz–Flory distributions.

Finally, Figure 5 shows the cloud-point curve (labeled Φ_1) and "shadow" curve (labeled $\Phi_1{}^A$), for the mixture CTBN(polydisperse)–E(382)(polydisperse). Horizontal lines tie pairs of points that coexist at the cloud point. The polydispersity of the epoxy monomer flattens the cloud-point curve in the critical-point region.

Conclusions

In a CTBN–epoxy mixture the consideration of the actual polydispersity of the system shifts the miscibility gap to lower χ values (i.e., the system becomes more incompatible than is expected by assuming monodisperse components). In spite of the low polydispersity index of liquid epoxies used

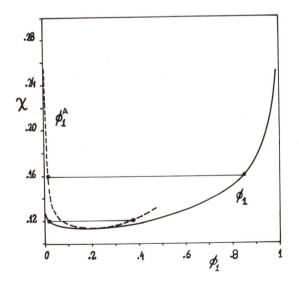

Figure 5. Cloud-point curve (labeled Φ_1) and "shadow" curve (labeled $\Phi_1{}^A$), for the mixture CTBN(polydisperse)–E(382)(polydisperse).

in industrial practice, for example E(382), the significance of its polydispersity on calculations is comparable to that of CTBN because the entropic contribution is very sensitive to changes in the distribution of species with small masses.

The effect of CTBN polydispersity is to produce a precipitation threshold at a low rubber volume fraction ($\Phi_{1th} = 0.07$–0.08), in agreement with experimental results reported in the literature (3). When the polydispersity of the epoxy monomer is also considered, a flattening of the cloud-point curve near the critical point is observed. Although the effect of polydispersity must be taken into account to fit experimental cloud-point curves, the monodisperse assumption gives a reasonable approximation when macroscopic separation is considered. This requires a fitting with experimental results at the critical point, implying a rescaling on the χ axis (i.e., a vertical shift so that the predicted critical point pertains to the experimental cloud-point curve).

The molecular weight distribution of CTBN segregated in the minority phase is shifted to high molecular weights. This effect is enhanced for low initial rubber concentrations.

References

1. *Rubber-Modified Thermoset Resins;* Riew, C. K.; Gillham, J. K., Eds.; Advances in Chemistry 208; American Chemical Society: Washington, DC, 1984, and references cited therein.
2. Visconti, S.; Marchessault, R. H. *Macromolecules* **1974**, *7*, 913.
3. Wang, T. T.; Zupko, H. M. *J. Appl. Polym. Sci.* **1981**, *26*, 2391.
4. Verchere, D. ; Sautereau, H.; Pascault, J. P.; Moschiar, S. M.; Riccardi, C. C.; Williams, R. J. *J. Polymer* **1989**, *30*, 107.
5. Williams, R. J. J.; Borrajo, J.; Adabbo, H. E.; Rojas, A. J. In *Rubber-Modified Thermoset Resins;* Riew, C. K.; Gillham, J. K., Eds.; Advances in Chemistry 208; American Chemical Society: Washington, DC, 1984; p 195.
6. Vazquez, A.; Rojas, A. J.; Adabbo, H. E.; Borrajo, J.; Williams, R. J. J. *Polymer* **1987**, *28*, 1156.
7. Roe, R. J.; Lu, L. *J. Polym. Sci., Polym. Phys. Ed.* **1985**, *23*, 917.
8. Rosen, S. L. *Fundamental Principles of Polymeric Materials;* Wiley: New York, 1982; p 130.
9. Gulino, D.; Galy, J.; Pascault, J. P.; Tighzert, L.; Pham, Q. T. *Makromol. Chem.* **1983**, *184*, 411.
10. Koningsveld, R.; Kleintjens, L. A. *J. Polym. Sci., Polym. Symp.* **1977**, *61*, 221.

RECEIVED for review March 25, 1988. ACCEPTED revised manuscript September 16, 1988.

Stabilizer Partitioning in Rubber-Modified Systems

Donald M. Kulich and Michael D. Wolkowicz

Technology Center, GE Plastics, Washington, WV 26181

Additives in rubber-modified systems partition between the elasto-meric phase and the thermoplastic phase. Because the rate and mech-anism of oxidation of each phase differ, the actual concentration of stabilizer in the elastomeric and thermoplastic phases can be an im-portant factor influencing overall polymer stability. The partitioning behavior of various antioxidants in model two-phase systems was studied by electron microscopy by using energy-dispersive X-ray analysis. Antioxidants used in the study included thiodipropionates, phosphites, and a phenolic antioxidant. Highly alkylated additives partitioned preferentially into the polybutadiene phase when the elas-tomer was dispersed in styrene–acrylonitrile copolymer. Effects of composition and relative solubility measurements of additives in polybutadiene and thermoplastics are described. The effects of ma-trix-phase composition on oxidative stability as determined by oxygen uptake and dynamic scanning calorimetry are shown to correlate with partitioning behavior.

RUBBER-MODIFIED SYSTEMS such as ABS (acrylonitrile–butadiene–sty-rene) and HIPS (high-impact polystyrene) are composed of an elastomeric component dispersed as a discrete particulate phase in a thermoplastic con-tinuous phase. Rubber-modified polymers are susceptible to oxidative deg-radation, and studies have been conducted to determine which structural feature is most sensitive to oxidation, the effects of oxidation on property changes, and the effects of additives on stability (1–6). The rate and mech-anism of oxidation of the rubber phase, usually polybutadiene (PB) or bu-tadiene copolymers, will differ significantly from that of the thermoplastic

0065–2393/89/0222–0329$06.00/0

phase, usually polystyrene (PS) or styrene–acrylonitrile (SAN) copolymer. Because additives can partition between phases, the antioxidant concentration in the elastomeric and thermoplastic phase can differ significantly from the average concentration. Partitioning of the additive within the polymer may, therefore, exert a controlling influence on the resulting effectiveness of the additive.

This chapter describes the partitioning behavior of various antioxidants as determined by scanning electron microscopy (SEM) by using X-ray energy-dispersive spectrometry (XEDS). Antioxidants used in the study included thiodipropionates, phosphites, and a phenolic antioxidant. Relative additive concentrations were determined in two-phase model systems by using XEDS, and results are compared with a determination of the relative solubility of the additives in PB and various thermoplastics.

Additive concentrations in polymers have been previously determined by microscopic, radiographic, thermogravimetric, spectroscopic, and extraction techniques (7–12). The use of SEM–XEDS to observe the redistribution of additives during polypropylene crystallization has been described (8). This chapter applies SEM–XEDS to analysis of additives in rubber-modified systems.

Experimental Details

Model systems were prepared for partitioning studies by mixing PB emulsion crumb with additive and the selected thermoplastic with a mixing bowl (Brabender) operating at 190 °C for 5 min at 60 rpm. A cross section of the compounded material was then prepared for SEM–XEDS analysis. The selected mixing time was sufficient to achieve the desired particle size and dispersion. Designated samples were thermally equilibrated under inert atmospheres at elevated temperatures.

Solubility measurements were carried out by placing weighed amounts of polymer together with excess additive in an ampoule, followed by evacuation and sealing. Samples were then equilibrated at the specified temperature for extended times to ensure that equilibrium was reached. The samples were quenched and additive concentrations were determined by SEM–XEDS analysis by using calibration curves. Measured concentrations are considered to be representative of additive concentrations attained after equilibration at the specified elevated temperature.

The cross-linked PB was prepared by emulsion polymerization to >90% conversion. The linear PB was Firestone Diene 35. The SAN, PS, and linear low-density polyethylene (LLDPE) were commercially available materials. The following additives were commercially available: dilauryl thiodipropionate (DLTDP); distearyl thiodipropionate (DSTDP); triphenyl phosphite; tris(2,4-di-t-butylphenyl) phosphite; and 2,4-bis(N-octylthio)-6-(4-hydroxy-3,5-di-t-butylanilino)-1,3,5-triazine (Irganox 565, Structure 1). The bis(4-nonylphenyl) thiodipropionate and bis(4-cyanophenyl) thiodipropionate were synthesized.

Oxidative stability measurements were determined on a dynamic scanning calorimeter (DSC, Du Pont). Samples analyzed were prepared by compounding 20% PB with experimental bulk SANs with styrene/acrylonitrile (S/A) ratios of 63:37, 75:25, and 80:20, and also with commercial PS. Stabilized samples were prepared

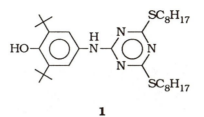

1

by adding 0.5 parts compound 1 during compounding. The samples were equilibrated at 190 °C in evacuated sealed ampoules for 60 h prior to analysis.

Oxygen-uptake measurements were conducted by using a previously described computerized method (*13*) on 40-mesh samples that had been compounded with a mixing bowl (Brabender).

Discussion

Thiodipropionates, phosphites, and hindered phenolic compounds are classes of compounds widely used as stabilizers against the thermal oxidation of rubber-modified systems such as HIPS and ABS. Sulfur and phosphorus stabilizers and some hindered phenolic compounds contain atoms of sufficient atomic number to permit detection by XEDS. Using model rubber-modified systems, relative antioxidant concentrations were thus determined from X-ray signal intensities in rubber and rigid areas. Rubber-particle sizes ranged from 10 to 100 μm, as shown in Figure 1.

As illustrated in Figure 2 for the system DLTDP in PB and SAN, the signal intensity emitted from the additive in the rubber phase is considerably greater than that from the additive in the SAN phase. This signal intensity indicates greater additive concentration in the rubber phase. The data in Table I demonstrate the significance of substituent effects on partitioning. An increase in substituent polarity reduces partitioning of the thiodipropionate into the rubber phase. The effects of the polymer matrix on partitioning are illustrated in Table II. Partitioning is strongly dependent on the nature of the matrix phase; increasing matrix polarity promotes concentration of the additive in the rubber phase. Significant differences exist between PB–PS and PB–SAN.

Differences in partitioning behavior are consistent with calculated solubility parameter differences. Solubility parameters (δ) were calculated from Small's equation (*14*)

$$\delta = \frac{d \, \Sigma G}{M} \tag{1}$$

where *d* is the density of the sample, *M* is the molecular weight of the additive or repeat unit in the polymer, and ΣG is the sum of the group molar

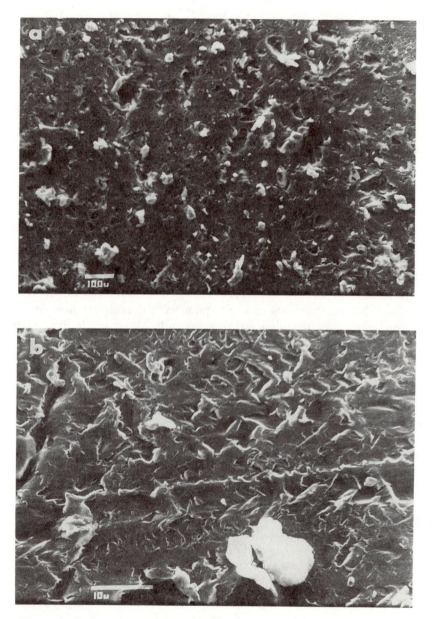

Figure 1. Cross section of PB–SAN blend, prepared by compounding on a mixing bowl. Key: a, ×120; b, ×2500.

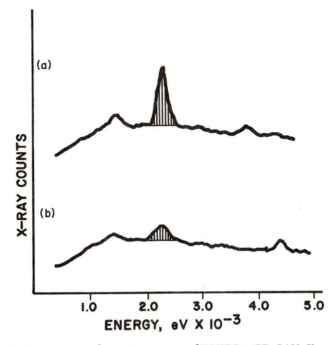

Figure 2. X-ray energy-dispersive spectra of DLTDP in PB–SAN. Key: a, sulfur (DLTDP) concentration in PB phase; b, sulfur (DLTDP) concentration in SAN phase.

attraction constants of all the chemical groups in the additive or polymer repeat unit. The solubility parameter of the SAN was calculated according to Krause (*15*)

$$\delta = \Sigma \delta_i \phi_i \tag{2}$$

where ϕ_i is the volume fraction of each component. Density values were calculated with equation 3

$$d = \frac{M}{\Sigma V_i} + V_r \tag{3}$$

where M is the molecular weight of additive or polymer repeat unit, V_i is the group contribution to the molar volume, and V_r is a residual volume factor (*16*).

As can be seen in Tables III and IV, the solubility parameters for the alkyl-substituted thiodipropionates are similar to the value for the PB component. However, polar substitution results in solubility parameter values of the additive closer to that of the PS or SAN phase. The solubility parameter

Table I. Substituent Effects on Partitioning for R in RO$_2$CCH$_2$CH$_2$SCH$_2$CH$_2$CO$_2$R

R	Additive, pph	Thiodipropionate, %	
		PB Phase	PSAN Phase[a]
$-C_{18}H_{37}$	2,3,5	100[b]	not detected
$-C_{12}H_{25}$	2,3,5	60–80[c]	20–40
$(C_6H_4)-C_9H_{19}$	3[d]	80	20
$-C_6H_5$	3[d]	70	30
$(C_6H_4)-CN$	3[d]	12	88

[a]Suspension with S/A = 2.2:1, w/w.
[b]Same results were obtained with mixing bowl.
[c]More heterogeneity was found with dilauryl vs. distearyl thiodipropionate.
[d]Thermally equilibrated samples at 190 °C.

Table II. Effect of Matrix on Partitioning for DSTDP

Matrix	PB Phase[a]	Matrix Phase
Styrene–acrylonitrile copolymer[b]	100	not detected
Polystyrene	50	50
Polyethylene[c]	33	67

NOTE: All results are percent DSTDP, with 5 parts per hundred of additive in 80:20 matrix–rubber system equilibrated at 190 °C after mixing bowl compounding.
[a]Emulsion rubber.
[b]S/A = 2.2:1, w/w.
[c]Linear low-density polyethylene.

Table III. Substituent Effect on Calculated Solubility Parameter for R in RO$_2$CCH$_2$CH$_2$SCH$_2$CH$_2$CO$_2$R

R	$\delta\ (cal/cm^3)^{1/2}$
$-C_{18}H_{37}$	8.4
$-C_{12}H_{25}$	8.5
$(C_6H_4)-C_9H_{19}$	8.9
$-C_6H_5$	10.4
$(C_6H_4)-CN$	11.7

Table IV. Calculated Solubility Parameters for Various Polymers

Polymer	$\delta^a\ (cal/cm^3)^{1/2}$
Polybutadiene	8.1
Polyethylene	8.2
Polystyrene	9.0
Styrene–acrylonitrile copolymer[b]	10.7

[a]Calculated from group molar attraction constants according to Small.
[b]S/A = 2.2:1, w/w.

rankings in Table III correlate with the partitioning trends in Table I. However, a relatively higher concentration of the diphenyl derivative was found in the PB phase than expected on the basis of the solubility parameter values calculated with Small's equation. These values do not take into account the reduction in the polar component's contribution to the solubility parameter caused by identical polar groups present in a symmetrical position. This change would increase the expected concentration of the additive in the less-polar phase.

SEM–XEDS analyses were also used to determine relative antioxidant solubilities in PB alone and in various thermoplastics. Solubility measurements were conducted to provide comparisons with previous partitioning data and to provide comparative data on phosphites and a hindered phenolic antioxidant. Concentrations of DLTDP and DSTDP in quenched samples of various thermoplastics equilibrated with excess additives are shown in Table V. DLTDP and DSTDP were found to be miscible with PB at elevated temperatures. Differences in behavior between DLTDP and DSTDP were observed, with DLTDP exhibiting slightly greater solubility in SAN and PS.

Increasing alkyl chain length in DSTDP (C_{18}) versus DLTDP (C_{12}) reduces solubility in SAN and PS versus PB. Solubility behavior thus correlates with partitioning behavior. Figure 3 illustrates that these solubility rankings remain unchanged over a wide temperature range.

Phosphite esters represent another class of widely used antioxidants. Most commercially available phosphite antioxidants are isomeric mixtures. Triphenyl phosphite and tris(2,4-di-*t*-butylphenyl) phosphite were selected for investigation to avoid the presence of structural isomers. The triaryl phosphite was shown to exhibit high solubility in SAN, as well as in PS and PB (*see* Table VI). Alkyl substitution again significantly reduces solubility in SAN and favors solubility in PB.

Most hindered phenolic antioxidants do not contain atoms sufficiently high in atomic number to permit detection by X-ray analysis. However,

Table V. Solubility of Thiodipropionates in Various Polymers at 157 °C

Polymer	DLTDP[a]	DSTDP[a]
PB (linear)[b]	>95	>95
PB (cross-linked)[c]	48	59
Polyethylene[d]	>95	>95
Polystyrene	>95	32
Styrene–acrylonitrile copolymer[e]	4	2

[a]Additive concentration, w/w %.
[b]Solution polymerized with 35:55 *cis–trans* and 10% vinyl.
[c]Prepared by emulsion polymerization.
[d]Linear low-density.
[e]S/A = 2.2:1, w/w.

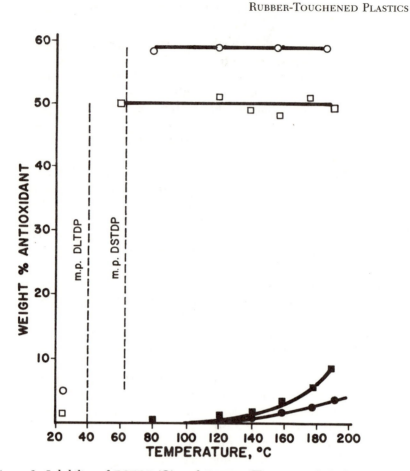

Figure 3. Solubility of DSTDP (○) and DLTDP (□) in cross-linked PB and DSTDP (●) and DLTDP (■) in SAN, with 2.2:1 S/A (w/w) as a function of temperature.

Table VI. Solubility of Phosphites in Various Polymers at 193 °C

Polymer	$P{+}O{-}\langle\bigcirc\rangle]_3^a$	$P{+}O{-}\langle\bigcirc\rangle{+}]_3^a$
PB (linear)[b]	>95	>95
PB (cross-linked)[c]	52	95
Polystyrene	>95	94
Styrene–acrylonitrile copolymer[d]	>95	8.5

[a]Additive concentration, w/w %.
[b]Solution polymerized with 35:55 *cis–trans* and 10% vinyl.
[c]Prepared by emulsion polymerization.
[d]S/A = 2.2:1, w/w.

compound 1 is soluble in various polymers (191 °C) at the following w/w additive concentrations:

- cross-linked PB, prepared by emulsion polymerization, 60
- PS, >95
- SAN (3:1 w/w), 20
- SAN (2.2:1 w/w), 11

Compound 1 shows low solubility in SAN, despite the presence of polar functional groups. Consistent with previous results, decreasing acrylonitrile comonomer content significantly increases additive solubility.

The ratio of styrene to acrylonitrile (S/A) of the matrix phase was changed in the PB–SAN model system, and the effects on partitioning and oxidative stability were determined by oxygen uptake and also by DSC. The relative stabilities of the unstabilized polymeric components are shown in Figure 4. The PB component oxidizes at a significantly higher rate than the SAN matrix components. No significant differences in stabilities were evident in the SAN components alone under the test conditions. Thus, stability measurements on the PB–SAN model system reflect primarily the stability of the rubber component.

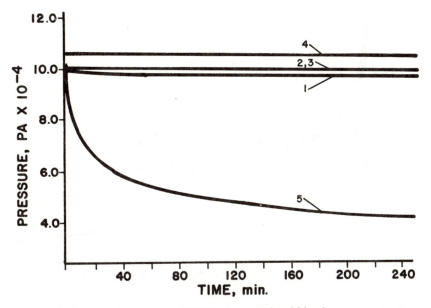

Figure 4. Oxygen absorption at 161 °C of unstabilized blend components. Key: 1, SAN with S/A of 63:37; 2, SAN with S/A of 75:25; 3, SAN with S/A of 80:20; 4, PS; 5, PB.

No significant differences between unstabilized PB–SAN blends were detected, as indicated by the negligible induction times obtained at the test conditions selected (*see* Figure 5). However, in the presence of 0.5 parts of compound **1**, induction times increase with increasing acrylonitrile content of the matrix phase (127 min at 100:0; 179.5 min at 80:20; 206.4 min at 75:25; and 241.5 min at 63:37) (*see* Figure 6).

Stability differences determined by DSC correlate with oxygen absorption measurements. Thus, time to onset of exotherm using programmed temperature rise or under isothermal conditions increases with increasing acrylonitrile content in the matrix phase (*see* Table VII). Figure 7 shows that increasing percent antioxidant in the rubber phase (due to differences in additive partitioning caused by changing S/A in the matrix phase) increases time to onset of exotherm. Each sample contained the same total loading of antioxidant.

Summary

The partitioning of additives between the rubber and rigid phases of rubber-modified systems was studied by SEM–XEDS. The effects of partitioning on additive concentrations were determined by direct measurements in model systems and assessed indirectly through solubility measurements.

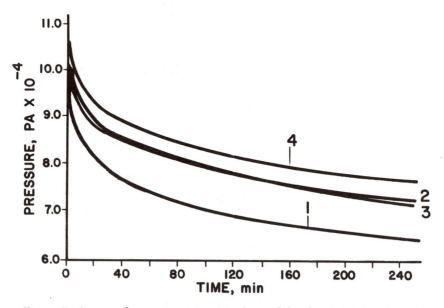

Figure 5. Oxygen absorption at 161 °C of unstabilized 20% PB blends with various rigid-phase compositions. Key: 1, SAN with S/A of 63:37; 2, SAN with S/A of 75:25; 3, SAN with S/A of 80:20; 4, PS.

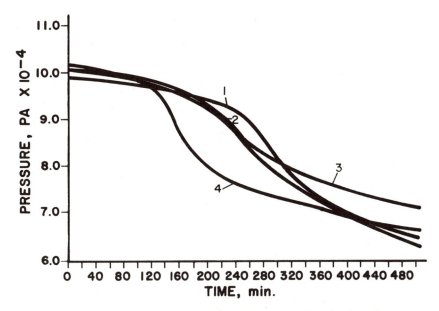

Figure 6. Oxygen absorption at 161 °C of stabilized 20% PB blends with various rigid-phase compositions. Key: 1, SAN with S/A of 63:37; 2, SAN with S/A of 75:25; 3, SAN with S/A of 80:20; 4, PS.

Table VII. Effect of Matrix S/A on Stability by DSC

Matrix, S/A	Isothermal[a] DSC, min	Dynamic[b] DSC, °C
100:0	5.4	209
80:20	6.3	208.5
75:25	9.9	217
63:37	17.3	226

[a]At 190 °C.
[b]At 10 °C/min.

Significant differences in additive concentration exist in each phase, depending upon the composition of the additive and that of the polymeric components. Solubility measurements correlate with partitioning experiments.

The behavior of various thiodipropionates, phosphites, and a phenolic antioxidant is described. In general, with PB-modified styrenics, increasing alkyl substitution in the additive strongly favors partitioning of the additive into the PB phase. Increasing acrylonitrile content in the styrenic portion also significantly promotes the concentration of these additives into the PB component. Oxidative stability differences correlate with partitioning behavior in model systems.

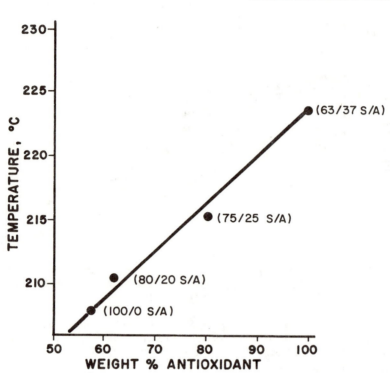

Figure 7. Effect of weight percent antioxidant in the PB phase of 20% PB blends with various rigid-phase compositions on temperature to exotherm in air by DSC. Each blend contains 0.5 parts total antioxidant loading.

Acknowledgment

The authors thank the technical staff of GE Plastics for their support.

References

1. Shimada, J.; Kabuki, K.; Ando, M. *Rev. Electr. Commun. Lab.* **1972**, *20*, 564.
2. Gesner, B. D. *J. Appl. Polym. Sci.* **1965**, *9*, 3701.
3. Kelleher, P. G. *J. Appl. Polym. Sci.* **1966**, *10*, 843.
4. Wolkowicz, M. D.; Gaggar, S. K. *Polym. Eng. Sci.* **1981**, *21*, 571.
5. *Developments in Polymer Stabilization—1*; Scott, G., Ed.; Applied Science: London, 1979; Chapter 9.
6. Ghaemy, M.; Scott, G. *Polym. Degrad. Stab.* **1981**, *3*, 233.
7. Roe, R.; Blair, H. E.; Gieniewski, C. *J. Appl. Polym. Sci.* **1974**, *18*, 843.
8. Billingham, N. C.; Calvert, P. D.; Manke, A. S. *J. Appl. Polym. Sci.* **1981**, *26*, 3543.
9. Klein, J.; Briscoe, B. *J. Polymer* **1976**, *17*, 481.
10. Kwei, T. K.; Zupko, H. M. *J. Polym. Sci. Part A2* **1969**, *7*, 867.

11. Cicchetti, O.; Dybini, M.; Parrini, P.; Vicario, G. P.; Bua, E. *Eur. Polym. J.* **1968,** *4,* 419.
12. *Stabilization and Degradation of Polymers;* Allara, D. L.; Hawkins, W. L., Eds.; Advances in Chemistry 169; American Chemical Society: Washington, DC, 1978.
13. Wozny, J. In *Polymer Additives;* Kresta, J., Ed.; Polymer Science and Engineering, Vol. 26; Plenum: New York, 1984; pp 111–126.
14. *Polymer Handbook*, 2nd ed.; Brandrup, J.; Immergut, E. H., Eds.; Wiley: New York, 1975; pp IV–339.
15. Krause, S. In *Polymer–Polymer Compatibility in Polymer Blends;* Paul, D.; Newman, S., Eds.; Academic: New York, 1978; Vol. 1, Chapter 2.
16. Van Krelvelen, D. V.; Hoftyzer, P. J. *Properties of Polymers*, 2nd ed.; Elsevier: New York, 1976; Chapter 4.

RECEIVED for review February 11, 1988. ACCEPTED revised manuscript September 15, 1988.

Synthesis, Characterization, and Evaluation of Telechelic Acrylate Oligomers and Related Toughened Epoxy Networks

Ajit K. Banthia, Prakash N. Chaturvedi, Vandana Jha, and Veera N. S. Pendyala

Materials Science Centre, Indian Institute of Technology, Kharagpur 721 302, India

Telechelic elastomeric carboxyl end-capped acrylates have been synthesized and thoroughly characterized. The effect of initiator [4,4'-azobis(4-cyanovaleric acid)] and chain-transfer agent (dithiodiglycolic acid) concentration, polymerization technique and temperature, and nature of solvent(s) on the gelation, molecular weight, and carboxyl functionality of the resulting oligomer has been explored in detail. Acrylate oligomers ($\overline{M}_n \leq 15,000$) exhibited extremely good miscibility with the conventional epoxy resin (diglycidyl ether of bisphenol A), but precipitated as a distinct (single–mixed) phase during network formation. Incorporation of the reactive elastomer (4–10%) significantly enhanced the impact strength of the epoxy network and correlated well with energy absorption of the relaxation process. Carboxyl-terminated telechelic ethylhexyl acrylate oligomers were found to be a potential elastomeric toughening agent for epoxy resin.

TOUGHNESS IS THE ABILITY OF A MATERIAL TO ABSORB ENERGY and undergo large permanent set without rupture. For many engineering applications, toughness is often the deciding factor. Plastics, because of their inherent brittleness, are an important candidate for toughening studies, with special emphasis on the brittleness of cross-linked glassy polymers. Epoxy resin (*1*), a versatile glassy network, exhibits excellent resistance to corrosion and solvents, good adhesion, reasonably high glass-transition temperatures (T_g), and adequate electrical properties. However, its poor fracture toughness has been the subject of intense investigations throughout the world.

Approximately 5–20% of a low-molecular-weight reactive liquid elastomeric oligomer, initially miscible and homogeneously dispersed in the host resin, may provide a multiphase toughened network (*2*) on curing. To optimize toughening, parameters such as modifier structure, molecular weight, solubility, and elastomer concentration must be clearly defined. These parameters are responsible for the dynamics of multiphase morphology.

For effective toughening of the epoxy resin, the dispersed elastomeric phase must be grafted, at least to a certain extent, to the matrix resin (*3–5*). However, excessive grafting of the elastomeric phase may lead to the formation of a single-phase morphological system (*6*), with disastrous results. These factors suggest the use of various reactive oligomers, such as carboxyl–amine end-capped acrylonitrile–butadiene oligomer (*7–10*), carboxyl-terminated polyisobutylene (*11*), and functionalized siloxane oligomers (*12–14*). Some commercially important oligomers used for toughening have a built-in unsaturation in the backbone that enhances their thermal and oxidative degradation. Acrylate-based reactive oligomers may be an ideal choice for the toughening of epoxy resin because of its comparatively better oxidative and thermal stability and its complete miscibility with diglycidyl ether of bisphenol A (DGEBA). Simultaneous interpenetrating network (SIN) has been used to improve the impact strength of epoxy resin by simultaneously polymerizing it with *n*-butyl acrylate (*15*). This chapter deals with synthesis, characterization, and evaluation of 2-ethylhexyl acrylate (EHA) based carboxyl-terminated oligomers as a toughening agent for DGEBA.

Experimental Details

Materials.   The monomers ethyl acrylate (EA), butyl acrylate (BA), and EHA were obtained commercially. They were purified by washing twice with aqueous sodium hydroxide solution (10% w/v) to remove inhibitor and then twice with distilled water. They were then dried over anhydrous calcium chloride for 48 h and distilled under reduced pressure. 4,4′-Azobis(4-cyanovaleric acid) (ABCVA) (**1**) and azobisisobutyronitrile recrystallized from ethanol were used as free-radical initiators and stored in the dark at –25 °C.

DGEBA liquid epoxy resin (Dobeckot 520F, epoxy equivalent weight 190 g) was used. The amine curing agent, bis(4-aminocyclohexyl)methane (PACM-20), was used as supplied.

Synthesis of Carboxyl-Terminated Acrylate Oligomers.   The following is a typical synthesis of carboxyl-terminated telechelic EHA oligomer by bulk polymerization. Approximately 20 g of EHA monomer was placed in a 100-mL two-necked glass reactor fitted with a stirrer, thermometer, and gas inlet. The requisite amount of ABCVA was added. After the system was well purged with an inert gas, the reaction was rapidly brought to the desired temperature and allowed to continue for 10–30 min. The reaction mixture was diluted with a suitable solvent such as aromatic hydrocarbons and immediately quenched to room temperature. Unreacted ABCVA was precipitated overnight and subsequently filtered. Unreacted monomer, if any, and the solvent were removed under vacuum on a rotary evaporator until a constant weight was obtained.

A typical preparation by a solution polymerization technique uses inhibitor-free monomer (0.1 mol), 2-propanol (3.0 mol) and ABCVA (0.015 mol). The materials were introduced into a flask, and nitrogen was slowly bubbled through them for 15 min to remove dissolved oxygen. The mixture was heated and stirred at the desired temperature for ≃ 12 h. The solution was cooled, filtered, and evaporated to dryness. Unreacted ABCVA was removed as previously described.

Isothermal free-radical bulk polymerization of acrylic acid esters was investigated at 100 °C with a differential scanning calorimeter (Perkin-Elmer DSC-2). ABCVA was added to initiate the polymerization. The enthalpies of the reaction and overall rate constants were calculated from the areas between the DSC curves and the baseline, which was obtained by back-extrapolation of the straight line recorded after the completion of polymerization. The DSC curves were calibrated with the melting enthalpy of indium.

The number-average molecular weight ($\overline{M}_n$) of the purified oligomers was determined in chloroform at 37 °C by using a vapor pressure osmometer (Knauer).

Carboxyl functionality of the oligomers was determined by potentiometric acid–base titration in nonaqueous medium, with alcoholic KOH as titrant (*14*). Indicators were also utilized to complement the potentiometric studies. The epoxy groups were titrated according to the procedure discussed in the literature (*14, 16*). The solubility parameters of the oligomers and DGEBA were determined by the recently developed method of Banthia et al. (*17*).

Epoxy Network Formation and Evaluation. *One-Stage Method.* The epoxy resin, the elastomer mixture (with 1% triphenylphosphine), and the curing agent were heated in separate beakers under vacuum (5 mm Hg) at 60 °C for 15 min to remove air bubbles. Then they were mixed, poured into hot aluminum molds, and cured at 100 °C for 2 h, followed by 160 °C for 2.5 h.

Two-Stage Method. The epoxy resin and the elastomer mixture were prereacted in the presence of 1% triphenylphosphine at 100 °C for 2 h. The prereacted epoxy–elastomer mixture and the curing agent were heated in separate beakers under vacuum (5 mm Hg) at 60 °C for 15 min to remove air bubbles. Then they were mixed, poured into hot aluminum molds, and cured at 160 °C for 2.5 h.

The glass-transition temperatures of the epoxy network were measured with a differential scanning calorimeter (Perkin-Elmer DSC-2). Dynamic mechanical spectroscopy was studied by using a dynamic mechanical thermal analyzer (Polymer Laboratories, PL–DMTA) in the temperature range –120 to 165 °C at 100 Hz, with a heating rate of 3 °C per min. Impact properties of some selected samples were determined with an instrumented falling-weight impact tester.

Results and Discussion

Synthesis and Characterization of Telechelic Acrylate Oligomers.   For higher acrylate monomers, termination of free-radical poly-

merization in bulk occurs predominantly via combination mode rather than diffusion-controlled termination. Furthermore, the onset of the gel effect is presumed to be delayed considerably as alkyl group length increases. Isothermal bulk polymerization of acrylate oligomers of interest provides a clue to the time of completion of free-radical polymerization reactions and to whether gelation took place in our systems at any stage of polymer or oligomer formation.

The course of the isothermal bulk polymerization of EA, BA, and EHA at 100 °C is presented in Figure 1. The polymerization trends of these acrylates are similar. After the initial brief period, a gradual increase in the exotherm is observed and a well-expressed gel effect sets in. However, with increasing length of the alkyl groups in the acrylic acid ester, the intensity of the gel effect decreases and the conversion at the onset of the gel effect increases. In the polymerization of EHA, the gel effect is not observed at any stage of polymer and oligomer formation.

Within the polymerization temperature range of our interest, 100% conversion takes place within 10–20 min. Similar observations of isothermal bulk polymerization of methacrylate have been reported by other workers (18). The same group reported that the gel effect in the polymerization of methyl acrylate starts immediately after the beginning of the reaction.

The shift of the onset of the gel effect to higher conversion and its ultimate suppression cannot be explained by the theory of diffusion-controlled termination in the highly viscous media, as a consequence of higher monomer viscosity. However, the shift can be explained by considering the fact that in the present system the increasing length of the alkyl group enhances the mobility of the polymer chains because of enhanced shielding of the carbonyl groups by the segmental rearrangement of these alkyl groups (18–22). Low-

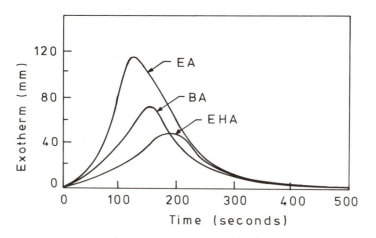

Figure 1. Isothermal bulk polymerization of EA, BA, and EHA at 100 °C in the presence of 1.0 × 10⁻² mol/L ABCVA.

ering of the glass-transition temperatures of polyacrylates (Table I) can also be explained in the same way.

The enthalpy of polymerization of these acrylates decreases with the length of the alkyl group (Table II). The polymerization enthalpies for the same monomers at different temperatures (80–120 °C) are within the limits of experimental error, estimated to be ±2%.

The molecular weight of EHA oligomers decreases with increasing temperature (Table III). The molecular weight decreases dramatically with the increasing polymerization temperature up to 100 °C in bulk (Table III). As expected, the increase in the initiator concentration resulted in a decrease of the molecular weight. However, the effect of initiator concentration is relatively small, as compared to the effect of polymerization temperature on the molecular weight.

Table I. Glass-Transition Temperatures of Some Representative Elastomeric Acrylates
$[CH_2-CH(COOR)]_n$

R	T_g, °C
C_2H_5	−25
n-C_4H_9	−55
C_5H_{11}	−60
C_6H_{13}	−57
C_8H_{17}	−45
$CH_2-CH(C_2H_5)-C_4H_9$	−55

Table II. Polymerization Enthalpies of EHA at Different Temperatures

Temperature (°C)	ΔH_p, kJ/mol
80	77.0
90	77.5
100	78.0
120	79.5

Table III. Number-Average Molecular Weight and the Functionality (f) of Carboxyl-Terminated EHA Oligomers Prepared by Bulk Polymerization

Sample No.	EHA, g	ABCVA, g	Temperature, °C	$\overline{M}_n$	f
1	20.0	2.0	120	10,231	2.07
2	20.0	2.0	100	11,564	2.01
3	20.0	2.0	90	12,325	2.02
4	20.0	2.0	80	15,026	2.04
5	20.0	2.5	100	10,078	1.99
6	20.0	3.0	100	10,196	2.01
7	20.0	4.0	100	9,525	2.03

Even these high-molecular-weight EHA oligomers were miscible with the conventional epoxy resin, DGEBA. This observation was supported by our studies pertaining to the determination of solubility parameters of polymers (17). The experimental solubility parameters of EA, BA, EHA oligomers, and DGEBA (Table IV) increase with increasing length of the alkyl group. This increase can be correlated with the segmental rearrangement and its effect on carboxyl group shielding. Presumably, the polymer–solvent interaction is enhanced by these factors. The close proximity of the solubility parameters of EHA oligomers (δ_T is 9.60 cal$^{1/2}$cm$^{3/2}$) and DGEBA resins (δ_T is 9.73 cal$^{1/2}$cm$^{3/2}$) ensures their complete miscibility up to $\simeq$ 20 wt % of the EHA oligomers in epoxy resin.

The measured functionality of carboxyl-terminated ethylhexyl acrylate (CTEHA) oligomers prepared by bulk polymerization technique was in the range of 1.99–2.07 (Table III). This characteristic carboxyl functionality value of CTEHA oligomer was found to be independent of the changes in the polymerization temperatures and initiator concentrations. Thus, these CTEHA polymers–oligomers are essentially telechelic in nature.

Dilute-Solution Polymerization of Acrylates. *Role of Disulfide Compounds as Chain-Transfer Agents.*

Solution polymerization of acrylic acid esters is better predicted and controlled, and dissipation of heat is easier than with bulk polymerization. Lack of heat dissipation could promote the onset of gelation for most of the lower acrylates. Several workers have studied the use of various symmetrical difunctional compounds (e.g., aromatic and aliphatic disulfide compounds) for the synthesis of telechelic polymers and oligomers by the free-radical polymerization technique (23–26).

Dithiodiglycolic acid (DTDGA) (2) was used as a chain-transfer agent. It can evidently react with the growing polymer radicals through cleavage of the sulfur–sulfur bond to provide a carboxyl end group. A representation of the tentative reaction mechanism is shown in Scheme I. This reaction will reduce the probability of chain termination of growing acrylate radicals by combination or disproportionation. In turn, disproportionation will be

Table IV. Solubility Parameters of Some Potential Oligomeric Acrylates and DGEBA

Oligomers	Theoretical[a]	Experimental[b]
Methyl acrylate	8.52	8.61
Ethyl acrylate	8.96	9.00
Butyl acrylate	9.32	9.30
Ethylhexyl acrylate	9.52	9.60
DGEBA	9.71	9.73

NOTE: All results for δ_T are given as the square root of calories per cubic centimeter at 25 °C.
[a]Small's method.
[b]Our results.

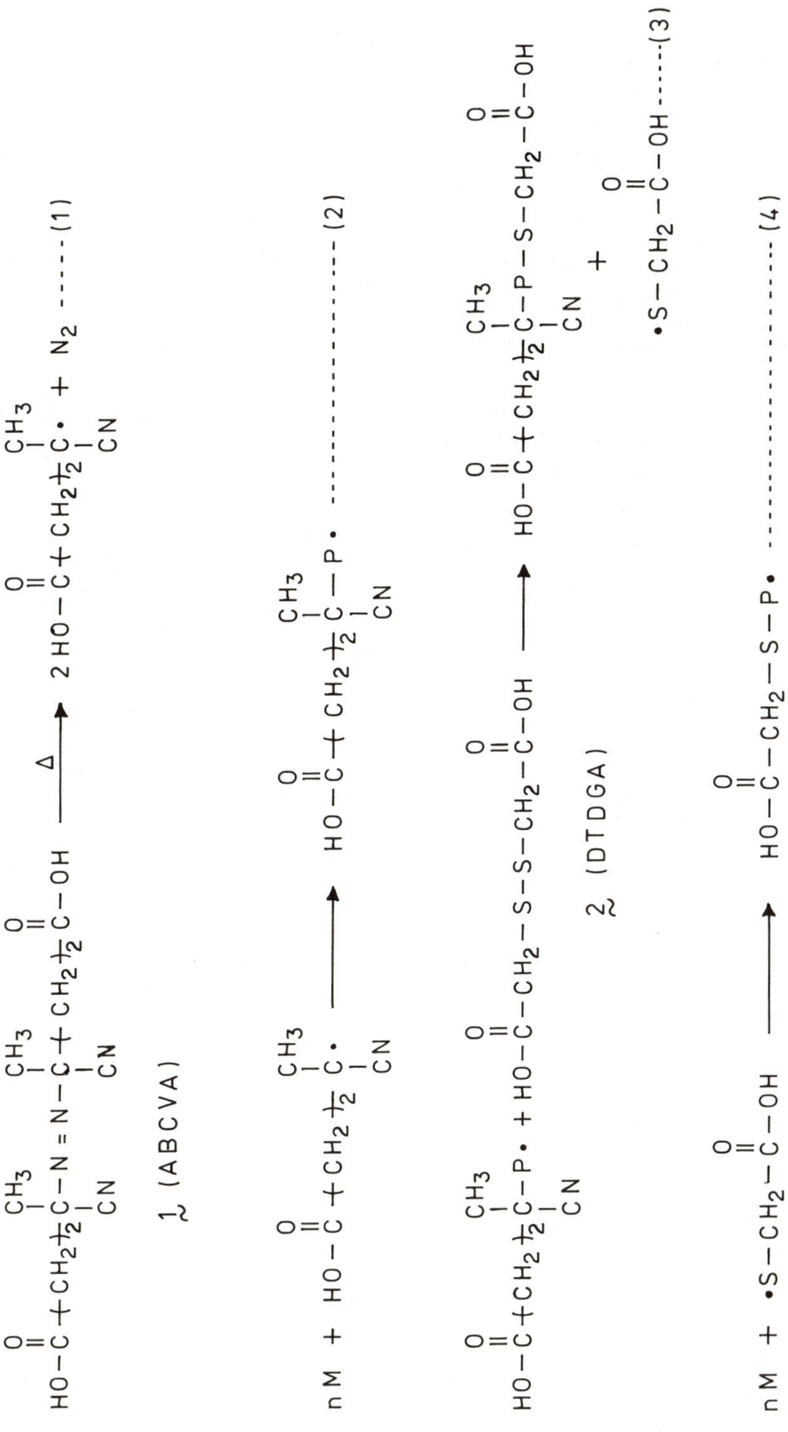

$$
\begin{aligned}
&\text{HO}-\overset{\overset{\text{O}}{\|}}{\text{C}}+\text{CH}_2\overset{\text{CH}_3}{\underset{\text{CN}}{\overset{|}{\text{C}}}}-\text{N}=\text{N}-\overset{\text{CH}_3}{\underset{\text{CN}}{\overset{|}{\text{C}}}}+\text{CH}_2\overset{\overset{\text{O}}{\|}}{\text{C}}-\text{OH} \\
&\qquad \underline{1}\ (\text{ABCVA}) \\
&\xrightarrow{\Delta}\ 2\ \text{HO}-\overset{\overset{\text{O}}{\|}}{\text{C}}+\text{CH}_2\overset{\text{CH}_3}{\underset{\text{CN}}{\overset{|}{\text{C}}}}\cdot\ +\ \text{N}_2\ \text{-----(1)}
\end{aligned}
$$

$$
n\,\text{M} + \text{HO}-\overset{\overset{\text{O}}{\|}}{\text{C}}+\text{CH}_2\overset{\text{CH}_3}{\underset{\text{CN}}{\overset{|}{\text{C}}}}\cdot\ \longrightarrow\ \text{HO}-\overset{\overset{\text{O}}{\|}}{\text{C}}+\text{CH}_2\overset{\text{CH}_3}{\underset{\text{CN}}{\overset{|}{\text{C}}}}-\text{P}\cdot\ \cdots\cdots\cdots (2)
$$

$$
\begin{aligned}
&\text{HO}-\overset{\overset{\text{O}}{\|}}{\text{C}}+\text{CH}_2\overset{\text{CH}_3}{\underset{\text{CN}}{\overset{|}{\text{C}}}}-\text{P}\cdot\ +\ \text{HO}-\overset{\overset{\text{O}}{\|}}{\text{C}}-\text{CH}_2-\text{S}-\text{S}-\text{CH}_2-\overset{\overset{\text{O}}{\|}}{\text{C}}-\text{OH} \\
&\qquad\qquad\qquad\qquad\qquad \underline{2}\ (\text{DTDGA}) \\
&\longrightarrow\ \text{HO}-\overset{\overset{\text{O}}{\|}}{\text{C}}+\text{CH}_2\overset{\text{CH}_3}{\underset{\text{CN}}{\overset{|}{\text{C}}}}-\text{P}-\text{S}-\text{CH}_2-\overset{\overset{\text{O}}{\|}}{\text{C}}-\text{OH} \\
&\qquad\qquad\qquad + \\
&\qquad\qquad\cdot\text{S}-\text{CH}_2-\overset{\overset{\text{O}}{\|}}{\text{C}}-\text{OH}\ \cdots\cdots(3)
\end{aligned}
$$

$$
n\,\text{M} + \cdot\text{S}-\text{CH}_2-\overset{\overset{\text{O}}{\|}}{\text{C}}-\text{OH}\ \longrightarrow\ \text{HO}-\overset{\overset{\text{O}}{\|}}{\text{C}}-\text{CH}_2-\text{S}-\text{P}\cdot\ \cdots\cdots\cdots (4)
$$

Scheme I

reduced considerably because of the decreased viscosity of the reaction media. Any decrease in the termination via disproportionation will enhance the probability of obtaining polymers with two carboxyl end groups per chain. Furthermore, better dissipation and removal of heat of polymerization from the reaction media will completely eliminate any probability of gelation, even in the case of the polymerization of lower acrylates (e.g., EA and BA).

The results of the effects of the concentration of the chain-transfer agent, DTDGA, on the molecular weight and the functionality of the resulting carboxyl-terminated acrylate oligomers are summarized in Table V. The molecular weight of the resulting oligomers decreases gradually with an increasing concentration of DTDGA. The measured carboxyl functionality of all these elastomeric acrylates was in the range of 1.98–2.04. Furthermore, no gelation was observed in any of these experiments. Thus, use of ABCVA and DTDGA as free-radical initiator and chain-transfer agent, respectively, will provide acrylate oligomers that are essentially telechelic.

Solvents as Chain-Transfer Agents.   The results of the dilute-solution polymerization of acrylate monomers in the presence of certain selected solvents are summarized in Table VI. As expected, increasing dilution of the polymerization media with these solvents decreases not only the molecular weight of the resulting acrylate oligomers, but also their carboxyl functionality. Both are attributable to extensive chain-transfer potentiality of these solvents. Furthermore, termination of the growing polymer chain through disproportionation may also contribute toward lowering the carboxyl

Table V. Number-Average Molecular Weight and the Functionality of Carboxyl-Terminated Acrylate Oligomers

Sample No.	Monomer	[DTDGA]	$\overline{M}_n$	f
CT-1	ethyl	0.00	11,564	2.01
CT-2		0.40	10,200	2.03
CT-3	hexyl	0.80	9150	1.99
CT-4		1.20	8035	2.04
CT-5	acrylate	2.00	6750	2.00
CT-6		3.00	6225	1.99
CT-7	butyl	0.40	9975	1.98
CT-8		0.80	9075	2.04
CT-9		1.20	8065	2.02
CT-10		2.00	7000	2.00
CT-11	acrylate	3.00	6320	1.98
CT-12	ethyl	0.40	11,250	2.02
CT-13		0.80	4200	2.03
CT-14		1.20	8125	2.04
CT-15		2.00	6900	1.99
CT-16	acrylate	3.00	6080	2.01

NOTE: Oligomers were prepared by free-radical polymerization at 100 °C in the presence of varied concentrations of dithiodiglycolic acid as chain-transfer agent. [monomer] = 20.0 g, [ABCVA] = 2.0 g.

Table VI. Number-Average Molecular Weight and Carboxyl Functionality of Acrylate Oligomers

Sample No.	Monomer	Solvent	Polymerization Temperature, °C	$\overline{M}_n$	f
ST–1	EA	IPA	83	1500	0.70
ST–2	BA	IPA	83	1350	0.73
ST–3	EHA	IPA	83	1250	0.75
ST–4	EA	TBA	83	3535	1.05
ST–5	BA	TBA	83	3670	1.15
ST–6	EHA	TBA	83	4020	1.19
ST–7	EHA	THF	65	1120	0.25
ST–8	EHA	acetone	65	6120	1.25
ST–9	EHA	CHCl$_3$	65	1105	0.30

NOTE: Oligomers were prepared by free-radical polymerization in the presence of various solvents. [monomer] = 20 g, [solvents] = 500 mL, [ABCVA] = 2 g.

functionality of these oligomers. As expected, *tert*-butyl alcohol exhibited lower chain-transfer potentiality than isopropyl alcohol under similar conditions of free-radical polymerization.

The results of the dilute-solution polymerization experiments suggested that in all these cases chain-transfer potentialities of these solvents were the major governing factor in lowering the molecular weight and the carboxyl functionality of the resulting acrylate oligomers. Furthermore, these transfer reactions generate potential free radicals, which do not carry carboxyl functionality. Therefore, the resulting polymer chains will have essentially no reactive terminal groups. It follows that use of solvents should be avoided in the synthesis of acrylate telechelic oligomers.

Epoxy Network Formation and Characterization Studies.
Carboxy materials are capable of interacting with epoxy groups to generate a hydroxy ester, as shown in Scheme II. This reaction will provide a route for chemically bonding the elastomeric acrylate (CTEHA) component to the rigid matrix. This chemically bonded system should have a relatively stable linkage between the modifier and the matrix, as required for the enhancement of toughening.

An idealized cross-linking reaction of DGEBA with a diamine is depicted in Scheme III. Ideally, four function points should be generated from each diamine molecule and thus form a relatively uniform network density in such systems. Any deviation will result in nonuniformity of the network profile. At present no such study has been reported in the literature for our control systems, which were prepared from DGEBA and PACM-20. We used PACM-20 exclusively for the present studies because the curing kinetics of the DGEBA–PACM-20 system had already been studied by Yilgor et al. (27). However, because of the absence of reaction kinetics studies pertaining to the epoxy–acrylate oligomer system, a two-step curing cycle was preferred.

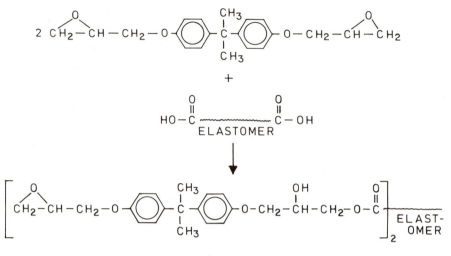

Scheme II

In general, this reaction of the modifier with DGEBA was allowed to proceed for 2 h at 100 °C in the presence of 1.0 wt % triphenylphosphine prior to the introduction of the curing agent. During this stage the CTEHA oligomers reacted with DGEBA to completion, as determined by the absence of carboxyl group. Presence of the theoretically calculated quantity of epoxy group was also confirmed by titration. The glass-transition temperature (T_g)

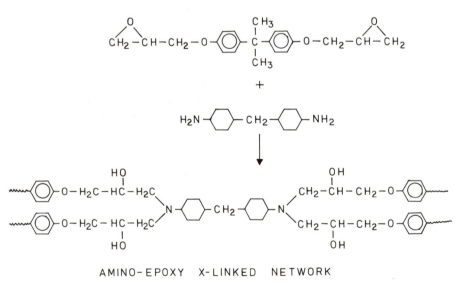

AMINO-EPOXY X-LINKED NETWORK

Scheme III

of the control and the modified network was always lower than 150 °C. Therefore, the chosen epoxy cross-linking temperature (160 °C) should be sufficient to provide adequate mobility for the chain ends to react and generate an optimized network structure. Furthermore, under the present cross-linking reaction conditions (160 °C for 2 h), DGEBA and PACM-20 react quantitatively (27).

All these modified networks have two distinctly separate glass-transition temperatures (Table VII). The upper transition temperature, around 150 °C, is attributable to the matrix resin network (DGEBA–PACM-20), whereas the lower transition, around –51 °C, is caused by the elastomeric acrylate modifier. Increasing acrylate concentrations (4–10 wt %) have little effect on the glass-transition behavior of the epoxy networks.

Furthermore, glass-transition behavior of these networks is independent of the modifier molecular weight within the range studied ($\overline{M}_n \sim 9500$–15,000). The proximity of these glass-transition values and the presence of two distinct transition temperatures confirms the formation of a multiphase morphology. DGEBA–PACM-20–CTEHA (one-step cured) and DGEBA–PACM-20–EHA (nonreactive, two-step cured) exhibited similar transition behavior. The nonreactive EHA oligomer was prepared by using azobisisobutyronitrile instead of ABCVA as polymerization catalyst. The number-average molecular weight of this sample was ~15,515.

Dynamic mechanical spectroscopy of the representative networks also confirms the presence of multiphase morphology, as two different β-relaxation peaks were observed (Figures 2 and 3). The location of the low- and high-temperature relaxation peaks at –48 and 148 °C does not change up to 10% concentration of the modifier. However, significant change occurs in the magnitude of these peaks, as is evident by broadening and the increase in the β-relaxation peak areas. The emergence of a new broad relaxation between the two pure components covering the initial plastics–elastomeric phase at higher modifier concentration can be attributed to the epoxy–EHA

Table VII. Glass-Transition Temperature of Carboxyl-Terminated EHA Oligomer Modified Epoxy Networks

Sample	EHA $\overline{M}_n$	Oligomer wt%	T_g, °C
1	9,525	10	149, –52
2	10,200	10	149, –51
3	11,564	10	150, –52
4	12,325	10	150, –52
5	15,026	10	149, –50
6	11,564	8	149, –51
7	11,564	6	149, –51
8	11,564	4	150, –52

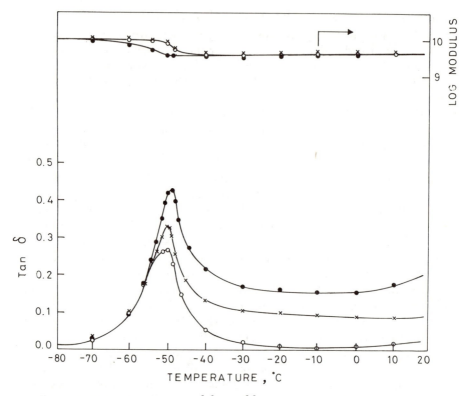

*Figure 2. Dynamic storage modulus and loss tangent, tan δ, vs. temperature
(–80 to 20 °C) curves for DGEBA–PACM-20 network and CTEHA-modified
DGEBA–PACM-20 networks for various CTEHA–DGEBA compositions.
Key: ●, 40–100; ×, 10–100; ○, 5–100.*

oligomer adduct, which precipitates as a distinct phase within the epoxy
matrix during curing. This type of attachment is important for toughening
as compared to simple inclusion of a nonreactive elastomer, which does not
improve toughness.

   Impact behavior of some typical representative samples is shown in
Figure 4. The results show that more than a 50% increase in impact strength
was obtained by using as little as 8 wt % of CTEHA ($\overline{M}_n$ 11,564) with respect
to an unmodified control network. Similar behavior was obtained earlier by
Pircivanoglu et al. (28), who used amine-terminated acrylonitrile–butadiene
elastomer-modified (ATBN 21, $\overline{M}_n$ 5200) epoxy networks prepared under
comparable reaction conditions. Furthermore, very insignificant increase in
the impact strength of the nonreactive EHA oligomer-modified epoxy network also confirms the fact that a chemically bonded elastomeric phase is
important for the enhancement of toughening. An attempt is being made to

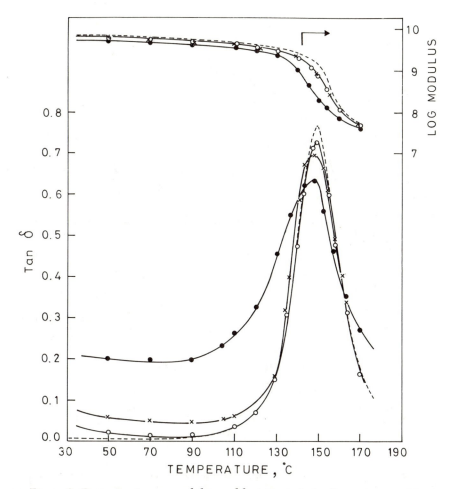

Figure 3. Dynamic storage modulus and loss tangent, tan δ, vs. temperature (30–180 °C) curves for DGEBA–PACM-20 network and CTEHA-modified DGEBA–PACM-20 networks for various CTEHA–DGEBA compositions. Key: ●, 40–100; ×, 10–100; ○, 5–100; - -, 0–100.

correlate the area under the relaxation peaks with the toughening behavior of epoxy–elastomer composites.

Conclusions

Carboxyl-terminated telechelic acrylate oligomers can easily be synthesized by bulk polymerization within the temperature range 80–120 °C by using ABCVA as free-radical initiator in the presence of DTDGA as chain-transfer agent.

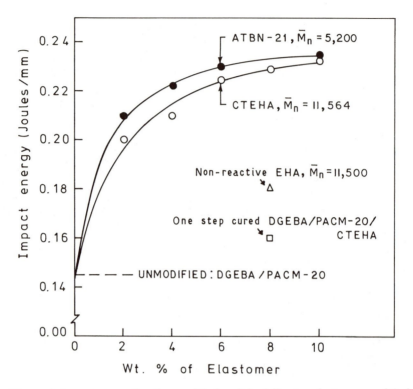

Figure 4. Impact strengths of unmodified and the following elastomer-modified DGEBA–PACM-20 networks. Key: ○, *2–10 wt % CTEHA* ($\overline{M}_n$ *11,564);* ●, *2–10 wt % ATBN-21* ($\overline{M}_n$ *5200);* △, *8% nonreactive EHA* ($\overline{M}_n$ *11,500);* □, *8% CTEHA (one-step cured).*

DSC curves of isothermal bulk polymerization of acrylates indicated that heat of polymerization decreases with increasing length of ester group. The chance of gelling decreases in the following order: EA > BA > EHA. Furthermore, nearly 100% conversion of acrylate monomers can be achieved.

The molecular weight of these acrylate oligomers can easily be controlled by polymerization temperature, initiator concentration, and chain-transfer agent concentration. Increasing initiator or chain-transfer agent concentration, as well as polymerization temperature, decreases the molecular weight.

Carboxyl-terminated EHA oligomers (δ_T 9.60 cal$^{1/2}$/cm$^{3/2}$, $\overline{M}_n$ < 15,000) were completely miscible with DGEBA (δ_T 9.73 cal$^{1/2}$/cm$^{3/2}$) at room temperature.

Severe chain transfer to solvents appeared to decrease the overall carboxyl functionality of the resulting acrylate oligomer in dilute-solution free-radical polymerization.

The presence of two distinct T_g and β-relaxation temperatures in all the cases under investigation indicates the multiphase nature of these systems.

There is a distinct shift in the β-relaxation temperature as the EHA concentration increases beyond 10 wt %. Furthermore, the formation of an epoxy–rubber reactive adduct may be the prime reason for the formation of a broad region of damping area covering both initial plastics–elastomeric phases. This damping area may be an important governing factor in the enhancement of toughening and may have synergistic effects. Similar effects have been observed in various other toughened systems.

Preliminary results indicated that impact strength of the CTEHA modifier epoxy network increases with increasing modifier concentration (2–10%). However, nonreactive EHA oligomer failed to enhance the toughening characteristic of the epoxy network.

Carboxyl-terminated telechelic ethylhexyl acrylate oligomers might be a potential elastomeric toughening agent.

Acknowledgments

A. K. Banthia expresses his appreciation to the Department of Science and Technology, Government of India, for the financial support needed to carry out the present investigation. The bis(4-aminocyclohexyl)methane sample was donated by E. I. du Pont de Nemours and Company.

References

1. Lee, H.; Neville, K. *Handbook of Epoxy Resins;* McGraw-Hill: New York, 1947.
2. Bucknall, C. B. *Toughened Plastics;* Applied Science: New York, 1977.
3. McGarry, F. J.; Willner, A. W. *Toughening of an Epoxy Resin by an Elastomeric Second Phase;* R 68-8, Massachusetts Institute of Technology: Cambridge, March 1968.
4. Bucknall, C. B.; Yoshii, T. *Br. Polym. J.* **1978**, *10*, 53.
5. Scarits, P. R.; Sperling, L. H. *Polym. Eng. Sci.* **1979**, *19*, 297.
6. Noshay, A.; Robeson, L. H. *J. Polym. Sci., Polym. Chem. Ed.* **1974**, *12*, 689.
7. Rowe, E. H.; Siebert, A. R.; Drake, R. S. *Mod. Plast.* **1970**, *49*, 110.
8. Riew, C. K.; Rowe, E. H.; Siebert, A. R. In *Toughness and Brittleness of Plastics;* Deanin, R.; Crugnola, A. M., Eds.; Advances in Chemistry 154; American Chemical Society: Washington, DC, 1976; p 326.
9. Riew, C. K. *Rubber Chem. Technol.* **1981**, *54*, 374.
10. Drake, R. K.; Egan, D. R.; Murphy, W. T. *Org. Coat. Appl. Polym. Sci. Proc.* **1982**, *46*, 392.
11. Slysh, R. In *Epoxy Resins;* Advances in Chemistry 92; American Chemical Society: Washington, DC, 1970; p 108.
12. Banthia, A. K.; Riffle, J. S.; Steckle, W. P.; Yilgor, I.; Wilkes, G.; McGrath, J. E. In *Proceedings of the International Conference on Structure—Property Relationship of Rubber;* Bhowmick, A. K.; De, S. K., Eds.; Indian Institute of Technology: Kharagpur, India, 1980; P-217.
13. Riffle, J.; Yilgor, I.; Banthia, A. K.; Wilkes, G. L.; McGrath, J. E. *Org. Coat. Appl. Polym. Sci. Proc.* **1982**, *46*, 397.
14. Riffle, J. S.; Yilgor, I.; Tran, C.; Wilkes, G. L.; McGrath, J. E.; Banthia, A. K.

In *Epoxy Resin Chemistry II;* Bauer, R. S., Ed.; ACS Symposium Series 221; American Chemical Society: Washington, DC, 1983; p 21.

15. Sperling, L. H.; Friedman, D. W. *J. Polym. Sci., Part A-2* **1969**, *7*, 425.
16. Standard Test Method for Epoxy Content Epoxy Resins, ASTM Designation: D 1652-73.
17. Pendyala, V. N. S.; Shareef, K. M. A.; Banthia, A. K., unpublished results.
18. Malavasec, T.; Osredkar, U.; Anzur, L.; Vizovisek, I. *J. Macromol. Sci. Chem.* **1986**, *A23*, 853.
19. Cardenas, J. N.; O'Driscoll, K. F. *J. Polym. Sci., Polym. Chem. Ed.* **1976**, *14*, 883; **1977**, *15*, 2097.
20. North, A. M.; Reed, G. A. *J. Polym. Sci., Part A* **1963**, *1*, 1311.
21. Benson, S. W.; North, A. M. *J. Am. Chem. Soc.* **1962**, *84*, 935.
22. Burkhart, R. D. *J. Polym. Sci., Part A* **1965, 3, 883.**
23. Pierson, R. M.; Costanza, A. J.; Weinstein, A. H. *J. Polym. Sci.* **1955**, *17*, 221, 319.
24. Athey, R. D.; Mosher, W. A.; Weston, N. W. *J. Polym. Sci.* **1977**, *15*, 1423.
25. Athey, R. D.; Mosher, W. A.; Weston, N. W. *Polym. Prepr. (Am. Chem. Soc. Div. Polym. Chem.)* **1979**, *20*, 20.
26. Athey, R. D. *Prog. Org. Coat.* **1979**, *7*, 289.
27. Yilgor, I.; Yilgor, E.; Banthia, A. K.; Wilkes, G. L.; McGrath, J. E. *Polym. Bull.* **1981**, *4*, 323.
28. Pircivanoglu, V. S.; Ward, T.; Yilgor, I.; Banthia, A. K.; McGrath, J. E. *Polym. Prepr. (Am. Chem. Soc. Div. Polym. Chem.)* **1981**, *22*, 174.

RECEIVED for review February 11, 1988. ACCEPTED revised manuscript December 27, 1988.

Ethylene–Propylene Rubber Bulk Functionalization

Roberto Greco[1], Pellegrino Musto[1], Fernando Riva[1], Gennaro Scarinzi[1], and Giovanni Maglio[2]

[1]Istituto di Ricerche su Tecnologia dei Polimeri e Reologia del Consiglio Nazionale delle Ricerche, 80072 Arco Felice (Napoli), Italy
[2]Dipartimento di Chimica, Università di Napoli, Naples, Italy

Ethylene–propylene random copolymers (EPR) were functionalized by inserting on their backbones a given amount of dibutyl maleate (DBM) by means of a bulk reactive process initiated by bis(α,α-dimethylbenzyl) peroxide (dicumyl peroxide, DCPO). The effect of temperature and relative DCPO amount in the reactant composition on the kinetics of the DBM grafting reaction was investigated. The kinetics of the degradation of the polymeric EPR substrate was monitored as well. A molecular model for the kinetics of both reactions was proposed and discussed with respect to the experimental data. An accurate analysis of the functionalized rubbers was performed by infrared spectroscopy, differential scanning calorimetry, and wide- and small-angle X-ray scattering. The DBM molecules were preferentially inserted on the ethylene sequences (C_2) of the EPR rubbers because of the steric hindrance of the methyl groups on the propylene (C_3) sequences.

POLYOLEFIN FUNCTIONALIZATION INITIATED BY ORGANIC PEROXIDES in solvent media has been carried out in recent years to improve properties such as dyeability and adhesion or to use the modified polymers for further reactions such as cross-linking and grafting (*1–8*). Ethylene(C_2)–propylene(C_3) random copolymers (EPR) were functionalized in solution by inserting onto their backbones maleic anhydride (MAH) molecules. The succinic anhydride (SA) groups, reacting with a polyamide (PA6), generated a graft copolymer, (EPR-g-SA)-g-PA6, which acted as an interfacial agent between PA6 and the rubbery dispersed phase (EPR or EPR-g-SA). The in situ formation of the (EPR-g-SA)-g-PA6 graft copolymer was obtained by melt-mixing of EPR-g-SA with molten PA6 and by hydrolytic polymerization of caprolactam (CL) in the presence of EPR-g-SA. The resulting binary (from PA6–EPR-g-SA) and ternary (from PA6–EPR–EPR-g-SA) blends both exhibited high impact resistance (9–14).

Other functional molecules, such as maleic and fumaric esters, were grafted on EPR copolymers as well. In these cases, however, the PA6 toughening was satisfactory only for blends obtained by the CL hydrolytic polymerization route (13). Melt-mixing residence times were too short for the ester groups, which react slowly with PA6, to reach a sufficient degree of conversion.

All these results were obtained by solution methods, which did not seem very suitable for large industrial applications. Therefore, some attempts at bulk functionalizations were made in our institute and by other authors (15–18). The MAH, however, is too volatile and toxic; it is quite incompatible with EPR and chemically aggressive toward metallic surfaces. Dibutyl maleate (DBM), though less reactive, does not show these shortcomings and hence was chosen as a model molecule for bulk reactions.

At first the reactants were put into the mixer simultaneously and analysis was carried out (19). Later we recognized that it was conceptually significant to divide the bulk process in two steps: the premixing of the reactants at conditions under which no reaction could take place and the reaction itself.

The present chapter is a brief review of the results of a systematic analysis of the model system EPR–DBM initiated by bis(α,α-dimethylbenzyl) peroxide (dicumyl peroxide, DCPO), in which the reaction temperature, the amount of DCPO, and the nature of the EPR substrate were varied (20–22). The data were compared with the predictions of a mathematical model derived from a proposed molecular mechanism of the functionalization reaction (23).

The unmodified EPR and the functionalized elastomers were investigated by infrared spectroscopy (IR), differential scanning calorimetry (DSC), and wide- (WAXS) and small-angle X-ray scattering (SAXS) to elucidate the influence of the degree of grafting on the structural and superreticular (long-range) order of the modified elastomers.

Experimental Details

The EPR were commercial polymers produced by Montedison with trade names Dutral Co054 (63% by mol of C_2; $\overline{M}_w$ = 1.8 × 10; [η] = 1.98 dL/g in tetrahydronaphthalene at 135 °C) and Dutral Co034 (72% by mol of C_2; intrinsic viscosity [η] = 1.95 dL/g). The DBM was reagent grade. The DCPO was crystallized from absolute ethanol and stored under vacuum on P_2O_5. EPR and DBM (with DCPO dissolved in it) were premixed at 90 °C in a batch mixer (Haake) for 5 min at a roller speed of 32 rpm. The real composition of the reactant mixture was measured by IR and NMR spectroscopy. The successive reaction was carried out at temperatures ranging from 140 to 180 °C, both in the same Haake mixer (dynamic mode) and in a thermostatic bath (static mode) (20).

At a fixed temperature (150 °C) the DCPO content was varied from 0.5 to 2 parts by weight, keeping EPR and DBM constant (100 and 10 parts, respectively). At the same temperature and composition (EPR–DBM–DCPO = 100–10–1) the substrate was changed to Co034 EPR (21).

The product of the reaction, the functionalized EPR, freed from the unreacted DBM, was analyzed by DSC, IR spectroscopy, and viscometry. Sheets 1.5 mm thick, obtained by compression molding at 160 °C, were used in the X-ray experiments.

WAXS profiles were carried out at (20 ± 0.5) °C by a powder diffractometer (PW 1050/71, Philips, Cu K_α, nickel-filtered radiation) in the reflection mode, continuously scanning the 2θ angle.

SAXS profiles were collected (by means of the same radiation and at the same temperature) by a step-by-step automated camera (Rigaku Denki), scanning the scattering 2θ angle at intervals of 0.01°; counting times were 1000 s per point. The intensities were corrected for parasitic scattering (background) and for absorption. WAXS and SAXS data were taken on the Co034 EPR and on the corresponding modified EPRs, whose degree of grafting ranged from 3.7 to 13.4% in weight of DBM (22).

Results and Discussion

General Considerations.   In bulk functionalizations, the mixing of the reactants in a highly viscous polymeric melt is not a trivial matter. In general, two cases are possible.

Case 1.   The functionalizing molecule (in which the radical initiator is usually dissolved) is sufficiently compatible with the polymeric substrate at the reaction conditions to ensure the achievement of the desired conversion in a homogeneous medium. In this case one must be sure that the mixing kinetics is much faster than that of the reaction. In other words, if t_R is the time needed to reach a given conversion and t_M is that necessary to get a molecular mixing, the following relation must be valid:

$$\frac{t_R}{t_M} \gg 1$$

If the two times are comparable, or even worse, if $t_R/t_M \ll 1$, the reaction will start to occur in a nonhomogeneous system, giving rise to nonuniform products. Therefore it is necessary to perform the whole process in two distinct steps: premixing of the reactants and reaction at the desired conditions. Because t_R and t_M are not easily measurable at the reaction conditions, it is generally advisable to adopt this double-step procedure. EPR and DBM fall in such a category with $t_R \gg t_M$. We followed the suggested double process for a careful kinetic analysis (20). Other authors, recognizing the problem for another kind of reaction, chose a modifier compatible with the substrate (15).

Case 2.   The compatibility between the modifier and the polymer substrate is quite low (case 1 is not fulfilled). The reactants will react mainly at the interface and only to a very limited extent in the homogeneous polymeric phase. Therefore the process must be designed case by case, depending on the thermodynamic compatibility of the components. However, some general suggestions can be given to improve the reactor performance. A continuous renewal of the interface must be provided by means of suitable shearing forces. A slow continuous (or stepwise) feeding of the modifier into the molten polymer can be used conveniently.

In the literature, some of the systems falling in this category are polyolefins (nonpolar substances) with (polar) anhydrides. In some cases the technique used has been a stepwise addition of reduced amounts of MAH to the molten polyolefin (16, 17). For our system (EPR–DBM–DCPO), which falls in case 1, the reaction in the second step was accomplished in both a dynamic and a static way (20).

Molecular Mechanism.   We proposed the following molecular mechanism in the literature (7, 16) for the EPR functionalizing process:

$$PH + R\cdot \rightarrow P\cdot + RH \tag{1a}$$

$$P\cdot + M \rightarrow PM\cdot \tag{1b}$$

$$PM\cdot + M \rightarrow PMM\cdot \tag{2}$$

$$PM\cdot + PH \rightarrow PM + P\cdot \tag{3}$$

where R, P, and M represent primary radicals, EPR, and DBM, respectively.

Termination Reactions:

$$P\cdot(PM\cdot) + R\cdot \rightarrow P\text{-}R\ (PM\text{-}R) \tag{4}$$

$$2P\cdot \rightarrow P\text{-}P \tag{5a}$$

$$2PM\cdot \rightarrow PM\text{-}PM \tag{5b}$$

$$P\cdot + PM\cdot \rightarrow P\text{-}PM \tag{5c}$$

$$2P\cdot \rightarrow PH + \wedge\!\wedge\!\wedge \tag{6a}$$

$$2\,PM\cdot \;\rightarrow\; PM \;+\; \text{\raisebox{0pt}{\rule{2cm}{0pt}}}$$
$$\underset{ROOC\;\;COOR}{}$$
(6b)

$$P\cdot +\; PM\cdot \;\rightarrow\; PM \;+\; \text{\raisebox{0pt}{\rule{2cm}{0pt}}}$$
(6c)

β-Scission Reactions:

$$\begin{array}{c} CH_3 \quad CH_3 \qquad\qquad\qquad\qquad CH_3 \\ \text{\small\textasciitilde}CH_2\text{-}\underset{\cdot}{C}\text{-}CH_2\text{-}CH\text{\textasciitilde} \;\rightarrow\; \text{\textasciitilde}CH_2\text{-}C\text{=}CH_2 \;+ \\ \qquad\qquad\qquad\qquad\qquad\qquad +\; \underset{\cdot}{C}H_2\text{\textasciitilde} \end{array}$$
(7)

This set of reactions, with certain simplifying assumptions, led to an analytical expression for the kinetics (23).

Kinetics of Functionalization.   At a constant composition of the reactant mixture (EPR–DBM–DCPO = 100–10–1) the values of the degree of grafting (DG) as a function of the reaction time, at four diverse temperatures (T_R = 140, 150, 160, and 180 °C), in the static mode, are reported in Figure 1a. An induction period (decreasing with increasing T_R) is observed, in which no DG is detectable. Then all the curves exhibit an initial linear trend, followed by a gradual diminution of the reaction rate and leveling off to a plateau region. The higher the T_R, the steeper the initial slopes of the curves. The plateau values are quite similar for the various T_R curves except the one relative to 140 °C, which is lower. The initial reaction rates, calculated from the curves, correlate well in a Arrhenius plot with an activation energy of 40 kcal/mol. Because the activation energy of the DCPO thermal decomposition is 37 kcal/mol, this could be the rate-determining step for the whole grafting process.

The intrinsic viscosity, measured at 135 °C in tetrahydronaphthalene, versus the reaction time is reported in Figure 1b, providing a rather qualitative picture of the deep molecular rearrangements that the EPR undergo. All the curves show a marked molecular degradation with elapsing time. A very surprising feature is observable, however, in some more or less pronounced maxima of [η] in the curves obtained at 140, 150, and 160 °C. Such an increase in molecular weight occurs at times corresponding to the induction period observed in Figure 1a, where no DG is detected. Therefore a different mechanism, other than that previously described, must govern the system. The DCPO thermal decomposition yields two primary radicals R·, which in turn give rise, by H abstraction, to two P· macroradicals. At first, because of the lack of molecular mobility in the absence of shear-

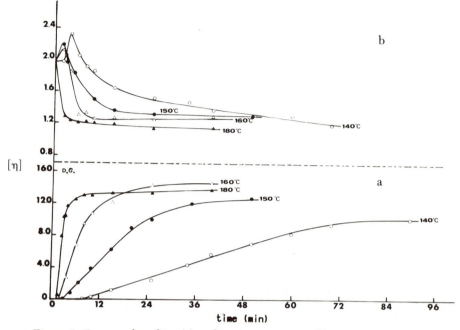

Figure 1. Degree of grafting (a) and intrinsic viscosity (b) versus reaction time for functionalization carried out at a fixed composition (EPR–DBM–DCPO = 100–10–1 by weight) and at temperatures as indicated. (Reproduced with permission from ref. 20. Copyright 1987 Wiley.)

ing forces, the formed P· radicals are trapped in a sort of cage. They tend to react together by coupling, increasing the EPR molecular weight in this way.

With elapsing time, however, a critical concentration of macroradicals is built up in the system. Radical sites start more and more to jump out of the cage onto neighboring chains by transfer and hence migrate throughout the polymeric bulk. For each of these escaped radical sites, a series of β-scissions can occur before their termination, generating a continuous molecular degradation.

Coupling prevails at the early stages of the reaction (initial [η] increase). It is rapidly balanced (at the maxima) by the starting β-scissions, which later begin to overcome the coupling (sharp [η] decrease). The final leveling of the curves indicates not only the exhausting of the radical sites and DBM concentration diminution as the grafting proceeds, but also the fact that [η] becomes scarcely sensitive to changes in the molecular characteristics as soon as the molecular weights become sufficiently low. The data given in Figure 1 seem to show some direct correlation between the grafting reaction and the EPR degradation. This correlation is indeed the case, as we can check by plotting [η] vs. DG. The data fall on a unique curve, independent of the temperature and the time of reaction (*20*). This result can be explained as follows:

In the absence of DBM, secondary P· macroradicals yield a coupling reaction (as in the case of high-density polyethylene, PE), and tertiary ones lead mainly to scissions (as for isotactic polypropylene, PP).

In the presence of DBM, grafting competes with these two reactions. It was found (*9*) that, under the same experimental conditions, the DG follows the order: PP $<<$ PE $\simeq$ EPR. Such considerations imply that, as grafting proceeds, the ratio C_3/C_2 of P· macroradicals increases, favoring β-scissions over coupling. In other words, the DBM is preferentially grafted on the C_2 longer sequences of the EPR chains, because of the steric hindrance of the C_3 methyl groups.

In a different kind of experiment, the DCPO amount was varied, at constant temperature (150 °C), from 0.5 to 2 parts by weight, keeping the EPR and the DBM contents constant. The DG values vs. time are reported in Figure 2. The trend of the curves is similar to that in Figure 1a, with an induction period and an initial linear portion of the curves (whose slopes and plateau values increase as the DCPO amount is enhanced). The initial reaction rates correlate linearly with the square root of the DCPO content, in accordance with the previously mentioned molecular model (*23*).

The [η] data versus time, relative to the same conditions as in Figure

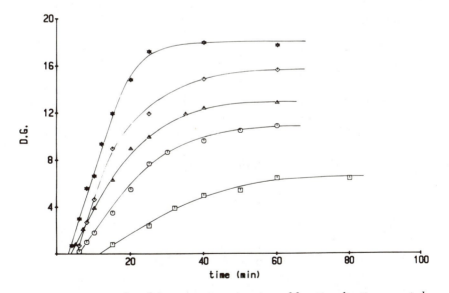

Figure 2. Degree of grafting versus reaction time of functionalizations carried out at a fixed temperature (150 °C) and with varying DCPO contents. Key: EPR–DBM–DCPO = 100–10–0.5 (□); 0.75 (○); 1.0 (△); 1.5 (◇); 2.0 (☆). (Reproduced with permission from ref. 21. Copyright 1989 Wiley.)

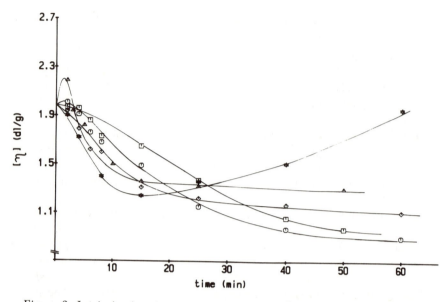

Figure 3. Intrinsic viscosity versus reaction time for EPR functionalizations carried out at conditions and compositions described in Figure 2. (Reproduced with permission from ref. 21. Copyright 1989 Wiley.)

2, are shown in Figure 3. The maxima, observed in Figure 1b, are less pronounced or absent here. The most striking feature is the up-turn at about 15 min of the curve indicating the highest DCPO content. The other curves show ever-decreasing values as the DCPO amount decreases. Some inversion can be caused by molecular rearrangements occurring during the induction period. These long-term features probably show DBM consumption at the end of the reaction, which leads to an inversion tendency, again favoring coupling reactions with respect to β-scissions. Such an effect is, of course, larger with higher radical site concentration in the system.

Structural and Superreticular Order Investigations. The WAXS powder spectra of Co034 EPR and of the corresponding functionalized EPRs are shown in Figure 4. All the compression-molded samples were annealed at 25 °C for 15 days to allow for complete development of the polyethylene crystallinity.

The profile of the unmodified elastomer (A) exhibits a crystalline reflection at about 21.5° of 2θ, identified as the (110) polyethylene strongest reflection (24–26). Its intensity decreases with enhancing DG. For the highest values (DG > 10%) only a very weak shoulder, superimposed to the amorphous halo, can be observed in the same 2θ region.

The crystallinity degree, X_c, was calculated from the WAXS profiles according to the Hermans–Weidinger method (27). The X_c values show a linear decrease with increasing DG contents, ranging from 9% (parent

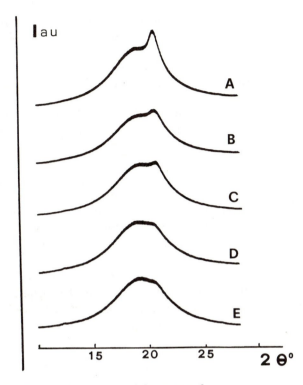

Figure 4. WAXS powder spectra of the Co034 elastomers at varying DG con-tents obtained after storage at 25 °C for 15 days. Key: A, 0% by weight; B, 3.7%; C, 5.1%; D, 10%; E, 13.4%. (Reproduced with permission from ref. 22. Copyright 1989 Wiley.)

EPR value) to about 1% for the sample with the highest DG percentage (13.4%).

These structural results, like those derived from IR and DSC investigations (22), indicate that DBM grafting involves, to a large extent, the ethylene sequences (C_2). Consequently, the residual crystallinity of EPR is strongly reduced by the functionalization process.

The shape of the SAXS profiles of the investigated elastomers (Figure 5a) shows a noticeable particle scattering and rather weak interference (28, 29) that rapidly decrease when the DG is enhanced.

According to the general theory of Hosemann (29), the intensities of the SAXS profiles are replotted in Figure 5b as u^2I versus u (I is the scattered intensity and $u = 2 \sin \theta/\lambda$, where 2θ is the diffusion angle and λ is the X-ray wavelength). By this method the superreticular (long-range) order of the examined EPRs is clearly shown. A sharp peak is observed for the unmodified EPR (A), whose height and sharpness gradually decrease with increasing DG (B, C, and D) up to a complete absence of interference for the E profile.

The location of the maximum, yielding the so-called "long period" (i.e., the main distance, d, between the centers of adjacent scattering particles), is independent of DG ($d \simeq 185$ Å). The polydispersity, instead, is directly related to the broadness of the peak and increases when DG is augmented. The results indicate that the significant degree of structure development at a superreticular level in the parent EPR is markedly diminished as the functionalization is enhanced. This diminished structure suggests that the paracrystalline superreticles become more and more distorted, gradually approaching a liquidlike system.

Conclusions

A molecular model and general considerations on the bulk functionalization of ethylene–propylene random copolymers by means of various functional molecules (compatible or incompatible with the EPR substrate) has been proposed and discussed.

A detailed analysis of the reaction kinetics has been carried out at constant composition with varying temperature and at constant temperature with changing composition on a commercial EPR (Dutral Co054). The results indicate that the functional molecule (DBM) is preferentially fixed on the

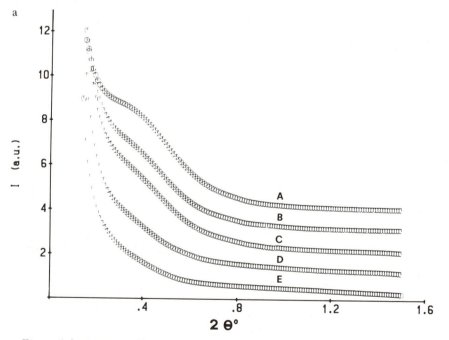

Figure 5. (a) SAXS profiles of the same EPRs as shown in Figure 4. (Reproduced with permission from ref. 22. Copyright 1989 Wiley.)

b

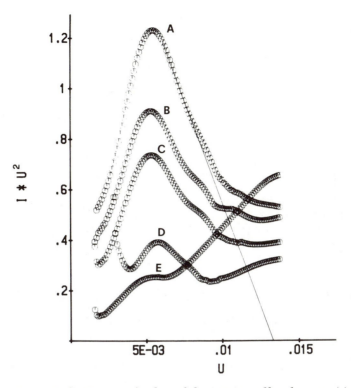

Figure 5. (b) Hosemann's plots of the SAXS profiles shown in (a).

longer ethylene (C_2) sequences. Such a finding has been confirmed by a parallel work (*21*), in which the EPR microstructure was varied. Larger C_2/C_3 ratios or more blocky C_2,C_3 sequences at equal C_2/C_3 ratios yield faster kinetics and higher final plateau values. Good agreement with a theoretical model is achieved for the data relative to the DCPO variation, but not for those relative to the temperature changes (*23*).

In another work (*22*) functionalized reaction products, relative to a different EPR (Dutral Co034), were characterized by IR, DSC, and X-ray diffraction patterns. Only the X-ray results have been reported here.

WAXS indicates a linear decrease of the C_2 residual crystallinity with increasing GD. The SAXS profiles, after Hosemann's treatment, indicate that the significant degree of structure at a superreticular level (long-range order) of the parent EPR is going to be lost more and more as the GD increases. Both the WAXS and SAXS profiles, as well as the DSC and IR data, depend on the amount of diminution of the C_2 sequences. This dependence confirms the hypothesis that steric effects (the presence of methyl groups of the C_3 sequences) cause the functional molecule to fix preferentially on the C_2 traits of the chain. The presence of the functional molecules disrupts the existing crystallinity and structure of the unmodified EPR.

References

1. Baldwin, F. P.; Ver Strate, G. *Rubber Chem. Tech.* **1972**, *45*, 834.
2. Jones, G. D. In *Chemical Reactions of Polymers;* Fettes, E. M., Ed.; Interscience: New York, 1964; p 247.
3. Minoura, Y.; Veda, M.; Minozouma, S.; Oba, M. *J. Appl. Polym. Sci.* **1969**, *13*, 1625.
4. Poreiko, S.; Gabara, W.; Kulesza, J. *J. Polym. Sci., Part A-1* **1967**, *5*, 1571.
5. Gabara, W.; Poreiko, S. *J. Polym. Sci., Part A-1* **1967**, *5*, 1547.
6. Avella, M.; Greco, R.; Lanzetta, N.; Maglio, G.; Malinconico, M.; Martuscelli, E.; Palumbo, R.; Ragosta, G. In *Polymer Blends: Processing, Morphology and Properties;* Martuscelli, E.; Palumbo, R.; Kryszewski, M., Eds.; Plenum: New York, 1981; Vol. 1, p 191.
7. De Vito, G.; Maglio, G.; Lanzetta, N.; Malinconico, M.; Musto, P.; Palumbo, R. *J. Polym. Sci., Polym. Chem. Ed.* **1984**, *22*, 1334.
8. Ide, F.; Hasegawa, A. *J. Appl. Polym. Sci.* **1974**, *18*, 963.
9. Avella, M.; Lanzetta, N.; Maglio, G.; Malinconico, M.; Musto, P.; Palumbo, R.; Volpe, M. G. In *Polymer Blends: Processing, Morphology and Properties;* Kryszewski, M.; Galeski, A.; Martuscelli, E., Eds.; Plenum: New York, 1984; Vol. 2, p 193.
10. Cimmino, S.; D'Orazio, L.; Greco, R.; Maglio, G.; Malinconico, M.; Mancarella, C.; Martuscelli, E.; Palumbo, R.; Ragosta, G. *Polym. Eng. Sci.* **1984**, *24*, 48.
11. Cimmino, S.; Coppola, F.; D'Orazio, L.; Greco, R.; Maglio, G.; Malinconico, M.; Mancarella, C.; Martuscelli, E.; Ragosta, G. *Polymer* **1986**, *27*, 1874.
12. Cimmino, S.; D'Orazio, L.; Greco, R.; Maglio, G.; Malinconico, M.; Mancarella, C.; Martuscelli, E.; Musto, P.; Palumbo, R.; Ragosta, G. *Polym. Eng. Sci.* **1985**, *25*, 193.
13. Greco, R.; Lanzetta, N.; Maglio, G.; Malinconico, M.; Martuscelli, E.; Palumbo, R.; Ragosta, G.; Scarinzi, G. *Polymer* **1986**, *27*, 299.
14. Ruggeri, G.; Aglietto, A.; Petragnani, A.; Ciardelli, F. *Eur. Polym. J.* **1983**, *19*, 863.
15. Bouillouxs, A.; Druz, J.; Lambla, M. *Polym. Proc. Eng.* **1986**, *4*, 235.
16. Gaylord, N. G.; Mishra, M. K. *J. Polym. Sci., Polym. Lett. Ed.* **1983**, *21*, 23.
17. Gaylord, N. G.; Metha, M. *J. Polym. Sci., Polym. Lett. Ed.* **1982**, *20*, 481.
18. Ide, F.; Kamada, K.; Hasegawa, A. *R. G. P. C.* **1970**, *47*, 487.
19. Greco, R.; Maglio, G.; Martuscelli, E.; Musto, P.; Palumbo, R. *Polym. Process Eng.* **1986**, *4*, 293.
20. Greco, R.; Maglio, G.; Musto, P.; Scarinzi, G. *J. Appl. Polym. Sci.* **1987**, *33*, 2513.
21. Greco, R.; Maglio, G.; Musto, P.; Scarinzi, G. *J. Appl. Polym. Sci.* **1989**, *37*, 777.
22. Greco, R.; Musto, P.; Riva, F.; Maglio, G. *J. Appl. Polym. Sci.* **1989**, *37*, 789.
23. Greco, R.; Musto, P., in preparation.
24. Maglio, G.; Milani, F.; Musto, P.; Riva, F. *Makromol. Chem., Rapid Commun.* **1987**, *8*, 589.
25. Ver Strate, G.; Wilchinsky, Z. W. *J. Polym. Sci., Part A-2* **1971**, *9*, 127.
26. Gilbert, M.; Briggs, J. E.; Omana, W. *Br. Polym. J.* **1979**, *11*, 81.
27. Hermans, P. H.; Weidinger, A. *Makromol. Chem.* **1961**, *44-46*, 24.
28. Guinier, A.; Fournet, G. *Small Angle Scattering of X-Rays;* Wiley: New York, 1970.

29. Hosemann, R.; Bagchi, S. N. *Direct Analysis of Diffraction by Matter*; North Holland: Amsterdam, 1962.

RECEIVED for review March 25, 1988. ACCEPTED revised manuscript September 14, 1988.

APPLICATIONS

Toughened Unsaturated Polyester Block Copolymers

Shen-Nan Tong, Chih-Chien Chen, and Peter T. K. Wu

Union Chemical Laboratories, Industrial Technology Research Institute, Hsinchu, Taiwan, Republic of China

Block copolymers of unsaturated polyester were prepared by cocondensation polymerization of hydroxyl- or carboxyl-terminated liquid rubbers or polyethylene glycols in the presence of maleic anhydride, phthalic anhydride, and propylene glycol. The resultant unsaturated polyesters were then diluted with styrene monomer and cured by using peroxide and heat. The mechanical properties of the cured block copolymers were determined and compared with those of a control resin or a corresponding toughened physical blend. In addition, the sheet-molding compounds made with these polyesters were obtained and investigated.

IMPROVEMENT OF THE FRACTURE TOUGHNESS and impact resistance of rigid unsaturated polyester has played an important role in the development of high-performance polymeric composites. Unsaturated polyesters can be modified by numerous chemical and physical methods (1–3). Chemical modifications by introducing either long-chain glycols (e.g., diethylene, dipropylene, or triethylene glycol) or long-chain saturated dibasic acids (e.g., adipic acid or sebacic acid) are common practice. However, a significant amount of such long-chain substances is needed to impart a sufficient improvement in toughness. This requirement, in turn, results in a large sacrifice of other mechanical properties.

Among physical modifications, the use of dispersed elastomers for enhancement of toughness in unsaturated polyester resins and molding materials has frequently been reported (4–17). These elastomers include both solid and liquid rubbers.

0065–2393/89/0222–0375$06.00/0
© 1989 American Chemical Society

The solid rubbers used contain 1,2-polybutadiene (18–20), 1,4-poly-butadiene (21), diene rubbers with carboxyl end groups (22), natural rubber (23, 24), butadiene–styrene rubbers (25–30), butadiene–acrylonitrile with amine end groups (31), halogenated butyl rubber (32), derivatives of acrylate rubber (33, 34), chlorosulfonated polyolefins (35), plasticized poly(vinyl chloride) (36), and thermoplastic grafted rubbers (37). Block copolymers such as butadiene–caprolactam styrene (38–40), ethylene–propylene–polystyrene rubber (41), and thermoplastic elastomers (42) were also reported.

The liquid rubbers used to modify unsaturated polyester resins are usually liquid polybutadienes (43–50) and liquid butadiene–acrylonitrile rubbers with reactive end groups (51–62). The use of organic siloxane elastomers (63), epichlorohydrin rubbers with hydroxyl end groups (29, 54, 57, 58, 64), unsaturated oligourethanes, oligourethanes with carboxyl groups (65–68), or other oligourethanes can also be found (8, 68).

In the modification by physical blending with elastomers, polyester is incompatible with most of the rubbers used, including those having hydroxyl or carboxyl terminal groups on the elastomer chain. There has been an effort recently to improve the bonding between the resin matrix and rubber phase. Vinyl-terminated liquid rubber was developed in the hope that the vinyl end group would react with the styrene monomer and become covalently bonded to the matrix.

Recently, block copolymers were developed that contain unsaturated polyester with polyurethanes (69–74), polyureas (75), saturated polyesters (e.g., polyethylene terephthalate and polybutylene terephthalate) (76–80), polysiloxanes (81–83), polyimide (84), polyoxazoline (85), or polyglycols (e.g., polyethylene glycols and polypropylene glycols) (86–93). Among them, poly-urethanes formed in a solution of unsaturated polyester resin do not form a separate phase, and are usually regarded as a resin-thickening agent. The block copolymers containing saturated polyester, polyimide, and polysilox-ane in chain form provide the main improvement in the thermal behavior. Improved impact strength of unsaturated polyester resins was reported in block copolymer systems containing polyoxazoline or polyglycols. However, the introduction of polyglycols into the chain increases the water absorption as it forms unsaturated polyester resins.

If the long-chain glycols or saturated dibasic acids used in conventional chemical modification are replaced by hydroxyl- or carboxyl-terminated liquid rubbers, the resultant unsaturated polyesters will have rubber segments in the main chain. If the molecular weight of the liquid rubber is significantly large, outstanding properties may be expected in these newly prepared unsaturated polyesters (94–96).

We prepared block copolymers of unsaturated polyesters containing rubber or polyethylene glycol segments on the polymer main chains. The properties of cured polyesters from these resins and from analogues modified by physical blending with elastomers and thermoplastics were investigated and compared. The purpose of this study is to evaluate mechanical properties of toughened unsaturated polyesters by various chemical and physical methods, and to develop a super-tough unsaturated polyester without sacrificing other mechanical properties.

Experimental Details

Reagent-grade styrene, phthalic anhydride (PA), maleic anhydride (MA), ethylene glycol (EG), diethylene glycol (DEG), propylene glycol (PG), hydroquinone, cobalt naphthenate, and methyl ethyl ketone peroxide (MEKPO) were used without further purification.

Elastomers used were copolymers of butadiene and acrylonitrile that have hydroxyl- (HTBN), carboxyl- (CTBN), amine- (ATBN) and vinyl- (VTBN) terminated functional groups. The acrylonitrile contents of these elastomers are 16.5% (except CTBN, which is 27.0%). Polyethylene glycols (PEG) are commercially available thermoplastics. The unsaturated polyester resins were prepared by the one-stage fusion process.

Control Resin. For example, PA (111.1 g; 0.75 mol), MA (73.8 g; 0.75 mol), and PG (120 g; 1.58 mol) were placed in a 1-L round-bottom four-neck flask that was equipped with an inlet tube for nitrogen gas, a thermometer, a stirrer, and a reflux condenser. The esterification reaction was carried out at 170–210 °C for 6 h, during which period 27 g of water was collected. After the reaction the system was cooled to 100 °C. Styrene (185 g) containing hydroquinone (0.07 g) was added to the contents of the reaction system, which was mixed thoroughly and cooled to room temperature (50% unsaturation).

Toughened Resin via Blending Method. The resins were prepared by blending the control resin in a mixer (Cowles) with various parts by weight of elastomers.

Toughened Resin via Chemical Modification. A given amount of long-chain glycol or reactive liquid rubber was used to replace short-chain glycol or dibasic acid. Treatment then proceeded as described for the control resin.

High-performance liquid chromatography (Waters Associates, 1000-, 500-, and 100-Å styrene copolymer columns) was used to study the molecular weight distribution of the resultant unsaturated polyesters. Tetrahydrofuran was the carrier solvent, and the detector was a differential refractometer (Waters). A 1.0-mg sample was injected each time, with a flow rate of 1 mL/min.

The resultant unsaturated polyesters were cured between two sheets of plate glass (12 × 12 in.) that were separated and sealed by a silicone rubber gasket. The unsaturated polyesters for casting were mixed with 0.2 phr (parts per hundred parts resin) cobalt naphthenate, then with 1 phr methyl ethyl ketone peroxide. The entire assembly was held together by a series of C clamps, with the clamp screws adjusted to prevent premature release during the gelation of unsaturated polyester. The inside surfaces of the plate glass were polished with a releasing agent before assembly. The catalyzed unsaturated polyester was heated at 80 °C for 1 h, at 100 °C for 1 h, and at 120 °C for 4 h.

Similarly, sheet-molding compound (SMC) was prepared with a composition of 100 phr unsaturated polyester resin (80% unsaturation; styrene content 40%), 80 phr $CaCO_3$, 60 phr glass fiber, 30 phr low-profile additive (polystyrene), 3 phr styrene monomer, 1 phr MgO, 3 phr zinc stearate, and 1.2 phr *tert*-butyl perbenzoate. The plate glass panels were molded at 150 kg/cm² at 135–140 °C for 2.5 min. These gave panels of 1/8-in. thickness.

Specimens used for mechanical properties testing were cut from 1/8-in.-thick unreinforced castings and SMC sheets. Tests were carried out according to the following specifications: tensile testing, ASTM (American Society for Testing and Materials) D638; flexural testing, ASTM D790; and impact testing, ASTM D256. The measurements of dynamic mechanical properties were performed on the torsion rectangular mode using a RDS rheometer (Rheometrics, Inc.) over the temperature range of –150 to 150 °C at a rate of 6.28 °C/min.

Results and Discussion

Unsaturated polyesters chemically modified with polyethylene glycols are homogeneous systems. The number-average molecular weights of the unsaturated polyesters thus formed were about 1100. Figure 1 shows gel permeation chromatograms of unsaturated polyesters based on PEG with different molecular weights. In contrast to a monomodal peak usually found in the analogous polyester derived from a low-molecular-weight glycol, bimodal peaks are observed when the molecular weight of PEG is greater than

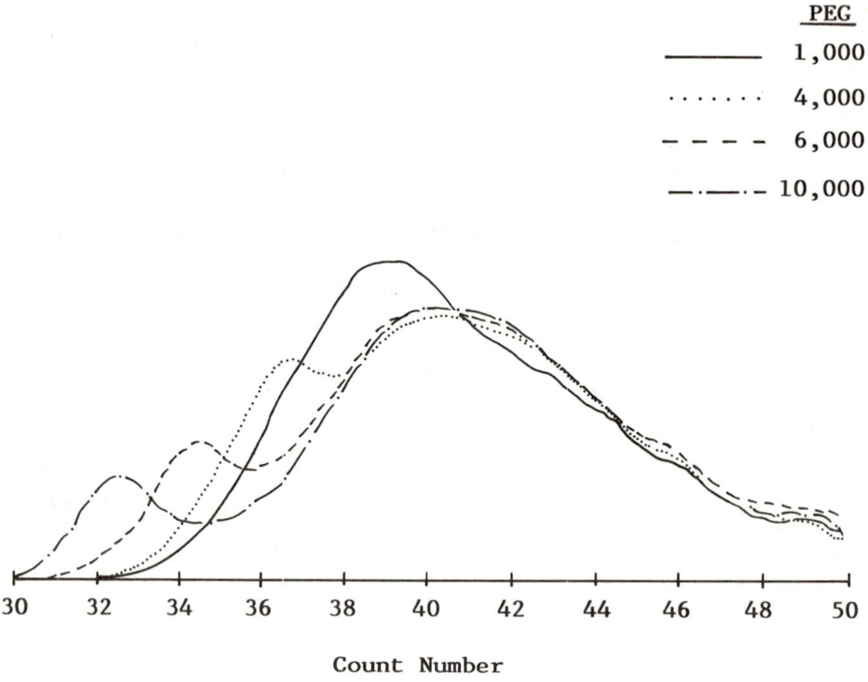

Figure 1. *GPC chromatograms of unsaturated polyesters based on different PEG molecular weights.*

1000. Obviously, this is a result of one-stage cocondensation consisting of both low- and high-molecular-weight glycols in the same system. In general, the long-chain glycols impart flexibility to the polyester chains and increase the elongation and impact strength of castings derived from them. However, they decrease the hardness, heat-distortion temperature, and other mechanical properties.

In contrast, unsaturated polyesters elastomer-modified by physical blending are two-phase systems containing dispersed rubber particles. This group includes polyester modified with vinyl-terminated liquid rubber. In principle, this polyester can cross-link with unsaturated polyester through its vinyl groups. Its particle diameter is usually in the range of 0.1–10 μm. Particle size and distribution depend strongly on the method and conditions of preparation.

Mechanical properties of unsaturated polyester castings with and without rubber modification are shown in Table I. The decrease in hardness, heat distortion temperature, and increase in elongation and toughness are as expected for rubber-toughened plastics.

Although the four rubbers possess different terminal functional groups, the mechanical properties of the blended systems are strongly dependent on the morphology of the dispersed rubber particles. The contribution of the rubber terminal functional groups to the mechanical properties of the rubber–polyester blends is only indirect. Different terminal groups have different effects on the miscibilities of the rubbers with the unsaturated polyester, and the result is different morphology of the dispersed rubber phase.

In rubber modification by chemical reaction, a small amount of HTBN or CTBN was added to a large amount of glycols or dibasic acids. In the resultant rubber-toughened block copolymer, each rubber molecule is linked to an ester segment at its two terminals. Because of the comparably large molecular weight of the liquid rubber ($\overline{M}_n$ = 3500–4000), only a few polyester molecules in the whole modified system contain centered rubber segments.

The mechanical properties of block-copolymer toughened systems are much improved over those of the systems toughened by physical blending, as shown in Table I. The properties are superior to those of a vinyl-terminated rubber system, in which reaction is expected to take place to some extent between the terminal vinyl group and the unsaturated polyester. Obviously, the improvement of mechanical properties cannot be explained satisfactorily by the formation of chemical bonds alone.

In Figure 2, the cured castings of the CTBN blend system show the characteristic mechanical damping peak of the rubber at −45 °C and the peak of the control resin matrix at 122 °C. Obviously, this system is heterogeneous and has good agreement with the results of the morphology study (*13*).

Table I. Properties of Rubber-Modified Resins and Their Casting Sheets

Resin and Property	Control	Amine	Carboxyl		Hydroxyl		Vinyl
Resin							
Rubber type	None	ATBN	CTBN		HTBN		VTBN
Rubber content (%)	0	5	5	5	5	5	5
Modification method	–	B[a]	B	R	B	R	B
Brookfield viscosity at 23 °C (cps)	200	500	290	360	400	420	420
Properties of casting sheet							
Barcol hardness	44	29	35	40	15	38	35
Unnotched Izod (ft-lb/in.)	1.2	1.0	1.1	2.0	1.4	2.2	1.3
Flexural strength ($\times 10^4$ psi)	1.18	1.02	1.20	1.34	1.27	1.62	1.27
Flexural modulus ($\times 10^5$ psi)	6.15	4.38	4.50	4.61	4.84	5.32	4.78
Tensile modulus ($\times 10^5$ psi)	5.30	4.22	4.31	4.67	4.44	5.09	4.44
Tensile strength ($\times 10^3$ psi)	6.04	4.70	6.02	6.40	6.19	8.36	5.65
Elongation (%)	0.9	1.5	1.6	1.6	1.5	1.9	1.6
Heat-distortion temperature (°C)	72.5	63.5	65.0	59.0	71.0	68.0	65.0
Toughness	1.00	1.53	1.55	1.96	1.85	2.66	1.90

[a] B: blending; R: reaction; unsaturation of resin: 50%.

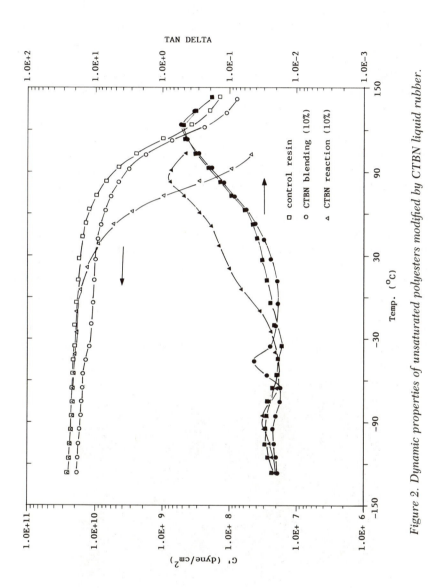

Figure 2. Dynamic properties of unsaturated polyesters modified by CTBN liquid rubber.

Photomicrographs of CTBN systems modified by chemical reaction show no evidence of large discrete domains. The dynamic mechanical properties suggest that the 10% CTBN material shows evidence of heterogeneity. In the cured castings of a chemically modified CTBN system, the mechanical damping maxima corresponding to the rubber and resin phases disappeared. A new single damping peak appeared at 87 °C. Obviously, it cannot fit the logarithmic mixture additivity rule. In the meantime, a shoulder centered around 20 °C is observed in the dynamic mechanical spectroscopy. It is assumed that a new phase is present. Microheterogeneous morphology is observed by TEM, as shown in Figure 3.

In the study of a 10% rubber-modified resin system, the toughness of the system combining 5% rubber incorporation into the main chain by chemical reaction and 5% rubber dispersed through the system by physical blending is 3.9 when the toughness of the control resin is arbitrarily taken as 1.0. In contrast, the toughness of the system containing 10% rubber dispersed through the system wholly by physical blending is only 3.0. Obviously, the coexistence of the micro- and macroheterogeneous phase shows a greater improvement in toughness. This evidence is also observed in the toughened epoxy resin system (54). The toughening theory of rigid resin can be used to describe this system (97).

Figure 4 shows the dynamic mechanical properties of a VTBN-modified system over the temperature range of –150 to 150 °C. Two mechanical damping peaks corresponding to resin matrix and rubber phase are obtained at 113 and –30 °C. Compared with the control system and VTBN rubber (130 and –39 °C), both damping peaks shift inward and become broad. These

Figure 3. Transmission electron micrograph of unsaturated polyester–rubber copolymer that was sliced to a thickness of 900 Å and stained with osmium tetroxide.

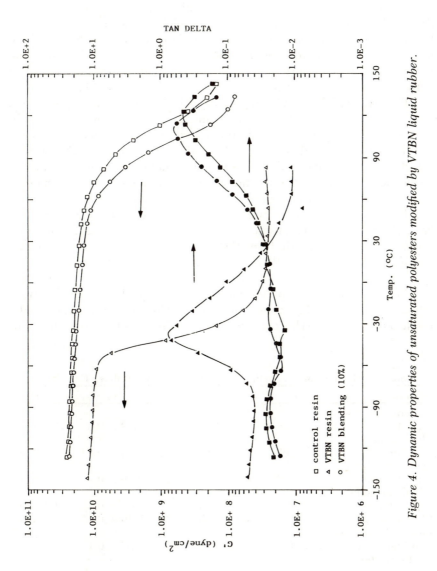

Figure 4. Dynamic properties of unsaturated polyesters modified by VTBN liquid rubber.

observations show the existence of interaction between the resin matrix and the dispersed rubber phase.

In general, the transition temperature of a chemical reaction system is lower than that of the corresponding physical blend. Figure 2 shows that the transition temperature of the CTBN blending system is very close to that of the control resin and that the chemically reacted system is much lower. This difference has also been observed in the VTBN system, as shown in Figure 3. Chemical reaction between the VTBN rubber and the polyester has occurred to some extent. Consequently, its transition temperature is between those of the CTBN reaction and the blending system.

The study of the toughening effect of highly unsaturated polyester is very significant, especially from the point of view of engineering. In Table II, the degree of unsaturation of unsaturated polyester is increased to 80% and the liquid rubber HTBN is used for chemical modification. Results show the effectiveness of molecularly incorporated rubber in the improvement of mechanical properties. The tensile, flexural, and impact strength and elongation increase with increasing rubber content, although modulus, hardness, and heat-distortion temperature show the reverse effect. The increase of flexural strength is two times; tensile strength, five times; elongation, seven times; and impact strength, three times. However, the decrease of modulus is only about 15%. This result indicates that rubber modification by chemical reaction is superior to the other methods.

Table III shows the application of this toughened resin for the preparation of a sheet-molding compound. The effectiveness of improvement of mechanical properties by chemical modification is still significant in the

Table II. Mechanical Properties of Casting Sheet vs. Rubber Content (%)

Properties	0	5	10	15
Flexural strength ($\times 10^3$ psi)	6.8 ± 0.4	11.6 ± 0.5	12.9 ± 0.6	15.6 ± 0.3
Flexural modulus ($\times 10^5$ psi)	4.92 ± 0.02	4.86 ± 0.02	4.36 ± 0.03	4.23 ± 0.08
Tensile strength ($\times 10^3$ psi)	1.8 ± 0.4	6.7 ± 0.3	7.6 ± 0.4	8.9 ± 0.1
Tensile modulus ($\times 10^5$ psi)	5.2 ± 0.3	5.0 ± 0.2	4.1 ± 0.3	4.2 ± 0.3
Elongation (%)	0.43 ± 0.05	1.55 ± 0.10	2.19 ± 0.06	3.14 ± 0.08
Impact strength (ft-lb/in.) unnotched Izod	0.7 ± 0.2	1.4 ± 0.1	1.7 ± 0.1	2.3 ± 0.4
Heat-distortion temperature (°C)	133 ± 2	89 ± 2	70 ± 2	85 ± 2
Hardness (Barcol)	54 ± 1	37 ± 1	26 ± 1	32 ± 1

NOTE: Toughener: HTBN by reaction. Styrene content: 40%. Curing agent: MEKPO, 1.0 phr; cobalt naphthenate, 0.2 phr. Unsaturation of resin: 80%.

Table III. Mechanical Properties of Sheet-Molding Compound

Properties	Control resin	Modified by chemical reaction
Flexural strength ($\times 10^4$ psi)	1.53 ± 0.16	3.53 ± 0.04
Flexural modulus ($\times 10^6$ psi)	0.97 ± 0.06	1.58 ± 0.05
Tensile strength ($\times 10^4$ psi)	0.83 ± 0.12	1.55 ± 0.10
Tensile modulus ($\times 10^6$ psi)	1.38 ± 0.12	1.68 ± 0.14
Elongation (%)	0.26 ± 0.10	1.63 ± 0.20
Impact strength (ft-lb/in.)		
notched Izod	4.94 ± 0.23	6.35 ± 0.55
Specific gravity	1.696 ± 0.146	1.828 ± 0.019
Water absorption	1.007 ± 0.12	1.425 ± 0.007
Heat-distortion temperature (°C)	220	220
Hardness (Barcol)	60–70	60–70

NOTE: Rubber content: control resin, 0%; tough resin, 15% by reaction.

presence of fillers and glass fibers. Therefore, this improved method can provide high-performance materials for engineering use.

Summary

Chemical modification by introducing long-chain glycols resulted in large sacrifices of most mechanical properties.

Rubber-modified unsaturated polyesters with ATBN, CTBN, HTBN, and VTBN liquid rubbers incorporated by physical blending showed that these were heterogeneous systems containing a dispersed rubber phase.

Block copolymers containing a rubber segment on the polymer main chains showed the existence of the microheterogeneous phase and superior properties. This material is developed as a high-performance polymeric composite for engineering applications.

Acknowledgments

The authors acknowledge the contributions of the following persons: J. M. Starita of Rheometrics, Inc., for his help in obtaining the dynamic properties data, and D. S. Chen, L. Z. Chung, and J. M. Perng of Union Chemical Laboratories for their assistance in specimen preparation. Acknowledgment is also due to T. K. Kwei for his helpful discussion.

References

1. Boenig, H. V. *Unsaturated Polyesters: Structure and Properties*; Elsevier: New York, 1970; Chapter 6.
2. Kostanski, L. K.; Krolikowski, W. *Int. Polym. Sci. Technol.* **1985**, *12*, T/13.
3. Anonymous *Toughening of Plastics*; Plastics & Rubber Institute: London, 1985.

4. Drake, R.; Siebert, A. *SAMPE Q.* **1975,** 6.
5. Carrol, B.; Collester, J. *35th Annu. Tech. Conf. SPI* **1980,** Sect. 16-A.
6. Golemba, F. *35th Annu. Tech. Conf. SPI* **1980,** Sect. 5-A.
7. Myers, F. A.; Nyo, H. *36th Annu. Tech. Conf. SPI* **1981,** Sect. 11-A.
8. Schulz-Walz, H. et al. *Kunstst. Plast.* **1981,** *28,* 26.
9. Fintelmann, C. et al. *Polym. Plast. Technol. Eng.* **1981,** *17,* 225.
10. Gardner, I. J. et al. *37th Annu. Tech. Conf. SPI* **1982,** Sect. 7-D.
11. Rizzi, M. A.; Kearney, M. R. *37th Annu. Tech. Conf. SPI* **1982,** Sect. 13-D.
12. Crosbie, G. A.; Phillips, M. G. *Mod. Plast. Int.* **1983,** *13,* 30.
13. Tong, S. N.; Chen, D. S.; Chung, L. Z.; Chen, C. C. *J. Chin. Inst. Chem. Eng.* **1983,** *14,* 333.
14. Yokohama Rubber Co. Jpn. Patent 8389629, 1983; *Chem. Abstr.* **1984,** *100,* 35465r.
15. Hesse, A; Buensch, H.; Czauderna, B. DE Patent 3332019, 1985; *Chem. Abstr.* **1985,** *103,* 38149p.
16. Takano, H.; Kunitomi, T.; Okada, S.; Awaji, T.; Shindo, K.; Atobe, D. DE Patent 3512791, 1985; *Chem. Abstr.* **1986,** *104,* 110808t.
17. Tonoki, S.; Tsunemi, H.; Nishimura, A.; Abe, K. EP Patent 186165, 1985; *Chem. Abstr.* **1986,** *105,* 173764s.
18. Bonnington, M. G. DE Patent 2260480, 1973; *Chem. Abstr.* **1973,** *79,* 67411j.
19. Teijin, Ltd. FR Patent 2139194, 1973; *Chem. Abstr.* **1974,** *80,* 4403h.
20. Matsushita Electric Works, Ltd. Jpn. Patent 8117259, 1981; *Chem. Abstr.* **1981,** *95,* 63184V.
21. Nowak, R. M.; Ginter, T. O. DE Patent 2014774, 1969; *Chem. Abstr.* **1971,** *74,* 13729x.
22. South, A., Jr. U.S. Patent 4020036, 1977; *Chem. Abstr.* **1977,** *86,* 190933c.
23. Willinger, E. J. DE Patent 1951261, 1969; *Chem. Abstr.* **1970,** *73,* 4504x.
24. Ibata, J.; Kobayashi, H.; Toyomoto, K.; Suzuoki, K.; Nogami, S. DE Patent 2313535, 1973; *Chem. Abstr.* **1974,** *80,* 28191p.
25. South, A., Jr.; Werkman, R. T. *32nd Annu. Tech. Conf. SPI* **1977,** Sect. 16-D.
26. South, A., Jr., et al. *33rd Annu. Tech. Conf. SPI* **1978,** Sect. 19-A.
27. Fintelmann, C. et al. *36th Annu. Tech. Conf. SPI* **1981,** Sect. 18-C.
28. Crain, D. L.; South, A., Jr. *Plast. Compd.* **1982,** *5,* 129, 132, 134.
29. Zaske, O. C. et al. *37th Annu. Tech. Conf. SPI* **1982,** Sect. 13-C.
30. Asahi Chemical Industry Co. Jpn. Patent 8308718, 1983; *Chem. Abstr.* **1983,** *99,* 23571v.
31. Ito, B.; Tanaka, H.; Iitani, H. Jpn. Patent 76128293, 1976; *Chem. Abstr.* **1977,** *86,* 107395u.
32. Ito, B.; Matsumoto, M.; Tanaka, K.; Iitani, H. Jpn. Patent 7575685, 1975; *Chem. Abstr.* **1975,** *83,* 194321n.
33. Kageyama, A.; Iwami, E.; Uchigasaki, I. Jpn. Patent 76131588, 1976; *Chem. Abstr.* **1977,** *86,* 91209g.
34. Diekie, R. A.; Newman, S. DE Patent 2163464, 1972; *Chem. Abstr.* **1972,** *77,* 102630n.
35. Ito, B.; Matsumoto, M.; Tanaka, K.; Iitani, H. Jpn. Patent 7533282, 1975; *Chem. Abstr.* **1975,** *83,* 115889c.
36. Adachi, T.; Funakisako, S.; Furuya, A.; Uchida, M. Jpn. Patent 8021474, 1980; *Chem. Abstr.* **1980,** *93,* 8990p.
37. Trostyanskaya, E. B.; Babaevskii, P. G.; Kulik, S. G.; Pavlguchenko, V. N.; Stepanova, M. I.; Kholodnova, L. V. SU Patent 992545, 1983; *Chem. Abstr.* **1983,** *98,* 161856x.
38. Dainippon Ink and Chemicals, Inc. Jpn. Patent 8301713, 1983; *Chem. Abstr.* **1983,** *99,* 6515n.

39. Uber, R. F.; Childers, C. W.; Naylor, F. E. U.S. Patent 4287313, 1981; *Chem. Abstr.* **1981**, *95*, 204973r.
40. Naylor, F. E. U.S. Patent 4360643, 1982; *Chem. Abstr.* **1983**, *98*, 107988d.
41. Kubota, H. U.S. Patent 3882078, 1975; *Chem. Abstr.* **1975**, *83*, 165117v.
42. Brownbill, D. *Mod Plast. Int.* **1983**, *13*, 33.
43. Golemba, F. *35th Annu. Tech. Conf. SPI* **1980**, Sect. 5-A.
44. Ryan, P. W.; Thompson, R. E. DE Patent 2160722, 1972; *Chem. Abstr.* **1973**, *78*, 30861j.
45. Kajiura, A.; Aito, Y.; Tamura, H.; Sugiyama, A.; Watanabe, K. DE Patent 2225736, 1973; *Chem. Abstr.* **1974**, *80*, 84193c.
46. Kajiura, A.; Irie, S.; Sugiyama, A. Jpn. Patent 7305889, 1973; *Chem. Abstr.* **1973**, *78*, 137456h.
47. Kajiura, J.; Tamura, H.; Sugiyama, A. Jpn. Patent 7443389, 1974; *Chem. Abstr.* **1975**, *82*, 141113u.
48. Hasegawa, M.; Okamoto, J.; Kojima, K. Jpn. Patent 79138082, 1979; *Chem. Abstr.* **1980**, *92*, 148071u.
49. Hasegawa, M.; Okamoto, J.; Kojima, K. Jpn. Patent 79138083, 1979; *Chem. Abstr.* **1980**, *92*, 148070t.
50. Inoue, M.; Hayashi, O.; Ono, H. Jpn. Patent 8043151, 1980; *Chem. Abstr.* **1980**, *93*, 48147a.
51. McGarry, F. J.; Sultan, J. N. *24th Annu. Tech. Conf. SPI* **1969**, Sect. 11-B.
52. Rowe, E. H. et al. *30th Annu. Tech. Conf. SPI* **1975**, Sect. 1-E.
53. McGarry, F. J. et al. *32nd Annu. Tech. Conf. SPI* **1977**, Sect. 16-C.
54. Rowe, E. H.; Howard, F. H. *33rd Annu. Tech. Conf. SPI* **1978**, Sect. 19-B.
55. Rowe, E. H. *34th Annu. Tech. Conf. SPI* **1979**, Sect. 23-B.
56. Rowe, E. H.; McGarry, F. J. *35th Annu. Tech. Conf. SPI* **1980**, Sect. 18-E.
57. Lee, B. L.; Howard, F. H. *36th Annu. Tech. Conf. SPI* **1981**, Sect. 23-E.
58. Crosbie, G. A.; Phillips, M. G. *37th Annu. Tech. Conf. SPI* **1982**, Sect. 8-C.
59. Ito, B.; Matsumoto, M.; Tanaka, K.; Iitani, H. Jpn. Patent 7575687, 1975; *Chem. Abstr.* **1975**, *83*, 194319t.
60. Takiyama, E. Jpn. Patent 7623557, 1976; *Chem. Abstr.* **1977**, *86*, 17646p.
61. Sumitomo Bakelite Co., Ltd. Jpn. Patent 8278409, 1982; *Chem. Abstr.* **1982**, *97*, 145698w.
62. B. F. Goodrich Co. Jpn. Patent 82100117, 1982; *Chem. Abstr.* **1982**, *97*, 183407h.
63. Ito, K. Jpn. Patent 74132184, 1974; *Chem. Abstr.* **1975**, *82*, 112736u.
64. Riew, C. K. U.S. Patent 4256904, 1981; *Chem. Abstr.* **1981**, *94*, 176097s.
65. O'Conner, J. M.; Lickei, D. L.; Sessions, W. J. DE Patent 3007614, 1980; *Chem. Abstr.* **1980**, *93*, 221538d.
66. Hess, B; Schulz-Walz, H. J.; Von Harpe, H.; Peltzer, B.; Bottenbruch, L. DE Patent 2843822, 1980; *Chem. Abstr.* **1980**, *93*, 72947g.
67. O'Conner, J. M.; Lickei, D. L.; Rosin, M. L.; EP Patent 74746, 1983; *Chem. Abstr.* **1983**, *98*, 199315h.
68. Von Harpe, H.; Bottenbruch, L.; Peltzer, B.; Schulz-Walz, H. J. DE Patent 2800468, 1979; *Chem. Abstr.* **1979**, *91*, 124473e.
69. Hutchinson, F. G.; Cleveland, R. G.; Leggett, M. K. DE Patent 2040594, 1971; *Chem. Abstr.* **1971**, *74*, 126793g.
70. Shaw, B.; Lawton, V.; McAinsh, J. DE Patent 2255567, 1973; *Chem. Abstr.* **1973**, *79*, 54383b.
71. Henbest, R. G. C. DE Patent 2309084, 1973; *Chem. Abstr.* **1973**, *79*, 147112c.
72. Henbest, R. G. C. DE Patent 2423483, 1974; *Chem. Abstr.* **1975**, *82*, 87180y.
73. Dainippon Ink and Chemical, Inc. Jpn. Patent 8242712, 1982; *Chem. Abstr.* **1982**, *97*, 39852g.
74. Hsu, T. J.; Lee, L. J. *42nd Annu. Tech. Conf. SPI* **1987**, Sect. 18-C.

75. Mitani, T.; Ogasawara, Y.; Hiraishi, S. Jpn. Patent 8045742, 1980; *Chem. Abstr.* **1980,** *93,* 47907m.
76. Toshima, H. Jpn. Patent 7564382, 1975; *Chem. Abstr.* **1975,** *83,* 116076x.
77. Miyake, H.; Makimura, O.; Tsuchida, T. DE Patent 2506774, 1975; *Chem. Abstr.* **1975,** *83,* 206992x.
78. Makimura, O.; Miyake, H. Jpn. Patent 7809275, 1978; *Chem. Abstr.* **1978,** *89,* 111200x.
79. Krolikowski, W.; Nowaczek, W.; Pawlak, M. PL Patent 115598, 1982; *Chem. Abstr.* **1983,** *98,* 199288b.
80. Tong, S. N.; Chen, D. S.; Chung, L. Z. *Polymer* **1983,** *24,* 469.
81. Madec, P. J.; Marechal, E. *J. Polym. Sci., Polym. Chem. Ed.* **1978,** *16,* 3157.
82. Madec, P. J.; Marechal, E. *J. Polym. Sci., Polym. Chem. Ed.* **1978,** *16,* 3165.
83. Bridges, R. D.; Swincer, A. G.; Strudwick, K. R. *42nd Annu. Tech. Conf. SPI* **1987,** Sect. 18-B.
84. Iwata, K.; Ogasawara, M.; Yoshida, T; Inata, H. Jpn. Patent 7897089, 1978; *Chem. Abstr.* **1978,** *89,* 216285g.
85. Hefner, R. E., Jr.; Hefner, R. E. U.S. Patent 4485220, 1984; *Chem. Abstr.* **1985,** *102,* 132939x.
86. Zofia, K. -W. *Polimery (Warsaw)* **1975,** *20,* 4.
87. Zofia, K. -W. *Polimery (Warsaw)* **1975,** *20,* 61.
88. Zofia, K. -W.; Wladyslaw, K. *Polimery (Warsaw)* **1975,** *20,* 586.
89. Fradet, A.; Marechal, E. *Eur. Polym. J.* **1978,** *14,* 749.
90. Fradet, A.; Marechal, E. *Eur. Polym. J.* **1978,** *14,* 755.
91. Fradet, A.; Marechal, E. *Eur. Polym. J.* **1978,** *14,* 761.
92. Guertler, E.; Gaertner, K.; Mufrick, M. DD Patent 212043, 1984; *Chem. Abstr.* **1985,** *102,* 79725w.
93. Tong, S. N.; Chen, D. S.; Kwei, T. K. *Polym. Eng. Sci.* **1985,** *25,* 54.
94. Tong, S. N.; Chen, D. S. *Proc. IUPAC, Macromol. Symp., 28th* **1982,** 190.
95. Tong, S. N.; Wu, Peter T. K. *42nd Annu. Tech. Conf. SPI* **1987,** Sect. 11-F.
96. Brecksville, R. J. B.; Riew, C. K. U.S. Patent 4290939, 1981.
97. Bucknall, C. B. *Toughened Plastics;* Applied Science: London, 1977; Chapter 7.

Received for review February 11, 1988. Accepted revised manuscript September 16, 1988.

Mechanical and Fracture Properties

Elastomer-Modified Epoxy Model Adhesives Cured with an Accelerated Dicyandiamide System

Alan R. Siebert, C. Dale Guiley, and Arthur M. Eplin

BF Goodrich Company, Research and Development Center, Specialty Polymers and Chemicals Division, Brecksville, OH 44141

Mechanical and fracture properties were measured on elastomer-modified model epoxy paste adhesive formulations, cured with an accelerated dicyandiamide system. Two carboxyl-terminated butadiene–acrylonitrile (CTBN) copolymers of differing bound acrylonitrile content in the form of CTBN–epoxy prereacts were used to modify these epoxy adhesive formulations. Improvements were found in breaking elongation, fracture energy with some loss of modulus, tensile strength, and thermal properties. Results are compared with the same elastomers in admixed systems cured with piperidine. The same filled recipes showed increases in lap shear and T-peel strengths on both cold-rolled and galvanized steel substrates.

Fracture toughness and impact strength of both liquid and solid diglycidyl ether of bisphenol A (DGEBA) epoxy resins can be significantly improved by modification with carboxyl-terminated butadiene–acrylonitrile (CTBN) copolymers (*1*, *2*). These improvements occur with minimum losses in mechanical and thermal properties (e.g., glass-transition temperatures, T_g). Much of this work centered on admixed CTBN–liquid DGEBA epoxy systems. Tertiary amines or tertiary amine salts, such as tris(dimethyl-aminomethyl)phenol (DMP-30) and its tris(2-ethylhexanoate) salt (*1*), and secondary amines, such as piperidine (*2*), were used as the cure agent. The work of Bascom et al. (*3*) and Kinloch and Shaw (*4*) demonstrated significant improvements in the adhesive performance of these modified systems. All

0065–2393/89/0222–0389$06.00/0

of the CTBN-modified epoxy systems showed the presence of a second rubbery phase that separates during cure. Extensive work, recently reviewed by Bascom and Hunston (5), has been reported on the morphology and mechanisms by which this second phase toughens the modified DGEBA epoxy resins.

Drake and Siebert (6) and Pocius (7) reviewed the use of liquid CTBN and solid butadiene–acrylonitrile copolymers as modifiers for epoxy structural adhesives. These modified structural adhesives impart peeling resistance into one- and two-component adhesive systems. One-component systems, which use primarily accelerated dicyandiamide (ADDA) as the latent cure agent, have been studied extensively as pastes or films (prepregs*). There is, however, very little published work on measurement of fracture energy and mechanical properties on modified DGEBA epoxy resin systems that use latent (ADDA) cured systems. In these modified structural adhesives the CTBN is usually added in the form of an epoxy adduct (i.e., CTBN is prereacted with a DGEBA epoxy resin, through alkyl hydroxy esterification reactions, so that only epoxy reactivity is present). This pre-reaction is an important difference from the previously discussed admixed systems using piperidine as the cure agent.

In this work, fracture energy, mechanical properties, T_gs, and adhesive performance were measured for elastomer-modified one-part ADDA-cured DGEBA epoxy paste systems. We wanted to determine if the fracture energy of these modified systems increases, as was found in admixed piperidine systems. If so, can these increases be related to improvements in adhesive performance, especially peel strength? It would be interesting to have an independent measure of adhesive performance such as fracture energy, a material property, to use to evaluate new modified adhesives. There are always questions on the proper adhesive test to use, and there is considerable controversy on the value of lap shear as an adhesive test.

Experimental Details

Materials. Hycar CTBN is a registered trade name of a series of carboxyl-terminated butadiene–acrylonitrile copolymers (BF Goodrich Co.). They can be represented structurally as

$$HOOC-[(CH_2-CH=CH-CH_2)_x-(CH_2-CH)_y]_m-COOH$$
$$|$$
$$CN$$

where averaged $x = 5$, $y = 1$, and $m = 10$. The CTBN products used were 1300 × 8, containing 18% bound acrylonitrile, and 1300 × 13, containing 26% bound acrylonitrile, hereafter called CTBN X8 and CTBN X13). WC–8005 and WC–8006 are

*A prepreg is a continuous fiber or woven fabric impregnated with a B thermosetting or thermoplastic resin, by solvent or melt processing, that is used to make laminated composites.

commercial adducts of CTBN X13 and CTBN X8, respectively, reacted with a liquid DGEBA epoxy resin (Wilmington Chemical Co.). They both contain 40% CTBN. The DGEBA liquid epoxy is Epon 828 (Shell Chemical Co.). Bisphenol A (BPA) is the trade name of 4,4′-isopropylidene diphenol (Dow Chemical Co.). Piperidine was used as received (Fisher Scientific Co.), as were tabular alumina filler, T–64, 325-mesh, aluminum oxide (Alcoa), fumed silica thixotrope, TS–720 (Cabot Corp.) dicyandiamide (cyanoguanidine), 325-mesh (Omicron Chemicals. Inc.), 3-phenyl-1,1′-dimethyl urea (Fike Chemical Co.), and mill oil (Almag 1564, Texaco).

Procedures.   *Castings.*   For mechanical property measurements, formulations were mixed under vacuum by using a marine blade and then cast into 20- × 25- × 0.635-cm (8- × 10- × 0.25-in.) poly(tetrafluoroethylene)-coated (Teflon) aluminum molds and cured as follows: 16 h at 120 °C for recipes cured with piperidine and 1 h at 120 °C for adhesive formulations. To aid in casting void-free specimens for property measurements, no fumed silica thixotrope was used.

Tensile Tests.   Dumbbells were machined from the castings described and mechanical properties were measured by using ASTM D 638. Three specimens were tested for each recipe.

Fracture Energy.   Samples for fracture-energy measurements were machined from the castings. At least five specimens were used for each recipe. Fracture energy was determined by using ASTM E 399, with the compact specimen (*see* Figure A4.1 of that test). The dimensions of the specimen are distance from center of holes to end of sample $W = 25.4$ mm (1.0 in.) and thickness $B =$ nominal 6.35 mm (0.25 in.). The precrack is formed with a single-edge razor blade. The stress intensity factor or fracture toughness, K_{Ic}, is calculated from the maximum load P, crack length a, and thickness B. The critical strain energy release rate or fracture energy, G_{Ic}, is related to K_{Ic} by the following expression:

$$G_{Ic} = \frac{K_{Ic}^2}{E} (1 - \nu^2) \tag{1}$$

where E is the elastic modulus and ν is Poisson's ratio. The value for ν used in these calculations is 0.34. A computer program was written for a personal computer to perform these calculations.

Adhesive Formulations.   All of the model adhesive formulations were hand mixed and run two times through a three-roll ink mill. Lap shear was determined according to ASTM D 1002 on cold-rolled steel (CRS), electrogalvanized steel (EGS), and hot-dipped galvanized steel (HDG) substrates. The metal blanks were washed in acetone and wiped with mill oil (Almag 1564) prior to bonding. In all cases a 12.7-mm (0.5-in.) overlap was used. T-peel measurements were made according to ASTM D 1876 on the same oily substrates. Both ASTM adhesive tests were modified in that precut metal blanks were used. Bond thickness was obtained with nominal 0.5-mm (19-mil) stainless steel wire spacers.

Differential Scanning Calorimetry.   Glass-transition temperatures (T_gs) were determined by using a Mettler TA 3000 instrument equipped with a differential scanning calorimeter (DSC 30 measuring cell). Scans were made at a heating rate of 20 °C/min.

Electron Microscopy. Micrographs were taken of fracture surfaces from compact specimens by using a Joel scanning electron microscope (SEM).

Results and Discussion

Physical Properties of Piperidine-Cured Admixed Elastomer-Modified DGEBA Epoxy Systems.

Recipes for piperidine-cured admixed CTBN-X8-modified DGEBA epoxies are given in Table I. Mechanical properties and fracture energy of these cured systems are given in Table II. The fracture energy, G_{Ic}, results for recipe 4 in Table II show about a 13-fold increase for 15 parts CTBN X8, compared to the unmodified epoxy, recipe 1. This improvement is accompanied by some loss in tensile strength, modulus, and T_g. These G_{Ic} and mechanical property results are similar to results reported earlier on similar systems (2, 3).

Table I also shows recipe 5, modified with 5 parts of CTBN X8 and 24 parts of BPA. Fracture and mechanical property results for this recipe are given in Table II. Recipe 5 shows the highest G_{Ic} of all the samples tested (2.79 KJ/m^2), and it contains only 5 parts of CTBN X8. Riew et al. (2) attributed this high G_{Ic} to a two-particle size distribution. The two-particle size was later confirmed by Kinloch (8), who used the same recipe. Yee and Pearson (9) showed that, for CTBN-X13-modified systems, G_{Ic} increases with the molecular weight of the starting epoxy resin. Whether these high results for fracture energy result from the dual particle sizes or from a lower cross-link density, they certainly show a much more efficient use of rubber, as only 5 parts are used. In addition, the T_g for this recipe is higher than that for the unmodified recipe.

Table I. Admixed CTBN X8-Modified Epoxy

Material	1	2	3	4	5
DGEBA liquid epoxy	100	100	100	100	100
CTBN 1300 X8	–	5	10	15	5
Bisphenol A	–	–	–	–	24
Piperidine	5	5	5	5	5

NOTE: All numbers are parts by weight.

Table II. Cured Properties of Admixed CTBN X8 Epoxy

Property Tested	1	2	3	4	5
Tensile strength, MPa	64.7	59.2	57.4	44.9	50.7
Elongation, %	6.44	7.29	5.79	9.50	8.49
Tensile modulus, GPa	2.68	2.50	2.24	2.11	2.37
K_{Ic}, MN/m$^{3/2}$	0.76	2.10	2.24	2.40	2.73
G_{Ic}, KJ/m^2	0.189	1.56	1.98	2.41	2.79
T_g, °C	91	89	92	84	98

NOTE: Cured for 16 h at 120 °C.

Tables III and IV are recipes and cured property data, respectively, for admixed epoxy systems modified with CTBN X13 (26% bound acrylonitrile). Property data for the control recipe with BPA are given in Table IV, recipe 1.

This CTBN-X13-modified system gives about half as much G_{Ic} (1.30 KJ/m^2) for 15 parts of elastomer as CTBN X8 (2.41 KJ/m^2). In addition to differences in G_{Ic}, there are also significant differences in tensile behavior, elongation in particular. The sample with 15 phr of CTBN X13 has an elongation at break of 16.7%, compared to 9.5 for the sample with 15 phr of CTBN X8, about a 75% increase. The sample with CTBN X13 is clear, a result that indicates very small particles or a match in index of refraction between the two phases.

Figure 1 is a SEM micrograph (3000×) of the fracture surface of CTBN X13 (15 phr) modified epoxy cured with piperidine. It shows very small particles in the range of 0.2–0.4 μm. Figure 2 is a similar micrograph of CTBN X8 (15 phr) modified epoxy. In this case the average diameter of the dispersed phase is about 10 times that for CTBN X13. Thus there is a very wide difference in the size of the dispersed phase between the two systems. This difference probably accounts for the lower G_{Ic} for the CTBN-X13-modified epoxy. The T_g was not lowered significantly for the CTBN-X13-modified recipe, a result that agrees with the elastomer being present as a second phase.

ADDA-Cured   Elastomer-Modified   DGEBA   Epoxy   Resin.

Table V gives recipes for filled and unfilled model adhesive recipes cured with 3-phenyl-1,1-dimethylurea accelerated dicyandiamide. These samples

Table III. Admixed CTBN X13-Modified Epoxy

Material	1	2	3	4	5
DGEBA liquid epoxy	100	100	100	100	100
CTBN 1300 X13	–	5	10	15	5
Bisphenol A	24	–	–	–	24
Piperidine	5	5	5	5	5

NOTE: All numbers are parts by weight.

Table IV. Cured Properties of Admixed CTBN X13 Epoxy

Property Tested	1	2	3	4	5
Tensile strength, MPa	56.1	57.9	47.1	43.9	49.3
Elongation, %	10.3	9.19	11.3	16.7	7.2
Tensile modulus, GPa	2.61	2.70	2.40	2.38	2.51
K_{Ic}, MN/m$^{3/2}$	0.78	1.45	1.55	1.84	1.94
G_{Ic}, KJ/m^2	0.21	0.69	0.88	1.30	1.33
T_g, °C	101	90	90	88	97

NOTE: Cured for 16 h at 120 °C.

*Figure 1. SEM micrograph (3000×) of fracture surface of CTBN X13 (15 phr)
modified model epoxy adhesive cured with piperidine.*

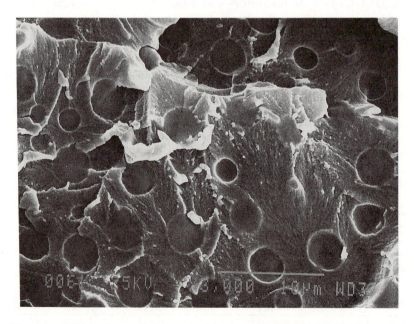

*Figure 2. SEM micrograph (3000×) of fracture surface of CTBN X8 (15 phr)
modified model epoxy adhesive cured with piperidine.*

are modified with 15 parts of CTBN X8 and CTBN X13 in the form of commercial adducts containing 40% CTBN. The 15-part elastomer level is used in this work because it gives the highest fracture energy while still retaining relatively high modulus and T_g properties. Recipes 4, 5, and 6 are filled with 40 parts of tabular alumina.

Cured properties for these recipes are given in Table VI. As expected, the unfilled control recipe has a very low G_{Ic}, indicating a very brittle system. The filled control sample has a G_{Ic} of 0.24 KJ/m^2, which is at least 2 times the value without filler and about the same as the control recipe from Table I with piperidine as the cure agent. Perhaps the increase in G_{Ic} in filled systems indicates a crack-pinning mechanism, as proposed by Lange (*10*). As expected, filled recipes show lower elongations and higher moduli than unfilled systems. Both filled and unfilled systems show significant increases in G_{Ic} for these elastomer-modified systems cured with ADDA.

For unfilled recipes we find about a 10-fold increase in fracture energy. This finding is not too different from increases found in elastomer-modified systems cured with piperidine. In these unfilled systems, both CTBN X8 and CTBN X13 give about the same value for G_{Ic}. The increases in G_{Ic} for the elastomer-modified filled systems were not as great, about 4 and 2.5 times for CTBN X13 and CTBN X8, respectively. However, the absolute value for the CTBN-X13-modified system is about the same in both filled and unfilled systems. We did not do any testing on filled systems with piperidine as the cure agent, as fillers would settle during the long cure cycles for these systems. The mechanical properties show a loss of tensile

Table V. Model One-Part Epoxy Adhesives Cured with ADDA

Material	1	2	3	4	5	6
DGEBA liquid epoxy	100	100	100	100	100	100
CTBN X13 DGEBA adduct	–	37.5	–	–	37.5	–
CTBN X8 DGEBA adduct	–	–	37.5	–	–	37.5
Tabular alumina	–	–	–	40	40	40
Fumed silica	–	–	–	3.5	3.5	3.5
Dicyandiamide	6	6	6	6	6	6
3-Phenyl-1,1-dimethylurea	2	2	2	2	2	2

NOTE: All numbers are parts by weight.

Table VI. Cured Properties of Model One-Part Epoxy Adhesives

Property Tested	1	2	3	4	5	6
Tensile strength, MPa	51.6	63.7	55.2	61.1	49.9	44.3
Elongation, %	1.90	5.19	3.13	1.76	2.23	1.93
Tensile modulus, GPa	2.98	2.38	2.28	4.01	3.26	2.52
K_{Ic}, MN/$m^{3/2}$	0.58	1.76	1.63	1.05	1.91	1.31
G_{Ic}, KJ/m^2	0.10	1.15	1.03	0.24	0.99	0.60
T_g, °C	137	122	131	125	117	127

NOTE: Cured 1 h at 120 °C.

strength and modulus for the modified recipes. There is some loss of T_g for the modified recipes.

The morphology for recipe 2, Table V (which is an unfilled system modified with CTBN X13) is given in Figure 3. It shows a very uniform distribution of particle sizes. All of the particles are about 1 μm. Figure 4 is a similar micrograph for an CTBN-X8-modified recipe. It has an average particle diameter of about 2 μm. This diameter is about twice the average diameter of the CTBN-X13-modified system. The two elastomers do not show the wide differences in particle size found in piperidine-cured systems. Figure 5 is the morphology for CTBN-X13-modified system that contains filler. It shows the same uniform distribution and average particle size. Such uniformity indicates that the presence of filler does not alter the second-phase formation or average size.

Thus these ADDA systems do show a second phase that is uniformly distributed and about 1 μm in average diameter. Even though the morphology appears to be similar for modified epoxy resins cured with either an ADDA or piperidine system, we cannot assume the composition of the second phases are the same. This question of composition of the second phase in elastomer-modified epoxies remains one of the more difficult problems to solve. Williams et al. (11) made some excellent theoretical calcula-

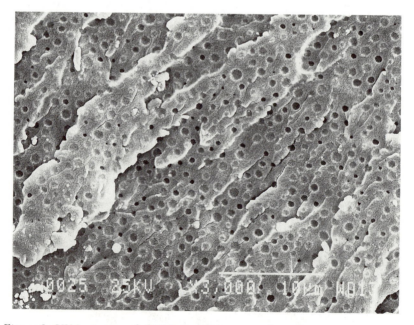

Figure 3. SEM micrograph (3000×) of fracture surface of CTBN X13 (15 phr) modified model epoxy adhesive cured with accelerated dicyandiamide system without filler.

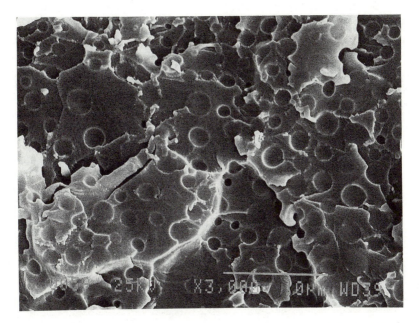

Figure 4. SEM micrograph (3000×) of fracture surface of CTBN X8 (15 phr) modified model epoxy adhesive cured with accelerated dicyandiamide system without filler.

tions on the composition of the second phase in CTBN-modified systems. However, we know of no method to verify these calculations experimentally.

Adhesive Properties of Model Adhesives.   Table VII gives lap shear and T-peel results for filled and unfilled recipes from Table V. Lap shear measurements are used here for comparison purposes only. For unfilled recipes there is a significant increase in both lap shear and T-peel for the CTBN-X13-modified systems compared to the unmodified control. Some improvement in adhesive properties of unmodified systems occurs with the addition of filler. This result is not unlike the increases found in G_{Ic} for these systems (*see* Table VI).

For filled recipes, a significant increase in adhesive performance occurred, but not quite as large an increase as was found in unfilled systems. If we compare filled and unfilled CTBN-X13-modified recipes, the filled recipes show the best adhesive performance. CRS CTBN-X13-modified recipes show better adhesive performance than CTBN-X8-modified recipes. This finding agrees with the G_{Ic} results from Table VI for filled recipes.

Table VIII shows the effect of cure temperature on the adhesive properties of filled systems for the control and CTBN-X13- and CTBN-X8-modified recipes on oily EGS metal substrates. Adhesive results when CRS is used are quite similar to those shown in Table VIII. Very modest increases

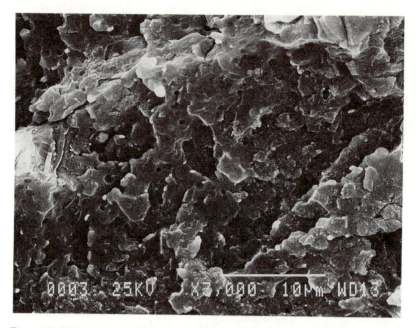

Figure 5. SEM micrograph (3000 ×) of fracture surface of CTBN X13 (15 phr)
modified model epoxy adhesive cured with accelerated dicyandiamide system
with filler.

Table VII. Lap Shear and T-Peel Strengths of Model One-Part Adhesives on Oily Cold-Rolled Steel

	Unfilled		Filled		
Property Tested	Control	X13	Control	X13	X8
Lap shear, MPa	5.51	9.72	7.33	11.0	8.86
T-Peel, Kg/cm	1.21	4.47	1.90	5.26	4.43

NOTE: Cured 1 h at 170 °C.

Table VIII. Effect of Cure Temperature on Adhesive Properties of One-Part Model Filled Systems on Oily Electrogalvanized Steel

Property Tested	Control	X13	X8
Lap shear,[a] MPa	8.10	12.3	10.2
Lap shear,[b] MPa	8.21	11.3	9.86
T-peel,[a] Kg/cm	2.49	7.00	5.00
T-peel,[b] Kg/cm	2.16	7.34	5.37

[a]Cured for 1 h at 170 °C.
[b]Cured for 20 min at 204 °C.

occurred in T-peel for higher-temperature cures. However, the most significant difference is that on EGS the CTBN-X13-modified recipes again give better adhesive properties than do CTBN-X8-modified recipes.

Table IX gives lap shear strengths as a function of cure temperature for unmodified, CTBN-X13-, and CTBN-X8-modified systems with an oily HDG metal substrate. On oily HDG, both modified systems show about the same lap shear strength, a decided increase over that for control recipes. Because the metals used in this work are relatively thin, about 13 mm, there is a considerable peel component when testing lap shear specimens. Thus, samples with lower peel strengths also give lower lap shear strengths.

Effect of BPA Addition on Adhesive Performance.   As stated earlier, the addition of BPA to piperidine-cured admixed epoxy systems gives significantly higher fracture energies. Thus we wanted to see what effect the addition of BPA to our model epoxy paste adhesives would have, particularly the effect on peel strength. Several problems are associated with the addition of BPA to a paste adhesive. First, these are one-part systems, and thus some degree of stability at room temperature is required. Mixtures of liquid DGEBA epoxy resins and BPA are known to advance at room temperature. Second, when BPA is added to liquid epoxies, a significant increase in viscosity occurs.

We examined adhesive performance at 3 and 6 parts of BPA, instead of the 24 parts used earlier, a decrease that should minimize these problems. Table X lists these adhesive formulations, as well as a recipe without BPA.

Table IX. Effect of Cure Temperature on Lap Shear Properties of One-Part Model Systems on Oily HDG Steel

Cure Time and Temperature	Control	X13	X8
1 h at 120 °C	4.21	8.16	8.87
1 h at 170 °C	6.50	10.8	10.5
20 min at 204 °C	7.21	12.7	12.6

NOTE: All results are in megapascals.

Table X. Model One-Part Epoxy Adhesive Recipes with BPA Additions, Cured with ADDA

Material	1	2	3
DGEBA liquid epoxy	77.5	77.5	77.5
CTBN X13 DGEBA adduct	37.5	37.5	37.5
Bisphenol A (BPA)	–	3	6
Tabular alumina	40	40	40
Fumed silica	3.5	3.5	3.5
Dicyandiamide	6	6	6
3-Phenyl-1,1-dimethylurea	2	2	2

NOTE: All numbers are parts by weight.

Table XI gives the adhesive properties for these recipes on oily galvanized metal. We find that the presence of BPA gives significantly higher lap shear and T-peel values than the recipe without BPA. In addition, recipes with BPA did not show a significant reduction in T_g. Thus the addition of BPA significantly increases adhesive properties on oily HDG steel.

We did not examine morphology of these adhesive formulations to see if a dual particle size exists for these model adhesives. Neither have we measured the mechanical properties for these recipes. In addition, it is doubtful if these small amounts of added BPA can significantly increase the molecular weight of the epoxy resin. Thus, additional work is needed to determine the mechanisms by which these systems improve adhesive performance.

Summary

Mechanical and fracture properties were measured for model epoxy paste adhesives modified with 15 parts of two CTBN liquid polymers with different acrylonitrile contents. These polymers were added in the form of prereacts with a liquid DGEBA epoxy resin. An ADDA latent one-part cure system was used. Improvements in fracture energy, G_{Ic}, were found with minimum losses in mechanical and T_g properties. Mechanical and fracture data showed that recipes modified with CTBN X13, containing 26% bound acrylonitrile, gave somewhat better performance than did CTBN-X8-modified recipes. This result is just the opposite of results obtained for model admixed formulations that used the same two polymers and cured with piperidine. In this work CTBN-X8-modified recipes gave significantly higher fracture properties than did those modified with CTBN X13.

Adhesive performance of these modified model adhesives was measured on CRS and EGS metal substrates. Improvements in both lap shear and T-peel strengths were found for both CTBN polymers. The adhesive data suggests that model formulations modified with CTBN X13 also give better adhesive performance than those modified with CTBN X8. Thus, it appears that fracture energy measurements of model recipes will give a good indication of their adhesive performance, peel strength in particular.

Table XI. Lap Shear and T-Peel Strengths of Model One-Part Adhesives with Added BPA on Oily HDG

Property Tested	1	2	3
Lap shear, MPa	12.4	14.6	15.0
T-peel, Kg/cm	6.44	7.80	9.00
T_g, °C	114	116	114

NOTE: Cured 1 h at 170 °C.

Acknowledgment

The authors thank Ron Smith, who did the SEM work.

References

1. McGarry, F. J.; Willner, A. M. *Org. Coat. Plast. Chem.* **1968**, *28*, 512.
2. Riew, C. K.; Rowe, E. H.; Siebert, A. R.; In *Toughness and Brittleness of Plastics;* Deanin, R. D.; Crugnola, A. M., Eds.; Advances in Chemistry 154; American Chemical Society: Washington, DC, 1976; pp 326–343.
3. Bascom, W. D.; Cottington, R. L.; Jones, R. L.; Peyser, P. *J. Appl. Polym. Sci.* **1975**, *19*, 2545.
4. Kinloch, A. J.; Shaw, S. J. *J. Adhes.* **1981**, *12*, 59.
5. Bascom, W. D.; Hunston, D. L. In *Rubber-Toughened Plastics;* Riew, C. K., Ed.; Advances in Chemistry 222; American Chemical Society: Washington, DC, 1989; Chapter 6.
6. Drake, R. S.; Siebert, A. R. *Adhes. Chem.* **1984**, *56*, 643–654.
7. Pocius, A. V. *Rubber Chem. Technol.* **1985**, *58*, 622–636.
8. Kinloch, A. J.; Hunston, D. L. *J. Mater. Sci. Lett.* **1986**, *5*, 909–911.
9. Yee, A. F.; Pearson, R. A. NASA Contract Report No. 3852, 1984.
10. Lange, F. F. *Philos. Mag.* **1970**, *22*, 983.
11. Williams, R. J. J.; Borrajo, J.; Adabbo, H. E.; Rojas, A. J. In *Rubber-Modified Thermoset Resins;* Riew, C. K.; Gillham, J. K., Eds.; Advances in Chemistry 208; American Chemical Society: Washington, DC, 1984; pp 195–214.

RECEIVED for review March 25, 1988. ACCEPTED revised manuscript November 22, 1988.

Polyblends of Polyvinyl Chloride with Ethylene–Vinyl Acetate Copolymers

Miscibility and Practical Properties

Rudolph D. Deanin, Sujal S. Rawal, Nikhil A. Shah, and Jan-Chan Huang

Plastics Engineering Department, University of Lowell, Lowell, MA 01854

In polyblends of polyvinyl chloride (PVC) with ethylene–vinyl acetate (EVA) copolymers, plots of properties versus the PVC/EVA ratio generally indicated two-phase behavior, growing softer with increasing EVA content in the blends and with increasing vinyl acetate (VA) content in the copolymers. Differential scanning calorimetry indicated that increasing the VA content in the copolymer lowered crystallinity, and that amorphous PVC plus amorphous EVA formed a new phase that behaved like plasticized PVC.

M ISCIBLE BLENDS ACT AS POLYMERIC PLASTICIZERS when a rubber is added to a rigid polymer. However, semicompatible blends can produce dramatic synergistic improvement of the impact strength of the rigid plastic (1–4). Conversely, when a rigid plastic is added to a rubber, it can produce useful increases in modulus and strength. These effects depend on the degree of molecular miscibility and practical compatibility.

Rigid polyvinyl chloride (PVC) is a fairly tough material, but for many applications it is toughened further by addition of 5–25% of rubbery "impact modifiers". Typically, these modifiers are graft copolymers of acrylonitrile, methyl methacrylate, and styrene on a backbone of butadiene rubber (5, 6). A particularly interesting family of impact modifiers comes from copolymers of ethylene with vinyl acetate (EVA). Increasing VA content in the EVA gradually converts them from crystalline, nonpolar, rigid materials into amorphous, polar, rubbery materials. In blends of PVC with EVA, there may be separate phases for crystalline PVC, amorphous PVC, crystalline polyeth-

0065–2393/89/0222–0403$06.00/0

ylene (PE) blocks, amorphous EVA, and one or two phases for molecular blends of amorphous PVC with amorphous EVA. Early studies indicated that EVA of 45–75% VA content produced maximum miscibility with PVC (7–10).

When PVC was blended with 0–25% EVA as impact modifiers, synergistic peak improvement occurred at 5–15% EVA (Figures 1 and 2). At higher VA content in the EVA, the peaks tended to move toward higher EVA content and become broader (11). Electron microscopy showed that increasing EVA content produced increasing EVA domain size, until phase inversion made EVA the continuous matrix phase and its properties inverted from a rigid high-impact plastic to a thermoplastic elastomer (12).

The present study was undertaken to explore the entire range of PVC/EVA ratios from 100:0 to 0:100, using EVAs of equal molecular weight and VA contents from 0 to 51% of the EVA. These were evaluated for mechanical and thermal mechanical properties and calorimetric characterization.

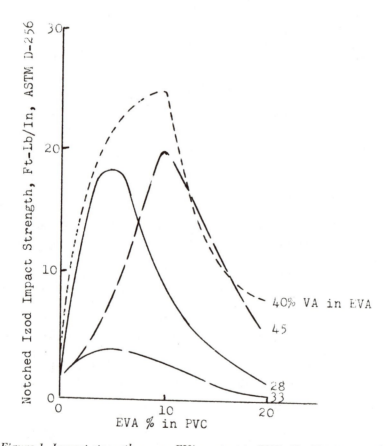

Figure 1. Impact strength versus EVA content in PVC; 28–45% VA in EVA.

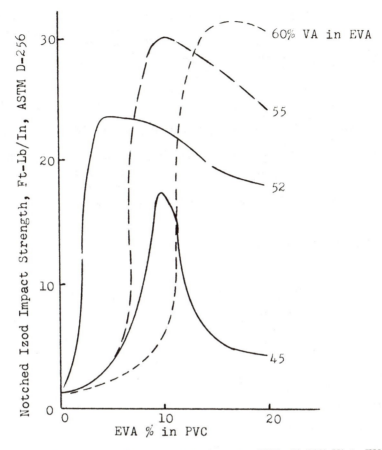

Figure 2. Impact strength versus EVA content in PVC; 45–60% VA in EVA.

Experimental Details

Medium-molecular-weight PVC (Goodrich Geon 30) was stabilized by 2 phr of organotin mercaptide (M&T Thermolite 31) and lubricated by 0.5 phr of stearic acid. It was blended with each of five EVAs (percent VA in parentheses):

- NA–171 (0)
- UE–637 (9)
- UE–621 (18)
- UE–645 (28)
- UE–904–25 (51)

in ratios of PVC/EVA of 100:0, 75:25, 50:50, 25:75, and 0:100, by mixing 5 min at 168 °C (335 °F) on a 6- × 12-in. (15.25- × 30.50-cm) differential-speed two-roll mill.

Milled sheets were pressed 0.125 in. (3.17 mm) thick for 3 min at 177 °C (350 °F) and cut into test specimens. The specimens were tested for tensile properties according to ASTM D 638 Type IV at 5.08 cm/min (2 in./min) and for Clash–Berg torsional modulus versus temperature according to ASTM D 1043. Thermal analysis was carried out on a differential scanning calorimeter (Perkin–Elmer DSC–2C) at 20 °C/min.

Mechanical and Thermal–Mechanical Properties

Modulus versus PVC/EVA ratio (Figure 3) gave S-shaped curves typical of two-phase polyblend systems. Increasing VA content in the EVA gave progressively softer curves. This configuration suggested that random copolymerization decreased the regularity and crystallinity of the EVA.

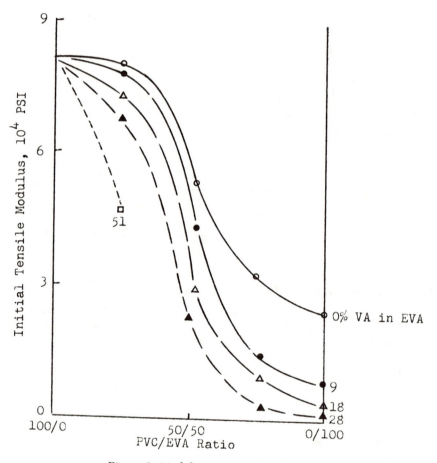

Figure 3. Modulus versus PVC/EVA ratio.

Ultimate tensile strength versus PVC/EVA ratio (Figure 4) gave sharply U-shaped curves typical of very sensitive properties in incompatible systems, where any heterogeneity can act as a stress concentrator to initiate premature failure.

Ultimate elongation increased with EVA content in the blends (Figure 5). The major factor was inversion to an EVA continuous matrix phase, with decreasing content of rigid PVC. Elongation also increased with VA content in the EVA because of increasing miscibility with PVC. This miscibility reduced the heterogeneities that could cause premature failure.

Clash–Berg torsional modulus versus temperature gave typical curves for the leathery region. For quantitative comparison, the temperature T_{lr} when torsional modulus approached 6035 psi was chosen as a measure of the leathery–rubbery transition. This value was read from the curves of all the blends (Figure 6). Transition temperature versus PVC/EVA ratio gave S-shaped curves typical of two-phase polyblend systems. Increasing VA content in the EVA gave progressively softer curves. This change suggested that

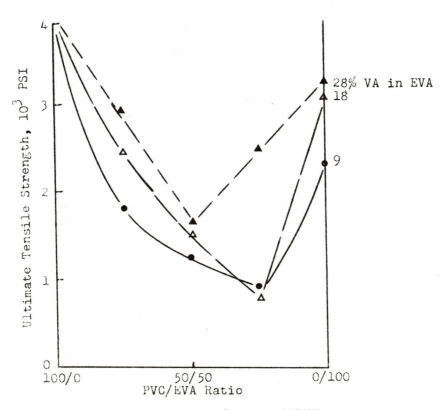

Figure 4. Tensile strength versus PVC/EVA ratio.

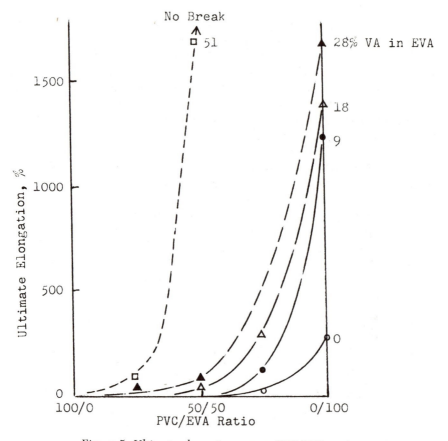

Figure 5. Ultimate elongation versus PVC/EVA ratio.

random copolymerization decreased the regularity and crystallinity of the EVA. The PVC plateau disappeared at 51% VA in the EVA. Apparently the EVA was miscible enough in the PVC matrix to act as a polymeric plasticizer.

Thermal Analysis

Differential scanning calorimetry suffered from overlap of the PVC glass transition versus the broad EVA melting range (Figures 7–11), but still yielded useful information. The crystalline melting point of low-density polyethylene (LDPE) at 108 °C (*13*) was lowered progressively by copolymerization with increasing amounts of VA (Table I). However, it was not further affected by blending with PVC. This stability indicated that the crystalline regions of PE blocks did not blend with the PVC. The enthalpy of melting was approximately proportional to the percent of EVA in the polyblends

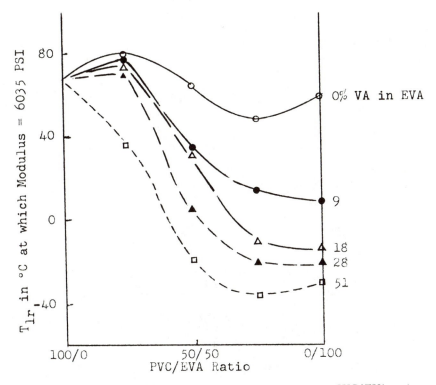

Figure 6. *Leathery–rubbery transition temperature versus PVC/EVA ratio.*

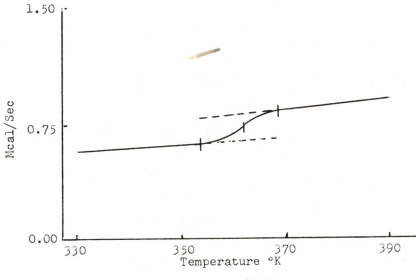

Figure 7. *DSC of PVC.*

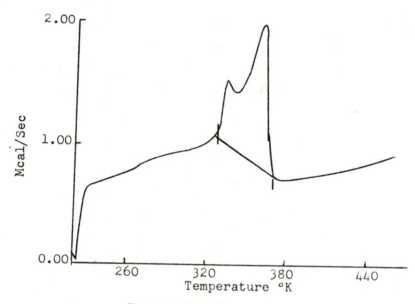

Figure 8. DSC of EVA; 18% VA.

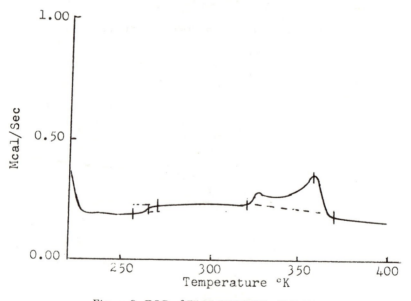

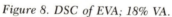

Figure 9. DSC of 75:25 PVC/EVA; 18% VA.

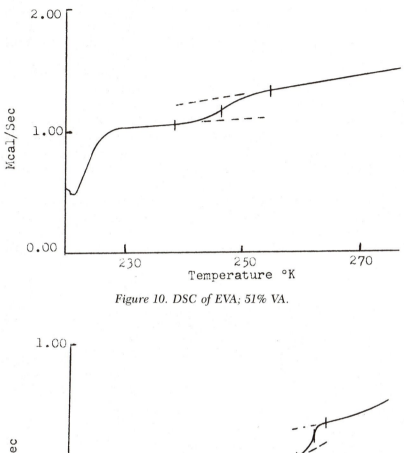

Figure 10. DSC of EVA; 51% VA.

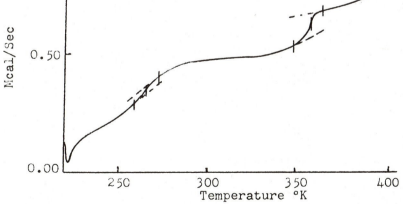

Figure 11. DSC of 75:25 PVC/EVA; 51% VA.

(Table II), a result that confirmed that the crystalline PE regions did not blend with the PVC.

Glass-transition temperatures (T_g) were more complex and sometimes too faint for the sensitivity of the instrument (Table III). Pure PVC was observed at 88 °C, denoting the maximum-use temperature of unmodified rigid vinyl plastics. High-VA-content EVAs were rubbery; their T_g values (−27 and −33 °C) represented the molecular flexibility of the random EVA copolymer molecule. In PVC–EVA blends, intermediate T_gs of −4 to −10 °C also appeared. These values reflect a new phase composed of a homogeneous blend of amorphous PVC with amorphous EVA copolymer. Thus the separate T_g values for PVC and for EVA would correspond to the plateaus in the property versus PVC/EVA ratio curves. Similarly, the T_g of the miscible polyblend phase would correspond to the more gradual linear property transitions.

In rigid vinyl plastics and thermoplastic EVA elastomers, the PVC and EVA plateaus would be of primary importance, and the miscible phase would serve mainly to provide interfacial adhesion. In highly miscible blends, the EVA would serve primarily as a polymeric plasticizer for flexible vinyl plastics and elastoplastics.

Table I. Effect of Copolymerization on Melting Points (° C) of PVC–EVA Blends

VA in EVA (%)	75:25	50:50	25:75	0:100
18	85	84	85	87
28	40–83	39–84	47–87	49–90
51	none	none	none	none

Table II. Effect of EVA Concentration in PVC–EVA Blends on Enthalpy of Melting (cal/g)

VA in EVA (%)	75:25	50:50	25:75	0:100
18	2.4	6.3	9.3	11.8
28	1.6	4.0	5.0	6.4

Table III. Effect of Concentration on Glass-Transition Temperatures (° C) of PVC–EVA Blends

VA in EVA (%)	100:0	75:25	50:50	25:75	0:100
0	88				
18		−8	−7	−8	
28		−10	−5	−9	−33
51		−8, 85	−25, −4, 86	−26, −9, 85	−27

Conclusions

Increasing VA content in EVA increases polarity closer to that of PVC, first producing stronger interfacial bonding between phases and later increasing miscibility of the two polymers to form one or more blended phases. Simultaneously, increasing VA content in EVA decreases polyethylene crystallinity, making the polymer more soft and flexible. These two effects together determine the efficiency of EVA as an impact modifier in rigid PVC, and also as a polymeric plasticizer in flexible PVC. Overall, this makes the EVA family a very versatile modifier for PVC.

References

1. Deanin, R. D. *Encycl. Polym. Sci. Technol. Suppl.* **1977**, *2*, 458.
2. Manson, J. A.; Sperling, L. H. *Polymer Blends and Composites;* Plenum: New York, 1976.
3. Paul, D. R.; Newman, S. *Polymer Blends;* Academic: New York, 1978.
4. Deanin, R. D. *Polym. Mater. Sci. Eng.* **1987**, *56*, 35; Seymour, R. B.; Mark, H. F. *Applications of Polymers;* Plenum: New York, 1988; p 53.
5. Ryan, C. F.; Jalbert, F. L. In *Encyclopedia of PVC;* Nass, L. I., Ed., Dekker: New York, 1977; Vol. 2, Chapter 12, pp 609–632.
6. Deegan, C. C. *Mod. Plast. Encycl.* **1985**, *62*, 154.
7. Hammer, C. F. *Macromolecules* **1971**, *4*, 69.
8. Marcincin, K.; Ramonov, A.; Pollak, V. *J. Appl. Polym. Sci.* **1972**, *16*, 2239.
9. Feldman, D.; Rusu, M. *Eur. Polym. J.* **1974**, *10*, 41.
10. Shur, Y. J.; Ranby, B. *J. Appl. Polym. Sci.* **1975**, *19*, 1337.
11. Deanin, R. D.; Shah, N. A. *Org. Coat. Plast. Chem.* **1981**, *45*, 290.
12. Deanin, R. D.; Shah, N. A. *J. Vinyl Technol.* **1983**, *4*, 167.
13. Brydson, J. A. *Plastics Materials;* Butterworth, London, 1982; p 199.

RECEIVED for review February 11, 1988. ACCEPTED revised manuscript September 7, 1988.

Blends of Unsaturated Polyesters with High-Molecular-Weight Elastomers Bearing Reactive Functional Groups

Richard F. Grossman

Synthetic Products Company, Stratford, CT 06497

Blends were prepared throughout the range of composition of a typical unsaturated polyester with acrylonitrile–butadiene copolymer elastomers (NBR), ethylene–propylene–diene–monomer copolymers (EPDM), carboxylated NBR, carboxylated EPDM, and amino-functional NBR elastomers. The elastomers bearing reactive functional groups show sufficient compatibility that their blends are suitable for further compounding. Blends with amino-functional NBR show improved fracture surface energy in glass-filled compositions.

BLENDS OF ELASTOMERS WITH THERMOSETTING RESINS have beer .n commercial use for many years in several applications. Blends with phenolic resins have been used to increase the hardness of rubber compounds. Blends with about equal weights of each component were popular many years ago in hard-rubber applications and, more recently, phenolic compounds have been modified with minor additions of various elastomers (*1*). The use of depolymerized rubber to modify the brittleness of epoxy resins is almost as old as the resins themselves (*2*). In addition, epoxy resins are common ingredients used by the rubber compounder, most often to improve the stability of halogenated elastomers (*3*). The toughening of epoxy resins by minor addition of liquid low-molecular-weight elastomer precursors has proven highly fruitful and led to considerable experimentation with elastomer modification of a variety of thermosets (*4*).

0065–2393/89/0222–0415$06.00/0

Elastomers in Unsaturated Polyesters

Similar results with unsaturated polyesters are desirable, as increased toughness would be of great commercial significance (5). Breakage of parts during demolding and finishing is a significant factor in the molding industry. In addition, availability of tougher compositions would enable wider use in automotive and other applications. Although some degree of success has been reported (6), and at least one low-molecular-weight elastomer is promoted in this area, improvements in toughness in actual working compounds are minor compared to those found with epoxy resins. Present commercial use is very small. For example, the most recent text on polyester-molding technology devotes a half page to the subject (7).

Explanations of the relative unresponsiveness of unsaturated polyesters to elastomer toughening have been diverse. One school of thought holds that matrix toughening is not significant in typical fiber-reinforced compositions (i.e., that failure occurs primarily at the resin–reinforcement interface). In this case, only toughening specifically at this locus would be effective (8). Others have suggested that overall matrix toughness probably is significant, but that poor elastomer–resin compatibility during the initial phase of cross-linking prevents development of a suitably distributed rubber phase of useful particle size (9).

With epoxy resins, considerable control of molecular architecture is possible. The molar mass, size, and mobility between cross-links are readily varied, as well as the same factors in the cross-link itself. The cross-linking reaction affects only certain groups present in the elastomer, whose numbers and frequency may also be varied. The importance of these considerations in developing toughened epoxy resins is well appreciated. The converse, that these controls are not readily available in the case of unsaturated polyesters, has not been stressed.

In theory, special-purpose polyester resins might be prepared with sites for peroxide-initiated free-radical cross-linking disposed so as to augment toughening with elastomers. Instead of a low-cost monomer, typically styrene, unsaturated oligomers or other specific cross-linking agents might be employed to enhance toughening. With epoxy resins, this type of approach would be normal compounding procedure. In the case of unsaturated polyesters, economics and past practice dictate against such an approach. The compounder in this field is normally limited to off-the-shelf polyester resins, low-cost monomers, and the range of fillers and additives promoted for this use. To date the inclusion of minor fractions of elastomers in this compounding framework has not generated improvements more significant than could be achieved by manipulation of the proportions of standard ingredients.

The present study has proceeded with two premises: that practices in the unsaturated polyester molding industry are likely to continue, and that, despite this, it may be possible to find elastomers that have a toughening effect. Search for such species was limited to elastomers of current commercial availability and those well along in development. Such an elastomer, it was thought, should have the following attributes:

1. The greater part of the elastomer should be thermodynamically incompatible with unsaturated polyesters. This feature is needed to prevent formation of a phase in which the elastomer acts as a plasticizer, thus to improve impact strength only by lending ductility, not through actual toughening, and consequently to reduce the physical and thermal properties of the composition.

2. Nonetheless, the elastomer should contain sufficient polar groups to permit solubility in the uncured resin–monomer mixture. These should be distributed so as to provide a large molar mass of nonpolar component between the polar groups so that there is sufficient mobility to form a distribution of discrete incompatible elastomeric particles during cross-linking. A necessary corollary is that the elastomer should have intrinsically high mobility (i.e., low glass-transition temperature).

3. The polar groups on the elastomer should have strong affinity not only for the unsaturated polyester matrix, but also for reinforcements such as treated glass fibers.

4. The rate of cross-linking of the elastomer via thermal or induced decomposition of peroxide initiators should be slow compared to that of the unsaturated polyester matrix, to permit formation of a distribution of discrete elastomer particles during the cross-linking phase.

5. The elastomer should be of relatively high molecular weight. It should possess elastomeric properties rather than be a liquid polymer that, under other conditions, could be chain-extended and cross-linked to yield a rubbery product. In respect to toughening brittle plastics in general, high molecular weight is not necessarily required, depending on whether the load-bearing ability of the elastomer particle enters into a specific toughening mechanism. In the case at hand, high molecular weight should be a positive factor because of its contribution of mass and size between polar groups, and because of its contribution to the work of adhesion needed to disrupt bonds to the matrix and to reinforcements.

Experimental Details

Unreinforced Compositions. Blends were made from 0–100% unsaturated polyester (Owens–Illinois OCF–RP–325), and the remainder elastomer. To each blend was added 50 parts per 100 total polymer of fine-particle amorphous silica, 35 parts of styrene monomer, 4 parts of zinc stearate, and 1.5 parts of dicumyl peroxide. No low-profile or other polymeric additives were used.

Rubber-rich blends were mixed in an internal mixer (Farrel Midget Banbury) and polyester-rich blends in a sigma blade mixer (Baker–Perkins). Compositions were press cured at 160 °C according to ASTM D 3182. Optimum cure time was

taken as the oscillating disk rheometer $t90$ value at this temperature, following ASTM
D 2804. Tensile properties were measured by using a tensile tester (Instron Model
1130) to perform ASTM D 638, with a crosshead speed of 5 mm/min on Type IV
samples, compression molded directly to dumbbell shape, as indicated previously.

Reinforced Compositions. Blends consisted of 0–100% unsaturated polyester
(Owens–Illinois OCF–RP–325), and the remainder elastomer. To each was added
200 parts per 100 total polymer of fine-particle calcium carbonate, 35 parts of styrene,
4 parts of zinc stearate, 1.5 parts of dicumyl peroxide, and 100 parts of 6-mm chopped
fiberglass strand (Owens–Corning 847GE).

Double-beam compact tension specimens were compression molded at 160 °C
according to the dimensions given by Kinloch (*10*). Cracks were initiated with a razor
blade and the samples loaded at a rate of 1 mm/min crosshead travel in a tensile
tester, to calculate stress-intensity factors and then fracture energies (*11*).

Specimens were also compression molded at 160 °C, 1.6 mm thick, 25 × 50
mm for measurement of flexural modulus, according to ASTM D 790, Method I, by
using a tensile tester with crosshead speed of 0.8 mm/min.

The elastomers amorphous ethylene–propylene–diene–monomer (EPDM, Epsyn 4506), carboxylated EPDM, (Epsyn DE203), acrylonitrile–butadiene copolymer
(NBR, Nysyn 33–3), and amino-functional NBR (Nysyn DN508–14A) were provided
by Copolymer Rubber Chemical Corp. Carboxylated NBR (Krynac 221) was provided
by Polysar, Inc.

Discussion

Amorphous EPDM was chosen as a control elastomer because it does not
contain polar groups and is known to have low affinity for both glass and
polyester. In blends with unsaturated polyester (Figure 1), tensile strength

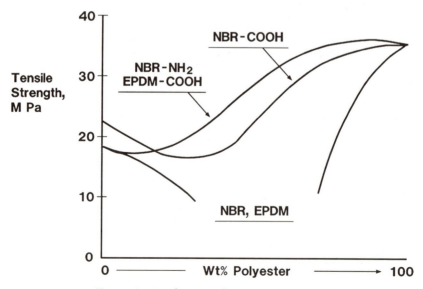

Figure 1. Tensile strength versus composition.

decreases rapidly as the composition curve is entered from either end, to levels far below the average of the individual components. Figure 2 shows that ultimate elongation falls sharply as polyester is introduced into the elastomer. In Figure 3, the tensile modulus of the polyester falls similarly sharply with introduction of the elastomer. This pattern of observations is typical of mixtures of incompatible polymers.

These experiments were repeated with carboxylated EPDM, essentially the same polymer except for the presence of a low level of pendant –COOH groups. This polymer is much closer to the criteria presented for useful toughening. The polar groups are of a useful type, widely disposed, and present at a level low enough not to interfere with overall mobility. Further, the rate of cross-linking with organic peroxides is very slow. It was found (Figures 1, 2, and 3) that tensile strength, elongation, and tensile modulus show relatively smooth curves from one extreme of composition to the other. A high level of elongation was obtained with approximately equal proportions of the two polymers.

Addition of 400 parts per 100 total polymer of calcium carbonate filler to the 50:50 blend produced a vulcanizate retaining ultimate elongation of 75–100%. Addition, instead, of 250 parts of hydrated alumina led to the same level of ultimate elongation. These compositions were easily compression molded and retained more than 50% of their original elongation after 7 days at 150 °C in a circulating-air oven. This level of heat-aging resistance is considerably better than that of typical hard-rubber compounds (1).

Acrylonitrile–butadiene rubber (NBR), 33% acrylonitrile (ACN) content,

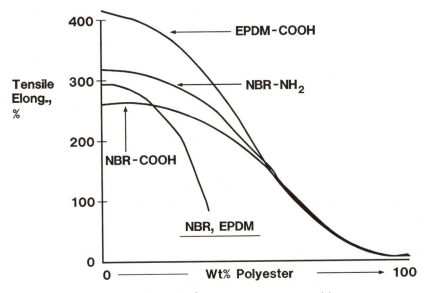

Figure 2. Ultimate elongation versus composition.

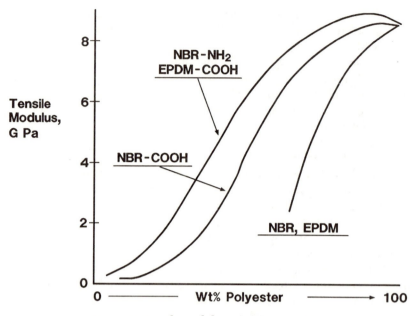

Figure 3. Tensile modulus versus composition.

yielded tensile, elongation, and modulus results in blends with unsaturated polyester very similar to those found with amorphous EPDM, as indicated in Figures 1, 2, and 3. This was the case despite a solubility parameter estimated to be similar to that of unsaturated polyesters (*12*). NBR may be considered another control polymer that does not meet the suggested criteria.

The experiments were repeated with a similar NBR containing a low level of pendant amino groups (Nysyn DN508–14A). This was considered another potential candidate to meet the criteria. Unlike the parent NBR, the cross-linking reaction with organic peroxides is very slow. The pattern of tensile, elongation, and modulus data was similar to that of carboxylated EPDM. When these data are plotted on the same scale as polyester, which has much greater tensile strength and modulus and much lower elongation than any of the elastomers, the differences tend to vanish because of distortions of scale. The same is true of the NBR and EPDM controls also given in Figures 1, 2, and 3. The scale needed to accommodate the pure polyester compound obscures the relatively minor differences between the two elastomers.

Compositions based on equal parts of unsaturated polyester and amino-functional NBR exhibited strong adhesion to polar substrates such as metal mill rolls and glass plates. Analogous compounds made with carboxylated EPDM, on the other hand, appeared to have easy release from polar substrates, compared to unmodified polyester compounds.

Carboxylated NBR (Krynac 221, medium ACN) was also investigated, despite its relatively rapid cross-linking by organic peroxides. This compound proved to be an intermediate case, yielding tensile strength at some points in the composition curve (Figure 1) lower than either component, but without the catastrophic loss noted with EPDM and NBR. The curves of tensile strength, elongation, and tensile modulus versus composition are similar to those of elastomer blends (e. g., NBR with EPDM) that are less than optimum but nonetheless often used in practice.

Although compositions containing equal fractions of carboxylated NBR and polyester have lower ultimate elongation than analogs that use carboxylated EPDM or amino-functional NBR, useful products may still result. For example, addition of 300–400 parts of hydrated alumina to the 50:50 test compound yielded vulcanizates having 10–20% elongation, with Izod impact strengths in the range of 5–10 J/cm. It was therefore decided to proceed with carboxylated NBR, as well as carboxylated EPDM and amino-functional NBR, in typical glass-reinforced polyester formulations.

Unmodified by elastomer, the glass-reinforced polyester compound yielded a flexural modulus in the range of 13.5–14 GPa. As shown in Figure 4, replacement of up to 20% of the polyester with carboxylated EPDM provided at most a 10% increase in fracture surface energy. At the 20% level, a flexural modulus of 13–13.5 GPa was retained. When the mix was increased to 30% carboxylated EPDM, the flexural modulus dropped to 8.5 GPa, indicating loss of the toughening effect and merely conversion from brittle failure.

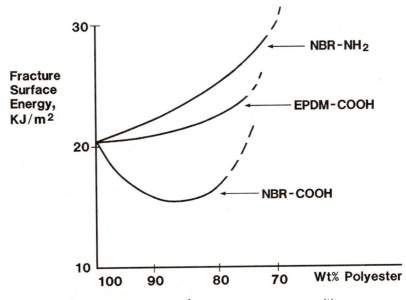

Figure 4. Fracture surface energy versus composition.

Replacement instead with amino-functional NBR led to more significant improvement in fracture surface energy. The original level of flexural modulus is maintained to about 25% rubber content with, as shown in Figure 4, an increase of almost 30% in fracture surface energy. Although not dramatic, this improvement is greater than most previous reports with glass-reinforced compounds. It is tempting to ascribe this result to increased interface adhesion provided by the amino-functional NBR. At levels above 25%, rubber flexural modulus again declines sharply. The remainder of the fracture surface energy versus composition curve no longer indicates toughening, but conversion from brittle failure.

Carboxylated NBR did not appear to provide any toughening effect. Minor additions resulted in a loss of fracture surface energy, as shown in Figure 4. Addition of 10% of this elastomer lowered the flexural modulus to 10.5 GPa, with further loss as the concentration was increased. It should be regarded as a reactive plasticizer for unsaturated polyesters. Improvements in brittleness will be attended by corresponding losses in properties.

Compounding polyesters with high-molecular-weight elastomers introduces several peripheral effects. As the fraction of elastomer is increased, there is a notable increase in apparent viscosity. Therefore thixotropic agents such as magnesium oxide must be reduced or eliminated. This viscosity increase is in contrast to the use of low-molecular-weight liquid elastomer precursors, which may provide the opposite effect (7). As the level of elastomer is increased, the proportion of thermoplastic modifier (low-profile additive) should be decreased, as part of its function will be provided. Compositions using a major fraction of elastomer may be conveniently processed by dry-rubber techniques. Such compositions, however, should be molded with tooling appropriate for solid thermosets rather than rubber, unless the elastomer content is very high.

Conclusions

The experiments described suggest that in typical glass-fiber-reinforced bulk-molding compounds (BMC) or sheet-molding compounds (SMC), elastomeric carboxylated NBR polymers will function only as reactive plasticizers, without true toughening. Carboxylated EPDM appears to have a minor toughening effect, but probably will not be of great use in BMC or SMC formulations because of poor adhesion to polar substrates. Amino-functional NBR elastomer exhibits toughening in reinforced compositions and appears to have good adhesion to polar substrates.

References

1. *Rubber Compounding Formulary;* Phillips Petroleum: Bartlesville, OK.
2. Flick, E. W. *Adhesive and Sealant Compound Formulations;* Noyes: Park Ridge, NJ, 1984.

3. Maynard, J. T.; Johnson, P. R. *Rubber Chem. Tech.* **1963**, *36*, 963.
4. *Rubber-Modified Thermoset Resins;* Riew, C. K.; Gillham, J. K., Eds.; Advances in Chemistry 208; American Chemical Society: Washington, DC, 1984.
5. Siebert, A. R.; Rowe, E. H.; Riew, C. K. *Toughening Polyester Resins with Liquid Rubbers;* 27th Ann. Conf. SPI, 1972.
6. McGarry, F. J.; Rowe, E. H.; Riew, C. K. *Polym. Eng. Sci.* **1978**, *18*, 2.
7. Meyer, R. W. *Handbook of Polyester Molding Compounds and Molding Technology;* Chapman and Hall: New York, 1987; p 111.
8. McGarry, F. J. *Rubber in Crosslinked Glassy Polymers;* Rubber Div., ACS, 1986, Spring Meeting, New York.
9. Korb, J. *Proc. Annu. Conf. SPE* **1984**, 660.
10. Kinloch, A. J.; Young, R. J. *Fracture Behavior of Polymers;* Applied Science: London, 1983.
11. Kinloch, A. J.; Shaw, S. J.; Tod, D. A. In *Rubber-Modified Thermoset Resins;* Riew, C. K.; Gillham, J. K., Eds.; Advances in Chemistry 208; American Chemical Society: Washington, DC, 1984.
12. Barton, A. F. M. *Handbook of Solubility Parameters;* CRC: Boca Raton, FL, 1983.

RECEIVED for review February 11, 1988. ACCEPTED revised manuscript September 1, 1988.

INDEXES

Author Index

Affiliation Index

Subject Index

Copy editing and indexing: Colleen P. Stamm

Production: Rebecca A. Hunsicker

Acquisitions Editor: Cheryl J. Shanks

Managing Editor: Janet S. Dodd

Typeset by Techna Type, Inc., York, PA
Front matter typeset by Hot Type Ltd., Washington, DC
Printing and binding by Maple Press, York, PA

Recent ACS Books

Biotechnology and Materials Science: Chemistry for the Future
Edited by Mary L. Good
160 pp; clothbound; ISBN 0-8412-1472-7

Practical Statistics for the Physical Sciences
By Larry L. Havlicek
ACS Professional Reference Book; 512 pp; clothbound;
ISBN 0-8412-1453-0

The Basics of Technical Communicating
By B. Edward Cain
ACS Professional Reference Book; 198 pp; clothbound;
ISBN 0-8412-1451-4

The ACS Style Guide: A Manual for Authors and Editors
Edited by Janet S. Dodd
264 pp; clothbound; ISBN 0-8412-0917-0

Personal Computers for Scientists: A Byte at a Time
By Glenn I. Ouchi
276 pp; clothbound; ISBN 0-8412-1000-4

Writing the Laboratory Notebook
By Howard M. Kanare
146 pp; clothbound; ISBN 0-8412-0906-5

Principles of Environmental Sampling
Edited by Lawrence H. Keith
458 pp; clothbound; ISBN 0-8412-1173-6

Phosphorus Chemistry in Everyday Living, Second Edition
By Arthur D. F. Toy and Edward N. Walsh
362 pp; clothbound; ISBN 0-8412-1002-0

Chemistry and Crime: From Sherlock Holmes to Today's Courtroom
Edited by Samuel M. Gerber
136 pp; clothbound; ISBN 0-8412-0784-4

Folk Medicine: The Art and the Science
Edited by Richard P. Steiner
224 pp; clothbound; ISBN 0-8412-0939-1

For further information and a free catalog of ACS books, contact:
American Chemical Society
Distribution Office, Department 225
1155 16th Street, NW, Washington, DC 20036
Telephone 800-227-5558